AF762837

MONOGRAPHS ON
STATISTICS AND APPLIED PROBABILITY

General Editors

D. R. Cox and D. V. Hinkley

Probability, Statistics and Time
M. S. Bartlett

The Statistical Analysis of Spatial Pattern
M. S. Bartlett

Stochastic Population Models in Ecology and Epidemiology
M. S. Bartlett

Risk Theory
R. E. Beard, T. Pentikäinen and E. Pesonen

Residuals and Influence in Regression
R. D. Cook and S. Weisberg

Point Processes
D. R. Cox and V. Isham

Analysis of Binary Data
D. R. Cox

The Statistical Analysis of Series of Events
D. R. Cox and P. A. W. Lewis

Analysis of Survival Data
D. R. Cox and D. Oakes

Queues
D. R. Cox and W. L. Smith

Stochastic Abundance Models
S. Engen

The Analysis of Contingency Tables
B. S. Everitt

Introduction to Latent Variable Models
B. S. Everitt

Finite Mixture Distributions
B. S. Everitt and D. J. Hand

Population Genetics
W. J. Ewens

Classification
A. D. Gordon

Monte Carlo Methods
J. M. Hammersley and D. C. Handscomb

Identification of Outliers
D. M. Hawkins

Generalized Linear Models
P. McCullagh and J. A. Nelder

Distribution-free Statistical Methods
J. S. Maritz

Multivariate Analysis in Behavioural Research
A. E. Maxwell

Applications of Queueing Theory
G. F. Newell

Some Basic Theory for Statistical Inference
E. J. G. Pitman

Statistical Inference
S. D. Silvey

Models in Regression and Related Topics
P. Sprent

Sequential Methods in Statistics
G. B. Wetherill

(Full details concerning this series are available from the Publishers)

Stochastic Modelling and Control

M. H. A. DAVIS
Department of Electrical Engineering
Imperial College
London

R. B. VINTER
Department of Electrical Engineering
Imperial College
London

LONDON NEW YORK
CHAPMAN AND HALL

First published 1985 by
Chapman and Hall Ltd
11 New Fetter Lane, London EC4P 4EE
Published in the USA by
Chapman and Hall
733 Third Avenue, New York NY10017
© 1985 M. H. A. Davis and R. B. Vinter
Softcover reprint of the hardcover 1st edition 1985

ISBN-13: 978-94-010-8640-0 e-ISBN-13: 978-94-009-4828-0
DOI: 10.1007/978-94-009-4828-0

All rights reserved. No part of this book may be reprinted, or reproduced or utilized in any form or by any electronic, mechanical or other means, now known or hereafter invented, including photocopying and recording, or in any information storage and retrieval system, without permission in writing from the publisher.

British Library Cataloguing in Publication Data

Davis, M. H. A.
Stochastic modelling and control.—
(Monographs on statistics and applied probability)
1. Mathematical models 2. Stochastic processes
I. Title II. Vinter, R. B. III. Series
519.2 QA402

Library of Congress Cataloging in Publication Data

Davis, M.H.A.
Stochastic modelling and control.

(Monographs on statistics and applied probability)
Includes bibliographies and index.
1. Control theory. 2. Discrete-time systems.
3. Stochastic systems. I. Vinter, R. B. (Richard B.)
II. Title. III. Series.
QA402.3.D39 1984 519.2 84-19924

To our friend and colleague
DAVID Q. MAYNE

Contents

Preface

This book aims to provide a unified treatment of input/output modelling and of control for discrete-time dynamical systems subject to random disturbances. The results presented are of wide applicability in control engineering, operations research, econometric modelling and many other areas.

There are two distinct approaches to mathematical modelling of physical systems: a direct analysis of the physical mechanisms that comprise the process, or a 'black box' approach based on analysis of input/output data. The second approach is adopted here, although of course the properties of the models we study, which within the limits of linearity are very general, are also relevant to the behaviour of systems represented by such models, however they are arrived at.

The type of system we are interested in is a discrete-time or sampled-data system where the relation between input and output is (at least approximately) linear and where additive random disturbances are also present, so that the behaviour of the system must be investigated by statistical methods. After a preliminary chapter summarizing elements of probability and linear system theory, we introduce in Chapter 2 some general linear stochastic models, both in input/output and state-space form. Chapter 3 concerns filtering theory: estimation of the state of a dynamical system from noisy observations. As well as being an important topic in its own right, filtering theory provides the link, via the so-called innovations representation, between input/output models (as identified by data analysis) and state-space models, as required for much contemporary control theory.

System identification – modelling from input/output data – is considered in Chapters 4 and 5. Most current techniques are based in one form or another either on least-squares or on maximum likelihood estimation and these procedures are described. A general approach to identification, due largely to L. Ljung and P. E. Caines, is the *prediction error* formulation, whereby a 'model' is thought of as an algorithm which generates one-step-ahead predictions of the output given past data. The corresponding model-fitting procedure is to choose that model within a specified class for which some measure of the average prediction error is minimized for the given data set. This gives a new slant on the idea of 'consistency': one asks, not whether the parameter estimates will converge to their 'true' values as the amount of available data increases – a question which is only relevant in the artificial case when the data was actually generated by some finitely-parametrized model – but rather whether one's identification procedure will succeed in giving the best available model within the prescribed model set to represent the data. Some general results along these lines have been provided by Ljung and we give a somewhat modified version of them in Chapter 5. In the last two chapters we turn to control topics. Chapter 6 covers the quadratic cost regulator theory for linear deterministic and stochastic systems. As is well known, the deterministic linear regulator is 'dual' to the Kalman filter in that the so-called *matrix Riccati equation* occurs in both contexts. The properties of this equation are studied in detail. The Kalman filter appears directly in the optimal stochastic linear regulator where state estimation is required as part of the control algorithm. We formulate the *separation* and *certainty-equivalence* principles which encapsulate this idea. In the final chapter, some topics in adaptive control are discussed. Adaptive control, that is, simultaneous identification and control of an initially 'unknown system', is a subject which is at the moment in a state of active development, and we restrict ourselves here to a discussion of the special but important topics of minimum-variance and self-tuning control. Conditions under which the self-tuning property is possible are investigated and one algorithm with guaranteed stability properties under well-defined conditions is presented.

Mathematical modelling and control are of course vast fields of enquiry and any single-volume treatment of them must necessarily be highly selective. In this book we do not enter into issues of practical data analysis such as are admirably covered in, for example, the influential book of Box and Jenkins. Neither do we discuss in any detail the numerical properties of the algorithms we present, although there has in fact been considerable recent research in this area. Rather, our objective has been to provide a cohesive account of the main mathematical methods and results underpinning most of the recent work in this area. The emphasis is on the *unity* of the subject, that is, on the fact that all the models are in some sense interchangeable and tend to appear in whatever guise is appropriate to the problem at hand, be it model fitting, prediction, regulation, or any other. In taking this point of view we make much more systematic use of linear system theory than is customary in 'time series analysis'.

This book is intended both to provide suitable material for postgraduate courses on the stochastic aspects of control systems, and to serve as a reference book for researchers in the field of stochastic systems. It has therefore been organized so that it can be read on several levels. A reader new to the field may wish to stick to the main body of the text, where intricate arguments are avoided; here certain results are merely stated (though we have made an effort in such cases to provide sufficient explanation that their significance can be appreciated). On the other hand, a reader with more experience should treat the appendices, where the more difficult proofs are to be found, as an integral part of the text.

We have tried to make our treatment as self-contained as possible. Our coverage of background topics is, however, brisk, and readers will undoubtedly benefit from some knowledge of probability, statistics, stochastic processes and linear system theory, as provided, for example, by the references at the end of Chapter 1.

This book grew out of our involvement in teaching and research in the Control Group at Imperial College, London. Our first debt of gratitude is to David Mayne, who has been largely responsible for creating, in the Control Group, an environment in which projects such as this can flourish, as well as for initiating the courses on which much of the material of this book was originally based. We would like to dedicate the book to him as a token of affection and esteem. We are indebted to Martin Clark and again to David Mayne for advice and discussions over the years, and to many other colleagues at Imperial College and elsewhere whose work has influenced our thinking. Of course, none of them can be blamed for the consequences. Doris Abeysekera has played a quite exceptional role in the creation of this book by typing, at great speed and often under considerable pressure, successive drafts of the various chapters, only to be confronted with irritating requests for additions and alterations. We are grateful to the Leverhulme Trust for a research grant to one of us (MHAD) which facilitated completion of the book. Finally, a word of thanks to David Cox for including this book in the Monographs on

Statistics and Applied Probability series under his editorship, and to our editors at Chapman and Hall for their collaboration and for tolerating what we modestly think must be a record-breaking series of missed deadlines.

M. H. A. Davis
R. B. Vinter
London,
September 1984

CHAPTER 1

Probability and linear system theory

This book is concerned with the analysis of discrete-time linear systems subject to random disturbances. This introductory chapter is designed to present the main results in the two areas of probability and linear systems theory as required for the main developments of the book, beginning in Chapter 2.

Section 1.1. on probability is divided into three subsections dealing with distributions and random variables, stochastic processes, and convergence of stochastic sequences. In the space available it is not possible to give a complete and self-contained account of these topics, which are in any case discussed at length in many other texts. The intention here is only to summarize the main ideas and results needed later in the book. Suggestions for further reading are contained in the Notes at the end of the chapter.

Section 1.2 covers the elements of linear system theory with particular emphasis on those aspects relevant to linear filtering and quadratic cost stochastic control. The section centres around the concepts of controllability and observability together with refinements of them in the form of stabilizability and detectability. The concepts are characterized and interrelated. Along the way there is discussion of pole assignment. The treatment is largely self-contained in that almost all results are proved in full, but the reader with little background in linear systems theory will probably none the less wish to consult the suggested references to complement the coverage here.

1.1 Probability and random processes

1.1.1 Distributions and random variables

A *random variable* X is the numerical outcome of some experiment the result of which cannot be exactly predicted in advance. Mathematically the properties of X are specified by a *distribution function*, F, which defines the probability that in a single trial the value of X will fall in a given interval of the real line. Symbolically,

$$F(a) = P[X < a] \tag{1.1.1}$$

so that

$$P[a \le X < b] = F(b) - F(a) \tag{1.1.2}$$

for arbitrary $a, b \in \mathbb{R}$. Thus F is a non-decreasing function with $F(-\infty) = 0, F(\infty) = 1$. It is *left-continuous* (this is due to the choice of $<$ rather than $\le$ in (1.1.1)), and the jump $F(a+) - F(a)$ is the probability that X takes exactly the value a. Two important special cases are the following.

(a) *Discrete random variables* Here X takes on one of a finite or countable number of values $x_1, x_2, \ldots$ with corresponding probabilities $p_1, p_2, \ldots$, which must satisfy

$$p_i \ge 0, \quad \sum_i p_i = 1.$$

The distribution function is

$$F(a) = \sum_{x_i < a} p_i$$

which is a piecewise-constant function with a jump of height p_i at x_i; see Fig. 1.1

(b) *Continuous random variables* These are random variables (r.v.s) whose distribution function F is absolutely continuous, i.e. can be written

$$F(a) = \int_{-\infty}^{a} f(x)\,\mathrm{d}x$$

for some function f, the *density function* of X. f must satisfy

$$f(x) \ge 0, \int_{-\infty}^{\infty} f(x)\,\mathrm{d}x = 1.$$

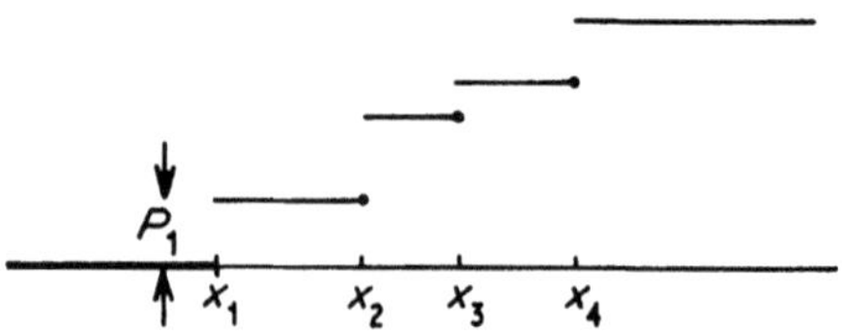

Fig. 1.1

In view of (1.1.2) we then have

$$P[a \le X < b] = \int_a^b f(x)\,dx. \tag{1.1.3}$$

Since F is continuous the probability that X takes exactly any particular value a is zero, so it is immaterial whether the endpoints of the interval $[a, b]$ are included or excluded in (1.1.3).

An important parameter of a random variable is its *expectation* or mean value EX. This is normally defined for discrete and continuous random variables respectively as follows:

$$EX = \sum_i x_i p_i \qquad \text{(discrete case)} \tag{1.1.4}$$

$$EX = \int_{-\infty}^{\infty} x f(x)\,dx \qquad \text{(continuous case)} \tag{1.1.5}$$

We can subsume these in a single formula as a *Stieltjes integral* with respect to the distribution function F. For positive-valued continuous functions g we define

$$\int_{-\infty}^{\infty} g(x)dF(x) := \lim_{n\to\infty} \sum_{k=-2^{2n}}^{2^{2n}} g(x_{k,n}^*)\left(F\left(\frac{k+1}{2^n}\right) - F\left(\frac{k}{2^n}\right)\right), \tag{1.1.6}$$

where $x_{k,n}^*$ is any minimizing point in the interval $[k/2^n, (k+1)/2^n]$, i.e. any point such that

$$g(x_{k,n}^*) \le g(x), \quad k/2^n \le x \le (k+1)/2^n.$$

The sum on the right is increasing as n increases and the limit may be finite or $+\infty$. For a general continuous function g we define

$$g^+(x) := \begin{cases} g(x) & \text{if } g(x) \ge 0 \\ 0 & \text{if } g(x) < 0 \end{cases}$$

$$g^-(x) := g^+(x) - g(x)$$

and

$$\int_{-\infty}^{\infty} g(x)\,dF(x) = \int_{-\infty}^{\infty} g^+(x)\,dF(x) - \int_{-\infty}^{\infty} g^-(x)\,dF(x)$$

as long as both integrals on the right are finite, which is the case if and only if

$$\int_{-\infty}^{\infty} |g(x)|\,dF(x) < \infty, \tag{1.1.7}$$

since

$$|g(x)| = g^+(x) + g^-(x).$$

It is easily seen that with the definition (1.1.6), the formula

$$EX = \int_{-\infty}^{\infty} x\,\mathrm{d}F(x) \tag{1.1.8}$$

agrees with (1.1.4) and (1.1.5) in those special cases, and this is our general definition of the expectation. In accordance with (1.1.7), for EX to be well defined we require that

$$\int_{-\infty}^{\infty} |x|\,\mathrm{d}F(x) < \infty.$$

Random variables whose distribution has this property are called *integrable*; thus only for integrable r.v.s X is the expectation EX well defined.

If g is a real-valued function and X is an r.v. then $g(X)$ is a r.v. whose expectation, if defined, is

$$Eg(X) = \int_{-\infty}^{\infty} g(x)\,\mathrm{d}F(x).$$

$g(X)$ is integrable if (1.1.7) is satisfied. It is not necessary for $g(\cdot)$ to be continuous for this to be valid but if g is not continuous (1.1.6) may require some modification. This technical point need not however detain us here.

The expectation measures the average value of X to be expected in a long series of trials. A measure of the spread around the mean value is given by the *variance*, defined by

$$\operatorname{var}(X) = E(X - EX)^2 = \int_{-\infty}^{\infty} (x - EX)^2\,\mathrm{d}F(x).$$

The *standard deviation* of X is

$$\sigma = \sqrt{\operatorname{var}(X)}.$$

This has the same units as X. The properties of $\operatorname{var}(X)$ are summarized in the following proposition.

Proposition 1.1.1

Suppose X^2 is integrable, i.e. $EX^2 < \infty$. Then:

(a) X is integrable, and hence $\operatorname{var}(X)$ is well defined; it is given by

$$\operatorname{var}(X) = EX^2 - (EX)^2.$$

We therefore say that X is a *finite variance* random variable if $EX^2 < \infty$.

(b) (Chebyshev inequality) For any positive constant a,

$$P[|X| > a] \leq (1/a^2)EX^2$$

(c) Define a function $v: \mathbb{R} \to \mathbb{R}$ by

$$v(b) = E(X-b)^2.$$

Then $v(b)$ takes its minimum at $b = EX$, and the minimum value is $\operatorname{var}(X)$.

(d) $EX^2 = 0$ if and only if $P[X=0] = 1$.

PROOF It is evident from (1.1.6) that if g, h are functions such that $h(x) \geq g(x)$ for all x then $Eh(X) \geq Eg(X)$. For part (a), take $g(x) = |x|$, $h(x) = 1 + x^2$ to conclude that $E|X| \leq 1 + EX^2 < \infty$. Thus X is integrable. For part (b), define $g(x) = 0$ for $|x| \leq a$ and $g(x) = a^2$ for $|x| > a$ and take $h(x) = x^2$. Then $h(x) \geq g(x)$ and $Eg(X) = a^2 P[|X| > a]$. The result follows. For any constant b we have

$$\begin{aligned} E[X-b]^2 &= \int_{-\infty}^{\infty} (x-b)^2 \, \mathrm{d}F(x) \\ &= \int_{-\infty}^{\infty} x^2 \, \mathrm{d}F(x) - 2b \int_{-\infty}^{\infty} x \, \mathrm{d}F(x) \\ &\quad + b^2 \int_{-\infty}^{\infty} \mathrm{d}F(x) \\ &= EX^2 - 2bEX + b^2. \end{aligned}$$

This last expression is minimized over b at $b = EX$; when $b = EX$ it is equal to $\operatorname{var}(X)$ and coincides with the expression given at part (a). Turning to part (d), to say that $P[X=0] = 1$ is equivalent to saying that the distribution function F of X is given by $F(a) = 0, a \leq 0$ and $F(a) = 1, a > 0$. It follows from (1.1.6) that $EX^2 = 0$ if X has this distribution. Conversely, if $EX^2 = 0$ then for any number $a \geq 0$

$$0 = \int_{-\infty}^{\infty} x^2 \, \mathrm{d}F(x) \geq \int_{a}^{\infty} x^2 \, \mathrm{d}F(x) \geq a^2 \int_{a}^{\infty} \mathrm{d}F(x) = a^2 P[X \geq a] \geq 0.$$

This shows that $P[X \geq a] = 0$ for any $a > 0$ and hence that $P[X > 0] = 0$.
A similar argument shows that $P[X < 0] = 0$; thus $P[X = 0] = 1$.

□

A *random n-vector* $X = (X_1, \ldots, X_n)^T$ is a collection of n random variables $X_1, \ldots, X_n$. To examine its probabilistic behaviour it is not sufficient to know the distribution of each X_i because this information does not specify how the components interact. In general one needs to know the *joint distribution function* $F(a_1, \ldots, a_n)$ which specifies the probabilities of events via the formula

$$P[X_1 < a_1, \ldots, X_n < a_n] = F(a_1, \ldots, a_n).$$

The random variables $X_1, \ldots, X_n$ are *independent* if

$$F(a_1, \ldots, a_n) = F_1(a_1)F_2(a_2)\ldots F_n(a_n)$$

where F_i is the distribution of X_i. This is the only case in which knowledge of $F_1, \ldots, F_n$ suffices to determine F. On the other hand, knowledge of F always determines the distribution of each X_i (the so-called *marginal distribution*) since, for example,

$$\begin{aligned} F_1(a_1) &= P[X_1 < a_1, X_2 < \infty, \ldots, X_n < \infty] \\ &= F(a_1, \infty, \ldots, \infty). \end{aligned}$$

$X_1, \ldots, X_n$ have a *joint density function* f if

$$F(a_1, \ldots, a_n) = \int_{-\infty}^{a_1} \cdots \int_{-\infty}^{a_n} f(x_1, \ldots, x_n)\, \mathrm{d}x_n \ldots \mathrm{d}x_1.$$

If the X_i are independent and X_j has density function f_j then

$$f(x_1, \ldots, x_n) = f_1(x_1)f_2(x_2)\ldots f_n(x_n).$$

If $g\colon \mathbb{R}^n \to \mathbb{R}$ is a continuous function then the expectation $Eg(X)$ can be defined using Stieltjes integrals in a way that agrees with the usual expression

$$Eg(x) = \int_{-\infty}^{\infty} \cdots \int_{-\infty}^{\infty} g(x_1, \ldots, x_n) f(x_1, \ldots, x_n)\, \mathrm{d}x_1 \ldots \mathrm{d}x_n$$

valid when X has joint density f. We give the definition for the bivariate case $n = 2$; for $n > 2$ it is similar but notationally cumbersome. For $n=2$ we have

$$P[a_1 \le X_1 < b_1, a_2 \le X_2 < b_2] = F(b_1, b_2) - F(b_1, a_2) - F(a_1, b_2) + F(a_1, a_2).$$

Let us denote this expression by $\Delta_n(i, j)$ when

$$a_1 = \frac{i}{2^n}, \quad b_1 = \frac{i+1}{2^n}, \quad a_2 = \frac{j}{2^n}, \quad b_2 = \frac{j+1}{2^n}.$$

Then we define

$$Eg(x) = \int_{-\infty}^{\infty} g(x)\,\mathrm{d}F(x) = \lim_{n\to\infty} \sum_{i,j=-2^{2n}}^{2^{2n}} g(x_{ij}^n)\Delta_n(i, j),$$

where x_{ij}^n is some point at which the function g attains its minimum in the rectangle $\{(x_1, x_2)\colon a_1 \le x_1 \le b_1,\ a_2 \le x_2 \le b_2\}$. As before, we require (1.1.7) to hold. It follows directly from the definition that if X_1 and X_2 are independent and $g(x) = g_1(x_1)g_2(x_2)$ then

$$\begin{aligned} Eg_1(X_1)g_2(X_2) &= \int_{-\infty}^{\infty} g_1(x_1)\,\mathrm{d}F_1(x_1) \int_{-\infty}^{\infty} g_2(x_2)\,\mathrm{d}F_2(x_2) \\ &= Eg_1(X_1)Eg_2(X_2) \end{aligned}$$

as long as all these expectations are well-defined.

Now let X_1, X_2 be any pair of finite variance random variables. Taking $g_i(x) = x_i - EX_i$, $i = 1, 2$ we obtain the *covariance* of X_1 and X_2:

$$\operatorname{cov}(X_1, X_2) := E[(X_1 - EX_1)(X_2 - EX_2)].$$

X_1 and X_2 are said to be *uncorrelated* if $\operatorname{cov}(X_1, X_2) = 0$. The properties of the covariance and some related results are summarized below.

Proposition 1.1.2

Let X_1, X_2 be finite-variance random variables, i.e. $EX_i^2 < \infty, i = 1, 2$. Then:

(a) $\operatorname{cov}(X_1, X_2)$ is well defined.
(b) If X_1, X_2 are independent then they are uncorrelated, but the converse is not generally true.

(c) (Schwarz inequality) $|\text{cov}(X_1, X_2)| \leq \sqrt{[\text{var}(X_1)\text{var}(X_2)]}$.
(d) $E[(X_1 - X_2)^2] = 0$ if and only if $P[X_1 = X_2] = 1$. In this case we say that $X_1 = X_2$ *almost surely* (a.s.).
(e) Define the *correlation coefficient* ρ as follows:

$$\rho := \frac{\text{cov}(X_1, X_2)}{\sigma_1 \sigma_2}$$

where $\sigma_i = \sqrt{(\text{var}(X_i))}$, $i = 1, 2$ (assumed non-zero). Then $|\rho| \leq 1$, and $|\rho| = 1$ if and only if there are constants c_1, c_2 such that

$$X_1 = c_1 X_2 + c_2 \quad \text{a.s.}$$

PROOF It is no loss of generality to suppose that $EX_1 = EX_2 = 0$ (otherwise, replace X_i by $X_i - EX_i$ throughout). Then $\text{cov}(X_1, X_2) = EX_1X_2$. For any numbers x, y,

$$|xy| \leq x^2 + y^2.$$

It follows that

$$E|X_1X_2| \leq EX_1^2 + EX_2^2 < \infty$$

and hence that $\text{cov}(X_1, X_2)$ is well-defined. If X_1, X_2 are independent then $EX_1X_2 = EX_1EX_2 = 0$, so that X_1, X_2 are uncorrelated. To see that uncorrelated random variables are not necessarily independent, consider a random variable X such that $EX = 0$ and $EX^3 = 0$ (for example, $X \sim N(0, 1)$; see below) and define $X_1 = X$, $X_2 = X^2 - EX^2$. Then $\text{cov}(X_1, X_2) = E[X(X^2 - EX^2)] = EX^3 - EXEX^2 = 0$, so that X_1, X_2 are uncorrelated; but they are generally not independent. To get the Schwarz inequality, take any number a and calculate

$$E[X_1 + aX_2]^2 = EX_1^2 + 2aEX_1X_2 + a^2EX_2^2. \qquad (1.1.9)$$

This expression takes its minimum value $EX_1^2 - (EX_1X_2)^2/EX_2^2$ at $a = -EX_1X_2/EX_2^2$. But this minimum value must be non-negative since $E[X_1 + aX_2]^2 \geq 0$ for any a. This gives (c). For part (d), note that (a) implies $E(X_1 - X_2)^2 < \infty$, i.e. $(X_1 - X_2)$ is a finite variance random variable. Applying Proposition 1.1.1(d) with $X = X_1 - X_2$ gives the result. Finally, turning to part (e), the fact that $|\rho| \leq 1$ is just a restatement of the Schwarz inequality. Rewrite (1.1.9) as

$$E[X_1 + aX_2]^2 = \sigma_1^2 + 2a\rho\sigma_1\sigma_2 + a^2\sigma_2^2.$$

If $\rho = \pm 1$ then the right hand side is $(\sigma_1 \pm a\sigma_2)^2$ and thus choosing $a = \mp \sigma_1/\sigma_2$ gives $E[X_1 + aX_2]^2 = 0$. In view of (d), this implies that

$X_1 = -aX_2$ a.s. Thus $c_1 = \pm\sigma_1/\sigma_2$. The constant c_2 is zero if $EX_1 = EX_2 = 0$; in general it takes the value $EX_1 \mp (\sigma_1/\sigma_2)EX_2$. Conversely, it is easy to check that $|\rho| = 1$ if $X_1 = c_1X_2 + c_2$ a.s. □

For a random n-vector $X = (X_1, \ldots, X_n)^{\mathrm{T}}$ the *mean* EX is the n-vector with ith element EX_i. The *covariance matrix* $\mathrm{cov}(X)$ is the $n \times n$ matrix with i,jth entry $\mathrm{cov}(X_i, X_j)$. One can check that

$$\mathrm{cov}(X) = EXX^{\mathrm{T}} - (EX)(EX)^{\mathrm{T}}. \tag{1.1.10}$$

Any covariance matrix is symmetric and non-negative definite, the latter property following from the fact that for any $a \in \mathbb{R}^n$,

$$0 \le E[a^{\mathrm{T}}(X - EX)]^2 = \sum_{i,j} a_i a_j E[X_i - EX_i)(X_j - EX_j).$$

An alternative way of specifying the distribution of a random vector (or random variable) is through its *characteristic function* defined for $u \in \mathbb{R}^n$ by

$$\phi_X(u) = Ee^{iu^{\mathrm{T}}X} = \int_{-\infty}^{\infty} \cdots \int_{-\infty}^{\infty} e^{iu^{\mathrm{T}}x}\,\mathrm{d}F(x).$$

This is always well-defined since $e^{iu^{\mathrm{T}}x} = \cos u^{\mathrm{T}}x + i \sin u^{\mathrm{T}}x$ and the trigonometric functions are bounded. There is a *one-to-one correspondence* between F and ϕ_X: if F has a density function f then ϕ_X is just the Fourier transform of f, and f can be recovered by the Fourier inversion formula (1.1.12) below. If F does not have a density then F can still be recovered uniquely from ϕ_X by a generalized inversion formula which it is not necessary to give here.

We shall have many occasions to consider linear transformations of a random vector X, i.e. random p-vectors of the form

$$Y = GX + b \tag{1.1.11}$$

where G is a $p \times n$ matrix and b a p-vector. The information we need is as follows.

Proposition 1.1.3

(a) If (1.1.11) holds and X is a finite-variance random vector then $EY = GEX + b$, $\mathrm{cov}(Y) = G\mathrm{cov}(X)G^{\mathrm{T}}$.

(b) If G is an $n \times n$ matrix then

$$E[X^{\mathrm{T}}GX] = (EX)^{\mathrm{T}}GEX + \mathrm{tr}[G\,\mathrm{cov}(X)].$$

(c) If Y is any finite variance random p-vector then there is a random

n-vector X for some $n \le p$ and a vector b such that $\operatorname{cov}(X) = I_n$ (the $n \times n$ identity matrix) and (1.1.11) holds.

(d) If ϕ_X, ϕ_Y are the characteristic functions of X and Y respectively, then

$$\phi_Y(u) = e^{iu^{\mathrm{T}}b}\phi_X(G^{\mathrm{T}}u).$$

(e) Suppose that $n = p$, that G is non-singular and that X has density function f_X. Then Y has density function f_Y, where

$$f_Y(y) = \frac{1}{|\det(G)|} f_X(G^{-1}(y-b)).$$

PROOF Part (a) is immediate from (1.1.10). For (b), suppose first that $EX = 0$. Then

$$\begin{aligned} E[X^{\mathrm{T}}GX] &= E\sum_{i,j=1}^{n} G_{ij}X_iX_j = \sum_{i,j=1}^{n} G_{ij}(\operatorname{cov}(X))_{ij} \\ &= \operatorname{tr}[G\operatorname{cov}(X)]. \end{aligned}$$

If $EX = M \ne 0$ then writing $\tilde{X} = X - M$ we have $E\tilde{X} = 0$ and hence

$$\begin{aligned} E[X^{\mathrm{T}}GX] &= E[(\tilde{X}+M)^{\mathrm{T}}G(\tilde{X}+M)] \\ &= E[\tilde{X}^{\mathrm{T}}G\tilde{X}] + E\tilde{X}^{\mathrm{T}}GM + EM^{\mathrm{T}}G\tilde{X} + M^{\mathrm{T}}GM \\ &= \operatorname{tr}[G\operatorname{cov}(X)] + M^{\mathrm{T}}GM. \end{aligned}$$

For (c), let $Q = \operatorname{cov}(Y)$. It is shown in Appendix C that Q can be factored in the form $Q = U\Lambda U^{\mathrm{T}}$ where U is orthogonal and Λ is diagonal with entries $\lambda_1, \ldots, \lambda_p$, the eigenvalues of a. Define

$$\Lambda^{1/2} = \begin{bmatrix} \sqrt{\lambda_1} & & & \bigcirc \\ & \cdot & & \\ & & \cdot & \\ & & & \cdot \\ \bigcirc & & & \sqrt{\lambda_p} \end{bmatrix}$$

and

$$G = U\Lambda^{1/2}.$$

Suppose for a moment that $\lambda_i > 0$ for all i; then G is non-singular. If we define $X = G^{-1}(Y - EY)$ then, by part (a), $EX = 0$ and $\operatorname{cov}(X) = I_p$, and $Y = GX + EY$. If only $p - n$ eigenvalues are non-zero then a similar construction applies but X has dimension n and is not determined as a *unique* linear combination of the Y_i.

Part (d) is immediate from the defintion, since

$$\begin{aligned}\phi_Y(u) &= Ee^{iu^TY}\\ &= Ee^{iu^T(GX+b)}\\ &= e^{iu^Tb}Ee^{i(G^Tu)^TX}\\ &= e^{iu^Tb}\phi_X(G^Tu).\end{aligned}$$

For (e) we use the Fourier inversion formula

$$\begin{aligned}f_Y(y) &= \frac{1}{2\pi}\int_{-\infty}^{\infty}\cdots\int_{-\infty}^{\infty} e^{-iu^Ty}\phi_Y(u)\,\mathrm{d}u_1\ldots\mathrm{d}u_p \qquad (1.1.12)\\ &= \frac{1}{2\pi}\int_{-\infty}^{\infty}\cdots\int_{-\infty}^{\infty} e^{iu^T(b-y)}\phi_X(G^Tu)\,\mathrm{d}u_1\ldots\mathrm{d}u_p\\ &= \frac{1}{2\pi|\det(G)|}\int_{-\infty}^{\infty}\cdots\int_{-\infty}^{\infty} e^{-iv^TG^{-1}(y-b)}\phi_X(v)\mathrm{d}v_1\ldots\mathrm{d}v_P\\ &= \frac{1}{|\det(G)|}f_X(G^{-1}(y-b)).\end{aligned}$$

□

Notice that in Proposition 1.1.3, part (d) is true with no restrictions on the distribution of X or on the dimensions n, p, whereas (e) holds only under special conditions, without which Y may not have a density at all. This is why the characteristic function is such a useful construction in dealing with linear combinations of random variables.

We now introduce the idea of the conditional distribution of a random variable X given another random variable Y. (In the following discussion X and Y are, for notational simplicity, taken as scalar but analogous results apply to the vector case.) Recall that for events A, B, the conditional probability of A given B is

$$P(A|B) = \frac{P(A \text{ and } B)}{P(B)}$$

if $P(B) > 0$, with arbitrary assignment if $P(B) = 0$. The obvious definition for the conditional distribution $F_{X|Y}(a;b)$ of X given Y would be

$$F_{X|Y}(a;b) = P[X < a|Y = b].$$

This is correct if Y is a discrete random variable taking values $b_1, b_2\ldots$

with positive probability, but not if Y is a continuous random variable since then the event $Y=b$ has probability 0 for all b. To circumvent this difficulty we adopt the following approach. Let $F(a,b)$ be the joint distribution function of X and Y, so that the marginal distribution of Y is $F_Y(b)=F(\infty,b)$. If $F_Y(b+\delta)-F_Y(b)>0$ for all $\delta>0$ then

$$\begin{aligned}P[X<a|b\leq Y<b+\delta] &= \frac{P[X<a \text{ and } b\leq Y<b+\delta]}{P[b\leq Y<b+\delta]}\\ &= \frac{F(a,b+\delta)-F(a,b)}{F_Y(b+\delta)-F_Y(b)}.\end{aligned}$$

We now define

$$F_{X|Y}(a;b)=\lim_{\delta\to 0}\frac{F(a,b+\delta)-F(a,b)}{F_Y(b+\delta)-F_Y(b)} \tag{1.1.13}$$

when this limit exists. If $F_Y(b+\delta)-F_Y(b)=0$ for some $\delta>0$ then $F_{X|Y}(a;b)$ is defined arbitrarily as $F_X(a)$. For each fixed b, $F_{X|Y}(a;b)$ is a distribution function in a.

This definition is still not completely general, but it does cover both discrete and continuous random variables. Indeed, it is easy to see that if X, Y have a continuous joint density function f then

$$F_{X|Y}(a;b)=\frac{\displaystyle\int_{-\infty}^{a} f(x,b)\,\mathrm{d}x}{\displaystyle\int_{-\infty}^{\infty} f(x,b)\,\mathrm{d}x}$$

if the denominator is positive, so that X has a *conditional density function*

$$f_{X|Y}(x;b)=\begin{cases}\dfrac{f(x,b)}{f_Y(b)} & f_Y(b)>0\\ f_X(x) & f_Y(b)=0\end{cases}$$

where f_X, f_Y are the marginal densities.

The *conditional expectation* of some function $g(X,Y)$ given Y is just the integral with respect to the conditional distribution, i.e.

$$E[g(X,Y)|Y]=\int_{-\infty}^{\infty} g(x,Y)\,\mathrm{d}F_{X|Y}(x;Y). \tag{1.1.14}$$

It is a function of the random variable Y. Conditional expectations have the following important properties. We state them for the vector case.

Proposition 1.1.4

Let X, Y be jointly distributed random vectors and g be a real-valued function such that $g(X)$ is integrable. Then

(a) If X and Y are independent then $E[g(X)|Y] = E[g(X)]$.
(b) If X is a function of Y, say $X = h(Y)$, then $E[g(X)|Y] = g(X) \quad (= g(h(Y)))$.
(c) $E[g(X)] = E[E[g(X)|Y]]$.
(d) $E[g(X)h(Y)|Y] = E[g(X)|Y]h(Y)$
for any function h such that $g(X)h(Y)$ is integrable.

REMARK The conditional distribution $F_{X|Y}$ exists for any random vectors X, Y and the above propositions hold. In fact, they hold even if Y has an infinite number of components. We give a partial proof here for the scalar case when the conditional distribution is defined by (1.1.13).

PROOF Part (a) follows from the fact that if X and Y are independent then the ratio in (1.1.13) is equal ot $F(a)$ for any δ. Thus the conditional distribution of X given Y is the same as the (unconditional) distribution of X. For (b), take first $a < h(b)$. Then $P[X < a \text{ and } b \leq Y \leq b + \delta] = P[h(Y) < a \text{ and } b \leq Y \leq b + \delta] = 0$ for sufficiently small δ (as long as h is continuous). Thus $F_{X|Y}(a; b) = 0$ if $a \geq h(b)$ and similarly $F_{X|Y}(a; b) = 1$ if $a \geq h(b)$. Thus $F_{X|Y}(a; b)$ is the distribution that puts probability 1 on the point $h(b)$ and hence $E[g(X)|Y = b] = g(h(b)) = g(X)$. Properties (c) and (d) follow immediately from the definitions (1.1.13) and (1.1.14) when the conditional density $f_{X|Y}$ exists. They also hold without this restriction but we do not give a proof here. □

Two further properties of conditional expectation will be required. The first of these relates to 'least-squares' estimation. Recall from Proposition 1.1.1 that the choice $a = E[g(X)]$ minimizes $E[g(X) - a]^2$ over constants a. One can regard $E[g(X)]$ as the 'best estimate' of $g(X)$ when *no information* about X (other than its distribution) is supplied. Now suppose we observe the random vector Y and base our estimate on the value of Y, that is we wish to choose a function $e(Y)$ so as to minimize $E[g(X)-e(Y)]^2$. This is the so-called *non-linear least-squares problem.*

Proposition 1.1.5

Let X, Y, g be as in Proposition 1.1.4. Then $E[g(X)-e(Y)]^2$ is minimized over functions e by the function $e(Y)=E[g(X)|Y]$.

PROOF Using Proposition 1.1.4(c) we can write

$$E[g(X)-e(Y)]^2=\int\int[g(x)-e(y)]^2\mathrm{d}F_{X|Y}(x;y)\mathrm{d}F_Y(y).$$

The double integral is certainly minimized if the inner integral is minimized pointwise for each y. But the inner integral is equal to $E[g(\tilde{X})-e(y)]^2$ where $\tilde{X}$ is a random vector with distribution $F_{X|Y}(x;y)$. It follows from Proposition 1.1.1 that the minimizing value of $e(y)$ is $E[g(\tilde{X})]=E[g(X)|Y=y]$. □

The final result states the rather natural property that if two random vectors Y and $\tilde{Y}$ are in one-to-one correspondence with each other then conditioning on Y is equivalent to conditioning on $\tilde{Y}$.

Proposition 1.1.6

Let X, Y, g be as in Proposition 1.1.4 and suppose $\tilde{Y}=\phi(Y)$ where ϕ is a one-to-one function. Then $E[g(X)|\tilde{Y}]=E[g(X)|Y]$ a.s.

PROOF Denote $e(Y)=E[g(X)|Y]$ and $\tilde{e}(\tilde{Y})=E[g(X)|\tilde{Y}]$. It is not hard to see, from Proposition 1.1.4(d), that $e(\cdot)$ is the unique function such that

$$E[h(Y)e(Y)]=E[h(Y)g(X)] \tag{1.1.15}$$

for all bounded functions $h(\cdot)$.† Similarly, $\tilde{e}$ is characterized by the property that

$$E[h(\tilde{Y})\tilde{e}(\tilde{Y})]=E[h(\tilde{Y})g(X)] \quad \text{for all } h$$

which we can write

$$E[h\circ\phi(Y)\tilde{e}\circ\phi(Y)]=E[h\circ\phi(Y)g(X)] \tag{1.1.16}$$

†It is unique up to equivalence, i.e. if $\hat{e}$ is a function such that $P[\hat{e}(Y)=e(Y)]=1$ then $E[g(X)|Y]$ can also be taken as $\hat{e}(Y)$.

where $h \circ \phi(Y) = h(\phi(Y))$, etc. But if j is any bounded function then $j = h \circ \phi$ where $h = j \circ \phi^{-1}$. Thus (1.1.16) is equivalent to

$$E[j(Y)\tilde{e} \circ \phi(Y)] = E[j(Y)g(X)] \quad \text{for all bounded } j(\cdot).$$

Comparing with (1.1.15) we see that

$$e = \tilde{e} \circ \phi$$

and hence that

$$E[g(X)|Y] = e(Y) = \tilde{e} \circ \phi(Y) = \tilde{e}(\tilde{Y}) = E[g(X)|\tilde{Y}]. \qquad \square$$

The normal distribution

This is probably the most important distribution in statistics and has many special properties. A random n-vector X has the *normal* or *gaussian* distribution if its characteristic function takes the form

$$\phi_X(u) = \exp(im^{\mathrm{T}}u - \tfrac{1}{2}u^{\mathrm{T}}Qu)$$

for some n-vector m and non-negative definite matrix Q. Then $m = Ex$ and $Q = \operatorname{cov}(X)$. We write $X \sim N(m, Q)$. In the special case $m = 0$, $Q = I_n$, X is said to be *standard normal*; it follows from Proposition 1.1.5 below that the components X_i are independent $N(0, 1)$ random variables (i.e. each component is normally distributed with zero mean and unit variance).

Any collection of r.v.s is said to be *jointly normal* if the vector r.v. containing those r.v.s as components has normal distribution.

Proposition 1.1.7

(a) Linear combinations of normal r.v.s are normal.
(b) If two jointly normal r.v.s are uncorrelated they are independent.
(c) Any normal vector can be expressed as a linear transformation of a standard normal random vector.
(d) If $Y \sim N(m, Q)$ and Q is non-singular then Y has density function

$$f_Y(x) = \frac{1}{(2\pi)^{n/2}(\det(Q))^{1/2}} \exp(-\tfrac{1}{2}(x - m)^{\mathrm{T}}Q^{-1}(x - m)).$$

(e) If X is a normal n-vector then the conditional distribution of $(X_1, \ldots, X_k)$ given $(X_{k+1}, \ldots, X_n)$ is normal. Its mean is an affine function of $(X_{k+1}, \ldots, X_n)$ and its covariance is constant (does not depend on $(X_{k+1}, \ldots, X_n)$).

PROOF (a) If $X \sim N(m, Q)$ and Y is given by (1.1.11) then, by Proposition 1.1.3(d),

$$\begin{aligned}\phi_Y(u) &= e^{iu^{\mathrm{T}}b}\phi_X(G^{\mathrm{T}}u)\\ &= \exp(iu^{\mathrm{T}}b + im^{\mathrm{T}}G^{\mathrm{T}}u - \tfrac{1}{2}u^{\mathrm{T}}GQG^{\mathrm{T}}u).\end{aligned}$$

This shows that $Y \sim N(Gm + b, GQG^{\mathrm{T}})$.

(b) If X_1, X_2 are uncorrelated and $Q = \mathrm{cov}(X)$ then

$$Q = \begin{bmatrix} v_1 & 0 \\ 0 & v_2 \end{bmatrix}$$

where $v_i = \mathrm{var}(X_i)$. Thus

$$\begin{aligned}\phi_X(u) &= \exp(im^{\mathrm{T}}u - \tfrac{1}{2}v_1u_1^2 - \tfrac{1}{2}v_2u_2^2)\\ &= \phi_{X_1}(u_1)\phi_{X_2}(u_2).\end{aligned}$$

This implies that X_1, X_2 are independent.

(c) This is immediate from part (b) of Proposition 1.1.3 together with (a) above.

(d) From part (c) we can write

$$Y = GX + m$$

where X is standard normal and G is non-singular. Now if $Z \sim N(0, 1)$ (scalar standard normal) then

$$\phi_Z(u) = e^{-u^2/2}.$$

and it follows from the Fourier inversion formula that the density is

$$f_Z(z) = \frac{1}{\sqrt{(2\pi)}}e^{-z^2/2}.$$

Therefore the density function for X is

$$f_X(x) = \frac{1}{(2\pi)^{n/2}}e^{-|x|^2/2}.$$

Applying part (e) of Proposition 1.1.3 we obtain the stated density function for Y.

(e) A full proof of this fact, and general expressions for the conditional mean and covariance, are contained in the section on linear estimation theory, Section 3.1. However, let us demonstrate it for the case $n = 2$, supposing that the covariance matrix $Q = \mathrm{cov}(X)$ is non-singular. Then $X = (X_1, X_2)$ has density function $f_X(x)$ as in

Part (d) and the conditional density of X_1 given X_2 is

$$f_{X_1|X_2}(x_1; x_2) = \frac{\exp(-\frac{1}{2}(x-m)^{\mathrm{T}}Q^{-1}(x-m))}{\int_{-\infty}^{\infty} \exp(-\frac{1}{2}(x-m)^{\mathrm{T}}Q^{-1}(x-m))\,\mathrm{d}x_1}.$$

This is a one-dimensional density function in x_1 for each fixed value of x_2. Note that the denominator does not depend on x_1 and is just an x_2-dependent 'normalizing constant'; denote it by $K_1^{-1}(x_2)$. Then if we denote $Q^{-1} = R = [r_{ij}]$,

$$\begin{aligned} f_{X_1|X_2}(x_1; x_2) &= K_1(x_2)\exp(-\tfrac{1}{2}(x-m)^{\mathrm{T}}R(x-m)) \\ &= K_1(x_2)\exp(-\tfrac{1}{2}\{(x_1-m_1)^2 r_{11} \\ &\quad + 2(x_1-m_1)(x_2-m_2)r_{12} + (x_2-m_2)^2 r_{22}\}) \\ &= K_1(x_2)\exp\left(-\tfrac{1}{2}r_{11}\left\{x_1 - \left(m_1 - (x_2-m_2)\frac{r_{12}}{r_{11}}\right)\right\}^2 \right. \\ &\quad \left. + K_2(x_2)\right) \end{aligned}$$

where $K_2(x_2)$ is a term not depending on x_1. We can write the last expression as

$$K_3(x_2)\exp\left(-\frac{1}{2\tilde{\sigma}^2}(x_1 - \tilde{m}_1)^2\right)$$

where

$$\tilde{m}_1 = m_1 - \frac{r_{12}}{r_{11}}(x_2 - m_2)$$

$$\tilde{\sigma}^2 = 1/r_{11}$$

and

$$K_3(x_2) = K_1(x_2)\exp(K_2(x_2)).$$

We know that this is a density function in x_1; it is clearly the density function $N(\tilde{m}_1, \tilde{\sigma}^2)$ and the normalizing constant $K_3(x_2)$ is therefore $1/\tilde{\sigma}\sqrt{(2\pi)}$ (it actually does not depend on x_2). Thus, as claimed, the conditional variance $\tilde{\sigma}^2$ does not depend on x_2 and the conditional mean $\tilde{m}_1$ is affine in x_2. To get the coefficients explicitly, note that

$$R = Q^{-1} = \frac{1}{q_{11}q_{22} - q_{12}^2}\begin{bmatrix} q_{22} & -q_{21} \\ -q_{21} & q_{11} \end{bmatrix}$$

where $Q=[q_{ij}]$. Using the fact that $Q=\operatorname{cov}(X)$ we see that

$$\tilde{m}_1 = m_1 + \frac{\operatorname{cov}(X_1, X_2)}{\operatorname{var}(X_2)}(X_2 - m_2)$$

$$\tilde{\sigma}_1^2 = (1-\rho^2)\operatorname{var}(X_1)$$

where ρ is the correlation coefficient. These agree with the general expressions given in Section 3.1. One notes in particular that $\tilde{\sigma}_1 = 0$ if $|\rho| = 1$ which is correct because then $X_1 = \pm X_2$ with probability 1. □

1.1.2 Stochastic processes

A *stochastic process* is just a collection $\{X_t, t \in T\}$ of random variables indexed by a set T. Generally T has the connotation of *time*: if it is an interval, say $[a, b]$, then $\{X_t\}$ is a *continuous-time* process, whereas if T contains only integer values then $\{X_t\}$ is a *discrete-time* process. The most commonly encountered time sets T for discrete-time processes are the integers $\mathbb{Z} = \{\ldots -1, 0, 1, \ldots\}$ and the non-negative integers $\mathbb{Z}^+ = \{0, 1, \ldots\}$. In this book we consider only discrete-time processes: they are mathematically simpler, and from the point of view of applications we must in any case discretize at some stage for digital computer implementation. The reader can consult Davis (1977) for an introduction to stochastic system theory in continuous time.

Time series which might be modelled by discrete-time processes arise in two ways:

(a) Series which are only available in discrete form, such as economic data.
(b) Series which are produced by sampling continuous data.

In the latter case, in addition to studying the time series itself, the relation between the series and the underlying continuous data needs to be considered: for example, one can ask what constitutes an appropriate sampling rate. Such questions are however beyond the scope of this book in that they cannot meaningfully be posed without bringing in the theory of continuous-time processes.

If $T = \{1, 2, \ldots, N\}$ then the process $\{X_t\} = \{X_1, X_2, \ldots, X_N\}$ is equivalent to a random vector and its probabilistic behaviour is specified by giving the joint distribution of the N random variables involved. In principle this covers all practical cases in that any data record is necessarily finite, but conceptually it is often useful to think of a process either as having started at some time in the distant past, or as continuing indefinitely into the future, or both, in which case T will be infinite. The probabilistic behaviour is then in principle specified by the *family of finite-dimensional distributions* of the process, i.e. by giving the joint distribution of $(X_{t_1}, \ldots, X_{t_n})$ for any arbitrary times $t_1, t_2, \ldots, t_n$. We say 'in principle' because giving an infinite set of distributions is a rather unwieldy way of specifying a process; usually it will be constructed in some well-defined way from some very simple process, and then the joint distributions can be calculated, if required. However, for the theory given in this book the complete distributions will rarely be required, analysis being generally carried out only in terms of means and covariances.

In this book we shall often consider vector processes $\{X_k, k \in T\}$, where each X_k is a random d-vector. The *mean* of such a process is the sequence of vectors $\{m(k), k \in T\}$ where

$$m(k) = EX_k.$$

The *covariance function* is the $d \times d$ matrix-valued function

$$R(k, l) = \operatorname{cov}(X_k, X_l) = E(X_k - m(k))(X_l - m(l))^{\mathrm{T}} \qquad k, l \in T.$$

In the scalar case $d = 1$ we usually denote the (scalar-valued) covariance function by $r(k, l)$. Note that these functions are defined in terms of the two-dimensional distributions, i.e. they can be calculated if one knows the distributions of all pairs of random vectors X_k, X_l. From the Schwarz inequality, Proposition 1.1.2(c), the mean and covariance functions are well-defined as long as the process has finite variance, i.e.

$$E|X_k|^2 < \infty \qquad \text{for all } k.$$

Since the mean is just a deterministic function, it is often convenient to assume that the process has mean zero, or equivalently to consider the *centred* process

$$X_k^c = X_k - m(k)$$

which has zero mean and the same covariance function as X_k.

While there are no restrictions on the form of the mean $m(k)$ this is not true of the covariance function $R(k, l)$. Indeed, pick n time instants $k_1, k_2, \ldots, k_n$ and d-vectors $a_1, \ldots, a_n$ and calculate

$$E\left(\sum_{i=1}^{n} a_i^{\mathrm{T}} X_{k_i}\right)^2 = \sum_{i,j} E a_i^{\mathrm{T}} X_{k_i} X_{k_j}^{\mathrm{T}} a_j$$
$$= \sum_{i,j} a_i^{\mathrm{T}} R(k_i, k_j) a_j.$$

Since the left-hand side is non-negative, it follows that

$$\sum_{i,j} a_i^{\mathrm{T}} R(k_i, k_j) a_j \geq 0 \tag{1.1.17}$$

for all possible choices of n, $k_1, \ldots, k_n$ and $a_1, \ldots, a_n$. A function with this property is said to be *non-negative definite*. R is also *symmetric* in that

$$R(k, l) = R^{\mathrm{T}}(l, k).$$

The process X_k is *normal* if all its finite-dimensional distributions are normal. In this case the finite-dimensional distributions are completely specified by the mean and covariance function. For the covariance matrix Q of the nd-vector random variable

$$X_{t_1,\ldots,t_n}^{\mathrm{T}} = (X_{t_1}^1, \ldots, X_{t_1}^d, X_{t_2}^1, \ldots, X_{t_n}^d)$$

is

$$Q = \begin{bmatrix} R(t_1, t_1) & R(t_1, t_2) \ldots R(t_1, t_n) \\ R(t_2, t_1) & R(t_2, t_2) \ldots & \\ \vdots & \vdots & R(t_n, t_n) \end{bmatrix}$$

which is a *bona fide* covariance matrix in view of condition (1.1.17). The mean is:

$$m = \begin{bmatrix} m(t_1) \\ \vdots \\ m(t_n) \end{bmatrix}$$

Thus the distribution of $(X_{t_1}, \ldots, X_{t_n})$ is specified by the characteristic function

$$\phi_{t_1 \ldots t_n}(u) = \exp(i m^{\mathrm{T}} u - \tfrac{1}{2} u^{\mathrm{T}} Q u).$$

This shows, among other things, that to every second-order process there corresponds a normal process having the same mean and covariance function. For if $\tilde{X}_k$ is an arbitrary (not necessarily normal) second-order process with mean $m(k)$ and covariance $R(k, l)$ then the above construction gives a normal process X_k whose mean and covariance coincide with those of $\tilde{X}_k$.

Stationary processes

A process $\{X_k, k \in T\}$ is said to be *stationary* (or *strict-sense stationary*) if its distributions do not vary with time, i.e. if for any $k_0, k_1, \dots, k_n$ the distribution of the n-vector random variable $(X_{k_1}, \dots, X_{k_n})$ is the same as that of $(X_{k_1+k_0}, \dots, X_{k_n+k_0})$. This means that the origin of time is irrelevant and the joint distributions of the random variables only depend on the time intervals separating them. Taking $n = 1$ we see in particular that all X_k have the same distribution – the distribution of, say, X_0. Thus if $EX_1^2 < \infty$ then $EX_k^2 < \infty$ for all k and the process has a well-defined mean $m(k)$ and covariance function $R(k, l)$. Since all X_k have the same distribution, $m(k) = m(0)$ for all k, i.e. *the mean of a stationary process is a constant*. Similarly, for any k_0, k, l, the joint distribution of (X_k, X_l) is the same as that of (X_{k+k_0}, X_{l+k_0}), so that

$$R(k, l) = R(k + k_0, l + k_0).$$

Take $k_0 = -l$; then

$$R(k, l) = R(k - l, 0).$$

Now define

$$\tilde{R}(m) = E[X_k X_{k-m}^{\mathrm{T}}] = R(m, 0).$$

Then we see that

$$\tilde{R}(-m) = \tilde{R}^{\mathrm{T}}(m)$$

and that

$$R(k, l) = \tilde{R}(k - l). \tag{1.1.18}$$

For a stationary process the term 'covariance function' usually refers to the one-parameter function $\tilde{R}$ defined as above. In the scalar case, where the (two-parameter) covariance function is denoted $r(k, l)$, we define $\tilde{r}(m) = r(m, 0)$; then $\tilde{r}(m) = \tilde{r}(-m)$ and

$$E[X_k X_l] = \tilde{r}(|k - l|).$$

Thus the covariance between X_k and X_l depends only on their distance apart in time.

The simplest form of stationary process is a sequence $\{X_1, X_2, \dots\}$ of independent identically distributed random variables. If F denotes their common distribution function then the distribution function of the random vector $(X_{t_1}, \dots, X_{t_n})$ is given by

$$F_{t_1\ldots t_n}(a_1,\ldots,a_n) = P[X^j_{t_1} < a^j_1,\ldots,X^j_{t_n} < a^j_n, j = 1,2,\ldots,d]$$
$$= \prod_{i=1}^{n} P[X^j_{t_i} < a^j_i, j = 1,\ldots,d]$$
$$= \prod_{i=1}^{n} F(a_i).$$

Thus the finite-dimensional distributions are completely determined by F. The mean and covariance are given by

$$m(k) = EX_1$$
$$R(k,l) = \begin{cases} \operatorname{var}(X_1) & k = l \\ 0 & k \neq l \end{cases}.$$

This process is, for reasons discussed below, sometimes known as a *white-noise sequence*. It plays a central role in the theory.

A finite-variance process X_k with constant mean and whose covariance function satisfies (1.1.18) for some function $\tilde{R}$ is said to be *weakly* or *wide-sense* stationary. As above, the one-parameter function $\tilde{R}(k)$ is known as the *covariance function* of the process. Not every wide-sense stationary process is strict-sense stationary: for example, let f_1, f_2 be two different density functions satisfying

$$\int_{-\infty}^{\infty} x f_i(x)\,dx = 0, \quad \int_{-\infty}^{\infty} x^2 f_i(x)\,dx = 1 \qquad i = 1, 2$$

and suppose $X_1, X_2, \ldots$ are independent random variables such that the density function of X_i is f_0 if i is odd and f_1 if i is even. Then $EX_i = m(i) = 0$ for all i and the covariance function is

$$r(k,l) = EX_kX_l = \delta(|k-l|)$$

where

$$\delta(i) = \begin{cases} 1 & i = 0 \\ 0 & i \neq 0 \end{cases}.$$

Thus X_k is wide-sense stationary, but it is not strict-sense stationary since X_k and X_{k+1} do not have the same distribution, for any k.

A *wide-sense white-noise sequence* $X_1, X_2, \ldots$ is a wide-sense stationary process with zero mean and a covariance function of the form

$$R(k) = Q\delta(k)$$

for some non-negative definite matrix Q. This merely stipulates that the random vectors X_i have the same mean and covariance and that X_i^k and X_j^l be uncorrelated for all k, l and $i \neq j$. Q can always be factored in the form $Q = AA^{\mathrm{T}}$ where A is a $d \times m$ matrix for some $m \leq d$. If (Y_k) is an m-vector weak-sense white-noise process with covariance $I_m\delta(k)$ (I_m is the $m \times m$ identity matrix) then $X_k := AY_k$ has covariance $Q\delta(k)$ so there is no real loss of generality in taking Q to be the identity matrix, in which case the components X_i^k and X_i^l are uncorrelated *at the same time* i for $k \neq l$.

In the analysis of wide-sense stationary processes a large rôle is played by Fourier series techniques, giving rise to the so-called *spectral theory* of stationary processes. We shall make occasional but not extensive use of spectral methods in this book. To introduce the ideas let us consider first a scalar zero-mean wide sense stationary process X_k with covariance function $r(k)$. Suppose that

$$\sum_{k=-\infty}^{\infty} |r(k)| < \infty. \tag{1.1.19}$$

Then we define the *spectral density function* $\Phi(\omega)$ for $-\pi \leq \omega \leq \pi$ by

$$\Phi(\omega) = \sum_{k=-\infty}^{\infty} r(k)e^{-i\omega k}. \tag{1.1.20}$$

Since $|e^{-i\omega k}| = 1$, condition (1.1.19) ensures that the sum converges for any ω and it is easily seen that $\Phi(\omega)$ is a *continuous function* of ω. It is also *real* and *non-negative*, due respectively to the symmetry and non-negative definiteness (1.1.17) of $r(k)$. Evidently, from the definition (1.1.20), $r(k)$ is the kth coefficient in the Fourier series expansion of $\Phi(\omega)$; it can therefore be recovered from Φ by the standard formula for calculating Fourier coefficients, namely

$$r(k) = \frac{1}{2\pi}\int_{-\pi}^{\pi} \Phi(\omega)e^{i\omega k}\,\mathrm{d}\omega$$

(the integral is certainly well-defined since Φ is bounded). In particular, the variance of the process is given by

$$\operatorname{var}(X_n) = r(0) = \frac{1}{2\pi}\int_{-\pi}^{\pi} \Phi(\omega)\,\mathrm{d}\omega.$$

Note that a scalar white-noise process with variance σ^2 has spectral density $\Phi(\omega) = \sigma^2$, i.e. a constant for all ω. This is the reason for the name 'white noise', by analogy with white light which has an approximately flat frequency spectrum.

Not every wide-sense stationary process has a spectral density function but each one has a *spectral distribution* function. A general result known as Bochner's theorem asserts that if $r(k)$ is the covariance function of some wide-sense stationary process with variance $r(0) = \sigma^2$ then $r(k)$ can always be represented in the form

$$r(k) = \frac{\sigma^2}{2\pi}\int_{-\pi}^{\pi} e^{i\omega k}\,\mathrm{d}F(\omega)$$

where F is a distribution function on $(-\pi, \pi)$, i.e. a monotone increasing function with $F(-\pi) = 0$, $F(\pi) = 1$. The integral is a Stieltjes integral as described earlier. The process has a spectral density Φ precisely when the spectral distribution F is absolutely continuous, and then

$$F(\omega) = \int_{-\pi}^{\omega} \Phi(\omega')\,\mathrm{d}\omega'.$$

Thus (1.1.19) is a sufficient condition for F to be absolutely continuous. Note that, since F is non-negative and monotone, $\Phi(\omega) \geq 0$ on $(-\pi, +\pi)$.

Analogous results hold for vector processes. The spectral density function now takes values matrices over the complex field. We summarize the results in the following proposition.

Proposition 1.1.18

Let $\{X_k, k \in \mathbb{Z}\}$ be a wide-sense stationary d-vector process with covariance $R(k)$ and suppose that

$$\sum_{k=-\infty}^{+\infty} \|R(K)\| < \infty$$

(the matrix norm $\|\quad\|$ here is, say, the spectral norm; see Appendix D.2). Then $\{X_k\}$ has a spectral density function $\Phi(\omega)$ given by

$$\Phi(\omega) = \sum_{k=-\infty}^{\infty} R(k) e^{-i\omega k}.$$

Φ has the following properties: $\Phi(-\omega) = \Phi^{\mathrm{T}}(\omega)$, $\Phi(-\omega) + \Phi(\omega)$ is real and $\Phi(-\omega) + \Phi(\omega) \geq 0$ for $\omega \in (-\pi, +\pi)$. The covariance function is given in terms of the spectral density by the inversion formula

$$R(k)=\frac{1}{2\pi}\int_{-\pi}^{\pi}\Phi(\omega)e^{i\omega k}d\omega.$$

1.1.3 Convergence of stochastic sequences

On many occasions in this book we shall wish to investigate questions such as whether a given process is asymptotically stationary, whether parameter estimates converge to their true values as the length of a data record increases, and so on. We need to know something about convergence of sequences of random variables in order to formulate such questions precisely.

First let us consider a non-random sequence $\{X_k\}=X_1,X_2,\ldots$ of real numbers. We say that $\{X_k\}$ converges to X, which we denote

$$X_k\to X \quad \text{as} \quad k\to\infty$$

or

$$\lim_{k\to\infty} X_k = X$$

if for any $\varepsilon>0$ there is an integer $k(\varepsilon)$ such that $|X_k - X|<\varepsilon$ for all $k>k(\varepsilon)$, i.e. if the distance between X_k and X is eventually arbitrarily small. $\{X_k\}$ is *bounded above* (resp. *below*) if there exists a number K such that $X_k\le K$ (resp. $X_k\ge K$) for all k; it is *bounded* if it is bounded above and below. Any sequence bounded above has a *least upper bound*, denoted $\sup_k X_k$, while any sequence bounded below has a *greatest lower bound* denoted $\inf_k X_k$. If $\{X_k\}$ is *not* bounded above (resp. below) we define $\sup_k X_k=+\infty$ (resp. $\inf X_k=-\infty$). Then $\sup_k X_k$ and $\inf_k X_k$ are well defined for *any* sequence $\{X_k\}$. It is clear that $\sup_k X_k\ge\inf_k X_k$ and that X_k is bounded if and only if $-\infty<\inf X_k<\sup X_k<+\infty$. $\{X_k\}$ is *monotone increasing* (resp. *decreasing*) if $X_{k+1}\ge X_k$ (resp. $X_{k+1}\le X_k$) for all k. A monotone increasing sequence always has a limit, namely $\sup_k X_k$, if we agree that '$X_k\to+\infty$' means that for any number M there is a number $k(M)$ such that $X_k>M$ for all $n\ge k(M)$. A monotone decreasing sequence has a limit also (the limit may possibly be $-\infty$).

For an arbitrary sequence $\{X_k\}$, define

$$y_n=\sup_{k\ge n} X_k$$

$$z_n=\inf_{k\ge n} X_k$$

Then y_n is monotone decreasing and z_n is monotone increasing, since the sup and inf are being taken over progressively fewer and fewer terms. We define

$$\limsup_{k\to\infty} X_k = \lim_{n\to\infty} y_n$$

$$\liminf_{k\to\infty} X_k = \lim_{n\to\infty} z_n$$

Thus lim sup X_k and lim inf X_k are well-defined for any sequence $\{X_k\}$; it is always the case that lim sup $X_k \geq$ lim inf X_k.

The lim sup operation describes the behaviour of 'large' values of the sequence in the following way.

Proposition 1.1.9

Let $\{X_k\}$ be any sequence such that $x^* := \limsup X_k < +\infty$. Then for any $\varepsilon > 0$ the statement $X_k > x^* + \varepsilon$ is true for only a finite number of indices k whereas the statement $X_k > x^* - \varepsilon$ is true for infinitely many k.

There is an analogous characterization of lim inf X_k.

Finally, a sequence $\{X_k\}$ is a *Cauchy sequence* if $|X_n - X_m| \to 0$ as $n, m \to \infty$, i.e. if for any $\varepsilon > 0$ there exists $n(\varepsilon)$ such that $|X_n - X_m| < \varepsilon$ for all $n, m \geq n(\varepsilon)$. Note that the definition of a Cauchy sequence refers only to the elements of the sequence themselves and does not involve any possible limit points.

We can formulate the idea of convergence in two alternative but equivalent ways using the above definitions.

Proposition 1.1.10

Let $\{X_k\}$ be any sequence of real numbers. Then the following statements are equivalent:

(a) $X_k \to X$ for some finite real number X.
(b) $\{X_k\}$ is a Cauchy sequence.
(c) $-\infty < \liminf_{k\to\infty} X_k = \limsup_{k\to\infty} X_k < +\infty$.

If any of these holds then

$$\lim_{k\to\infty} X_k = \limsup_{k\to\infty} X_k = \liminf_{k\to\infty} X_k.$$

Let us now turn to convergence of sequences of random variables or, equivalently, of stochastic processes $\{X_k, k \in \mathbb{Z}^+\}$. Then we have a different sequence of real numbers for every realization of the process. The most obvious way to define convergence would be to say that $X_k \to X$ as $k \to \infty$ for every realization of $\{X_k, X\}$, in the sense described above. Note that the limit X is in general a random variable, i.e. depends on the realization of $\{X_k\}$. This is known as *sure* convergence, but is not actually a very useful concept because it can be destroyed by trivial modifications of the process. Indeed, suppose $\{X'_k\}$ is another process such that $P[X_k = X'_k] = 1$ for all n; $\{X_k\}$ and $\{X'_k\}$ are then said to be *equivalent*. $\{X_k\}$ and $\{X'_k\}$ have exactly the same joint distributions and it is unreasonable to attempt to distinguish between them, yet it is quite possible that $\{X'_k\}$ converges surely and $\{X_k\}$ does not. We therefore make the following definition: $\{X_k\}$ converges *almost surely* (a.s.) to X if there exists an equivalent process $\{X'_k\}$ and a random variable X' such that $P[X = X'] = 1$ and $\{X'_k\}$ converges surely to X'. Similarly, we say that $\{X_k\}$ is a *Cauchy sequence a.s.* if every realization of some equivalent process $\{X'_k\}$ is a Cauchy sequence. We then have the following result.

Proposition 1.1.11

A process $\{X_k\}$ converges a.s. to some random variable X if and only if $\{X_k\}$ is a Cauchy sequence a.s.

Another approach to convergence of random variables is based on the following idea. In the case of sequences $\{X_k\}$ we know that $X_k \to X$ if and only if $d(X_k, X) \to 0$ where $d(X_k, X) = |X_k - X|$ is the *distance* between X_k and X. To apply this in the stochastic case we need some scalar measure of the 'distance' between two random variables. The most common such measure, used in most chapters of this book, is the *mean square deviation* $d_2(X_k, X) = E(X_k - X)^2$. Occasionally it is useful to replace the exponent 2 by some other number $p \geq 1$ giving the pth mean deviation $d_p(X_k, X) = E|X_k - X|^p$. In general we say that $X_k \to X$ *in* p*th mean* as $k \to \infty$ if $E|X_k|^p < \infty$ for all p and $E|X_k - X|^p \to 0$ as $n \to \infty$ (this will imply that $E|X|^p < \infty$). When $p = 2$ this is usually known as *quadratic mean convergence*.

These various modes of convergence are not equivalent. The standard example to demonstrate this is as follows: let U be a random variable uniformly distributed on $[0, 1]$ (i.e. with density function $f_U(x) = 1, 0 \leq x \leq 1, f_U(x) = 0$ elsewhere). Define

$$g_k(x) = \begin{cases} k & 0 \le x \le \frac{1}{k} \\ 0 & \text{elsewhere} \end{cases}$$

and

$$X_k = g_k(U).$$

Clearly $X_k \to 0$ a.s. since $g_k(U) = 0$ for all $k > 1/U$ but $EX_k^2 = E(X_k - 0)^2 = 1$ so X_k does not converge to zero in quadratic mean. Now define for $m = 1, 2, \ldots$ and $n = 0, 1, \ldots, 2^m - 1$,

$$h_{m,n}(x) = \begin{cases} 1 & \frac{n}{2^m} \le x \le \frac{n+1}{2^m} \\ 0 & \text{elsewhere} \end{cases}$$

and arrange these functions in a single sequence $\{h_{1,0}, h_{1,1}, h_{2,0}, \ldots, h_{2,3}, h_{3,0}, \ldots\}$. Let h_k denote the kth element of this sequence and define

$$Y_k = h_k(U).$$

Since $E[h_{m,n}(U)]^2 = 2^{-m}$ it is clear that $EY_k^2 \to 0$ so that $Y_k \to 0$ in quadratic mean; but almost sure convergence does not take place since for any $U \in (0, 1)$, $\limsup Y_k = 1$, $\liminf Y_k = 0$.

The following proposition summarizes the relationship between the various convergence concepts.

Proposition 1.1.12

Let $\{X_k, k \in \mathbb{Z}^+\}$ be a stochastic process. Then

(a) $X_k \to X$ in pth mean ($p \ge 1$) for some r.v. X such that $E|X|^p < \infty$ if and only if X_k is a Cauchy sequence in pth mean, i.e. $E|X_n - X_m|^p \to 0$ as $n, m \to \infty$.

(b) $X_k \to X$ in pth mean implies that $X_k \to X$ in rth mean for any r, $1 \le r \le p$.

(c) If $X_k \to X$ in pth mean, $p \ge 1$, then there exists a subsequence X_{k_m} such that $X_{k_m} \to X$ a.s. as $m \to \infty$.

As the name implies, a *subsequence* is a sequence $\{\tilde{X}_m, m \in \mathbb{Z}^+\}$ where $\tilde{X}_m = X_{k_m}$ for some increasing sequence of indices $k_1 < k_2 < k_3 < \cdots$. In the above example, for instance, it is clear that $h_{m,0}(U) \to 0$ a.s. as $m \to \infty$ and this is a subsequence of (Y_k).

All of the above discussion extends immediately to d-vector-valued processes. In this case $X_k \to X$ a.s. if and only if $X_k^i \to X^i$ a.s. for each $i = 1, 2, \dots, d$. The definition of pth mean convergence requires no change and all propositions are valid as stated.

Finally, we shall need the following *ergodic theorem.* It was stated earlier that EX is the 'average value of X in a long sequence of trials'. This is obviously what it *ought* to be but such properties are *results* of the theory rather than being built into the definitions. Ergodic theorems are the results which establish just such connections between sample averages and expected values. The one we are going to give depends on the so-called *Borel–Cantelli lemma.* We do not give a proof of this here.

Lemma 1.1.13 (Borel–Cantelli).

Suppose $\{A_k\}$ is a sequence of events, event A_k having probability PA_k. If

$$\sum_{k=1}^{\infty} PA_k < \infty$$

then $P[A_k$ occurs for infinitely many $k] = 0$.

Alternatively, one can say that if $\sum PA_k < \infty$ then with probability one there is some integer k_0 such that A_k does not occur for any k beyond k_0. This is very useful in proving almost sure convergence, as the next lemma illustrates.

Lemma 1.1.14

Let $\{X_k, k \in \mathbb{Z}^+\}$ be a vector process such that

$$\sum_{k=1}^{\infty} E|X_k|^2 < \infty.$$

Then $X_k \to 0$ a.s.

PROOF Fix $\varepsilon > 0$ and define

$$A_k = [|X_k| > \varepsilon].$$

By the Chebyshev inequality,

$$PA_k \le \frac{1}{\varepsilon^2} E|X_k|^2.$$

Therefore $\sum PA_k < \infty$ and from our alternative formulation of the Borel-Cantelli lemma this means that with probability one, $|X_k| \leq \varepsilon$ for all k greater than some k_0. Thus $|X_k| \to 0$. □

Here now is the main ergodic theorem. Note that, unlike many of its ilk, it does not require that the process $\{X_k\}$ be stationary.

Theorem 1.1.15

Let $\{X_k, k \in \mathbb{Z}^+\}$ be a scalar finite-variance process with covariance function $r(t,s)$. Suppose that there are numbers $c > 0$, $\lambda \in (0,1)$ such that

$$|r(k,l)| \leq c\lambda^{|k-l|} \quad \text{for all } k, l \geq 0. \tag{1.1.21}$$

Then

$$\lim_{N\to\infty} \frac{1}{N} \sum_{k=1}^{N} (X_k - EX_k) = 0 \quad \text{a.s.}$$

REMARK Suppose for example that the X_k are uncorrelated random variables with the same mean μ and variance σ^2; then the condition (1.1.21) is certainly satisfied and the theorem asserts that

$$\lim_{N\to\infty} \frac{1}{N} \sum_{k=1}^{N} X_k = \mu \quad \text{a.s.,}$$

i.e. sample averages converge to the mean value. This confirms our interpretation of the expectation as the average value in a long sequence of trials.

PROOF The theorem is true as stated if it is true when $EX_k = 0$, so we shall assume that $EX_k = 0$ for all k throughout. It is easily shown that for $\lambda \in (0,1)$ there exists a number K such that for all N, M,

$$\sum_{k=N}^{M} \sum_{l=N}^{M} \lambda^{|k-l|} \leq K|M-N|. \tag{1.1.22}$$

Define

$$\bar{X}_N = \frac{1}{N} \sum_{k=1}^{N} X_k.$$

Then

$$E\bar{X}_N^2 = \frac{1}{N^2} \sum_{k=1}^{N} \sum_{l=1}^{N} E[X_k X_l]$$

$$= \frac{1}{N^2} \sum_{k=1}^{N} \sum_{l=1}^{N} r(k,l) \leq \frac{Kc}{N}$$

where we have used condition (1.1.21) together with (1.1.22). Consider the subsequence $\bar{X}_{k(N)}$ where $k(N) = N^2$. Then $\sum_{N=1}^{\infty} E\bar{X}_{k(N)}^2 \leq Kc\sum_1^{\infty} N^{-2} < \infty$ so that by Lemma 1.1.14,

$$\bar{X}_{k(N)} \to 0 \quad \text{a.s.} \quad \text{as } N \to \infty.$$

To show that the entire sequence $\bar{X}_N$ converges and not just the subsequence $\bar{X}_{k(N)}$, it suffices to show that

$$Y_n \to 0 \quad \text{a.s.} \quad \text{as } n \to \infty$$

where

$$Y_n := \max_{k(n) \leq j \leq k(n+1)} |\bar{X}_j - \bar{X}_{k(n)}|.$$

Fix n and denote temporarily $p = k(n) = n^2, q = k(n+1) = (n+1)^2$. Then

$$\begin{aligned} Y_n &= \max_{p \leq j \leq q} \left| \left(\frac{1}{j} - \frac{1}{p}\right) \sum_{l=1}^{p} X_l + \frac{1}{j} \sum_{l=p+1}^{q} X_l \right| \\ &\leq \frac{q-p}{p^2} \sum_{l=1}^{p} |X_l| + \frac{1}{p} \sum_{l=p+1}^{q} |X_l|. \end{aligned}$$

Therefore

$$\begin{aligned} Y_n^2 &\leq \frac{2(q-p)^2}{p^4} \left(\sum_{l=1}^{p} |X_l| \right)^2 + \frac{2}{p^2} \left(\sum_{l=p+1}^{q} |X_l| \right)^2 \\ &= \frac{2(q-p)^2}{p^4} \sum_{l,m=1}^{p} |X_l X_m| + \frac{2}{p^2} \sum_{l,m=p+1}^{q} |X_l X_m|. \end{aligned}$$

On taking expectations and using (1.1.21) and (1.1.22) again, we find that for some constant K_1,

$$EY_n^2 \leq \frac{K_1}{n^3}.$$

It now follows from Lemma 1.1.14 that $Y_n \to 0$ a.s. This completes the proof. □

1.2 Linear system theory

System theory concerns the qualitative properties of devices whose responses depend on inputs applied to them and on the initial values of certain internal variables. Such devices are called systems. Issues connected with selection of inputs which give rise to desirable responses, extraction of information about the values of internal variables from the response and equivalent descriptions of the system equations are of primary interest. We shall look into some of these issues, laying special emphasis on aspects relevant to the study of filtering and control problems. As far as the problems studied in this book are concerned, system theory enters most explicitly when we come to the steady-state analysis of optimal estimators and controllers. Analysis is possible when certain hypotheses are made which involve the system-theoretic concepts of controllability, observability, stabilizability and detectability. We provide a largely self-contained, but rapid, coverage of the theory surrounding these concepts.

The systems we consider are discrete time, linear time-invariant systems. They are described by the equations

$$x_{k+1} = Ax_k + Bu_k \tag{1.2.1}$$

$$y_k = Hx_k \tag{1.2.2}$$

in which, A, B and H are $n \times n, n \times m$ and $r \times m$ matrices respectively.

In these equations the r-vector y_k is the output of the system, sampled at time k. (The time scale is assumed normalized so that sampling occurs at times $k = \ldots, -1, 0, +1, \ldots$). The m-vector u_k, the input (or control) at time k, summarizes the control action applied to the system during the interval of time $t, k \leq t \leq k+1$. The n-vector x_k, the state at time k, comprises variables which, loosely speaking, sum up the effect of past inputs and other influences on future outputs. Equation (1.2.1) is often called the state equation, and (1.2.2) the observation equation.

Notice that, given any time j, and given the state x_j at time j, x_j, and the inputs $u_j, u_{j+1}, \ldots$ at times $j, j+1, \ldots$, we can solve the system equations (1.2.1) and (1.2.2) for $x_k, y_k, k > j$, and obtain

$$x_k = A^{k-j}x_j + \sum_{i=j}^{k-1} A^{k-i-1}Bu_i \tag{1.2.3}$$

$$y_k = HA^{k-j}x_j + \sum_{i=j}^{k-1} HA^{k-i-1}Bu_i \tag{1.2.4}$$

(in these expressions A raised to the zeroth power is interpreted as the identity matrix).

The state has the following property: knowledge of x_j, the state at time j, in addition to knowledge of present and future inputs, namely

$u_j, u_{j+1}, \ldots$, suffices for calculation of future outputs $y_{j+1}, y_{j+2}, \ldots$. This is clear from (1.2.4). It is in this sense that the state contains all relevant information about the past history of the system for purposes of determining future outputs.

The discrete time system with description (1.2.1), (1.2.2) is called 'linear' because x_k and y_k depend linearly on x_0 and $u_0, \ldots, u_{k-1}$. It is called 'time-invariant' for the following reason. If we set an initial state at time 0 and apply an input sequence, then the state and output, x_k and y_k, at time k, coincide with the state and output $\tilde{x}_{k+j}$ and $\tilde{y}_{k+j}$ at some subsequent time $k+j$, which would result if the same initial state, previously set at time 0, is now set at time j, and the input sequence is delayed by the time interval j. These properties are obvious from the formulae (1.2.3) and (1.2.4). So the response of the system is invariant under time shifts.

1.2.1 Controllability and observability

Controllability

We first examine conditions under which we can change the state of the system at will by suitable choice of the input sequence. Systems having this property are called 'controllable systems'.

Definition 1.2.1

The system (1.2.1), (1.2.2) is *controllable* when, given any n-vectors x_a and x_b, there exist some non-negative integer j and inputs $u_0, \ldots, u_{j-1}$ such that x_j generated by the state equation

$$\begin{aligned} x_{k+1} &= Ax_k + Bu_k, \qquad k = 0, \ldots, j-1 \\ x_0 &= x_a \end{aligned}$$

satisfies $x_j = x_b$.

Notice that the definition of controllability involves only the state equation which is itself specified by the matrices A and B. For this reason we often say '(A, B) is controllable' in place of 'the system (1.2.1), (1.2.2) is controllable'.

We remark that variants of Definition 1.2.1 appear in the literature. Many authors reserve the terminology 'controllable' for systems which can be driven from an arbitrary initial state to zero, a notion of controllability which is strictly weaker than ours. (As an example of a system (A, B) which is not controllable in our sense but is controllable 'to the zero state' take A such that $A^k = 0$ for some k and $B = 0$). We could consider too systems which can be driven from the zero state to an arbitrary terminal state. Such systems are often called *reachable* systems. Actually reachability is equivalent to controllability in the sense of Definition 1.2.1.

A simple condition expressed directly in terms of the matrices A and B of the state equation (1.2.1) is available for testing controllability. This is Kalman's rank condition test, described in the following proposition.

Proposition 1.2.2

(A, B) is controllable if and only if

$$\operatorname{rank}[B \,\vdots\, AB \,\vdots\, \ldots \,\vdots\, A^{n-1}B] = n. \tag{1.2.5}$$

The $n \times nm$ matrix $[B \,\vdots\, AB \,\vdots\, \ldots \,\vdots\, A^{n-1}B]$ is called the *controllability matrix*. Since it has n rows the rank condition can be otherwise stated as: the controllability matrix has range all of $\mathbb{R}^n$. If $m = 1$, that is the input is scalar valued, then the controllability matrix is a square matrix and the rank condition reduces to the requirement that the controllability matrix be non-singular.

The validity of the rank condition test for controllability hinges on the Cayley–Hamilton theorem. For the moment we take A to be an arbitrary $n \times n$ matrix with characteristic polynomial $\alpha_0 + \alpha_1 s + \cdots + \alpha_{n-1}s^{n-1}$. The Cayley–Hamilton theorem tells us that A 'satisfies its own characteristic equation', by which is meant

$$A^n = -\alpha_{n-1}A^{n-1} - \alpha_{n-2}A^{n-2} - \cdots - \alpha_1 A - \alpha_0 I \tag{1.2.6}$$

(I is the $n \times n$ identity matrix). A consequence of this property is that, given any non-negative integer i, A^i satisfies

$$A^i = \beta_0 I + \beta_1 A + \cdots + \beta_{n-1}A^{n-1} \quad \text{for some scalars } \beta_0, \ldots, \beta_{n-1}. \tag{1.2.7}$$

In other words, A^i is some linear combination of the matrices $I, A, \ldots, A^{n-1}$. The representation (1.2.7) is obviously possible when $i = 0, \ldots, n-1$, and also when $i = n$, from (1.2.6). That it is possible for arbitrary i is now proved by induction; suppose that, given arbitrary $j \geq 0$, (1.2.7) is true whenever $i \leq j$. Under the induction hypothesis A^j can be expressed

$$A^j = \beta_0 I + \beta_1 A + \cdots + \beta_{n-1} A^{n-1}$$

for suitable coefficients $\beta_0, \ldots, \beta_{n-1}$. Premultiplying through by A we obtain

$$A^{j+1} = \beta_0 A + \beta_1 A^2 + \cdots + \beta_{n-1} A^n. \tag{1.2.8}$$

But each of the terms on the right-hand side of (1.2.8) is expressible as a linear combination of $I, A, \ldots, A^{n-1}$ since, as we have remarked, (1.2.7) is true for $i = 0, 1, \ldots, n$. It follows that A^{j+1}, given by (1.2.8) is also a linear combination of $I, A, \ldots, A^{n-1}$. This provides the required representation of A^{j+1} and the induction is complete.

We are now ready to establish the rank condition test.

PROOF OF PROPOSITION 1.2.2 Let us write W for the controllability matrix. Suppose first that W has rank n. Let x_a and x_b be arbitrary n-vectors. Under the assumption, W has range all of $\mathbb{R}^n$ and so there exists an nm-vector ξ (which we partition as a collection of m-vectors $u_0, \ldots, u_{n-1}$, thus $\xi = \text{col}\{u_0, \ldots, u_{n-1}\}^{\dagger}$) such that

$$x_b - A^n x_a = W\xi = [B \vdots AB \vdots \ldots \vdots A^{n-1}B]\,\text{col}\{u_{n-1}, \ldots, u_0\}.$$

This equation can be written in the form

$$x_b = A^n x_a + \sum_{j=0}^{n-1} A^{n-j-1} B u_j.$$

It is clear from (1.2.3) that the input sequence $u_0, \ldots, u_{n-1}$ drives the state x_a at time 0 to x_b at time n. We have shown that (A, B) is controllable.

Next suppose that W does not have rank n. This means that the rows of W are not linearly independent and so there exists a non-zero n-vector ξ such that

$$\xi^{\mathrm{T}}[B \vdots AB \vdots \ldots \vdots A^{n-1}B] = 0$$

or, otherwise expressed,

$$\xi^{\mathrm{T}} B = \xi^{\mathrm{T}} AB = \cdots = \xi^{\mathrm{T}} A^{n-1} B = 0. \tag{1.2.9}$$

†Given an ordered collection of matrices $\{F_1, \ldots, F_q\}$, each having the same number of columns, then $\text{col}\{F_1, \ldots, F_q\}$ denotes $[F_1^{\mathrm{T}} \vdots F_2^{\mathrm{T}} \vdots \ldots \vdots F_q^{\mathrm{T}}]^{\mathrm{T}}$.

It remains to show that (A, B) is not controllable. Equations (1.2.9) imply that

$$\xi^{\mathrm{T}} A^k B = 0 \qquad \text{for } k = 0, 1, \ldots \tag{1.2.10}$$

Indeed, for arbitrary k, A^k can be expressed as

$$A^k = \beta_0 I + \cdots + \beta_{n-1} A^{n-1}$$

for suitable coefficients $\beta_0, \ldots, \beta_{n-1}$, in view of our earlier remarks on the consequences of the Cayley–Hamilton theorem. But then

$$\xi^{\mathrm{T}} A^k B = \beta_0 \xi^{\mathrm{T}} B + \beta_1 \xi^{\mathrm{T}} AB + \cdots + \beta_{n-1} \xi^{\mathrm{T}} A^{n-1} B = 0.$$

We claim that there can exist no time k and input sequence $u_0, \ldots, u_{n-1}$ which drives the system from the origin at time 0 to ξ at time k; it would certainly follow that (A, B) is not controllable. If such a time k and input sequence did exist, we would have

$$\xi = \sum_{j=0}^{k-1} A^{k-j-1} B u_j.$$

Premultiplying through this equation by ξ^{T} we obtain

$$\xi^{\mathrm{T}} \xi = \xi^{\mathrm{T}} A^{k-1} B u_0 + \cdots + \xi^{\mathrm{T}} B u_{i-1}$$

which is a contradiction since the left-hand side is non-zero, and the right-hand side is zero by (1.2.10). We have shown that (A, B) is not controllable. □

A byproduct of our proof is the fact that, if (A, B) is controllable, then we can drive the system from one state to another in at most n time steps. What is an input sequence which achieves this transfer? Let x_a, x_b be arbitrary states. One input sequence $u_0, \ldots, u_{n-1}$ which transfers x_a at time 0 to x_b at time n is provided by the formula

$$\begin{bmatrix} u_{n-1} \\ u_{n-2} \\ \vdots \\ u_0 \end{bmatrix} = W^{\mathrm{T}} (W W^{\mathrm{T}})^{-1} (x_b - A^n x_a). \tag{1.2.11}$$

(To apply the formula we need to know that (WW^{T}) is non-singular: let ξ be any non-zero n vector. Since (A, B) is controllable, $\xi^{\mathrm{T}} W \neq 0$. But then $\xi^{\mathrm{T}} W W^{\mathrm{T}} \xi = (\xi^{\mathrm{T}} W)(\xi^{\mathrm{T}} W)^{\mathrm{T}} \neq 0$ and so, certainly, $W W^{\mathrm{T}} \xi \neq 0$, i.e. WW^{T} is non-singular.) We check that if the system is at state x_a at time 0 and the input sequence $u_0, \ldots, u_{n-1}$ defined by (1.2.11) is applied, then the state at time n (see (1.2.3)) is

$$A^n x_a + \sum_{j=0}^{n-1} A^{n-j-1} B u_j = A^n x_a + W \operatorname{col}\{u_{n-1}, \ldots, u_0\}$$
$$= A^n x_a + WW^{\mathrm{T}}(WW^{\mathrm{T}})^{-1}(x_b - A^n x_a) = x_b$$

as required.

As an example of a system which is not controllable, consider one involving the state equation

$$x_{k+1} = Ax_k + Bu_k$$

in which the matrices can be partitioned as follows:

$$A = \begin{bmatrix} \overbrace{A_{11}}^{\tilde{n}} & A_{12} \\ 0 & A_{22} \end{bmatrix} \}\tilde{n}, \quad B = \begin{bmatrix} B_1 \\ 0 \end{bmatrix} \}\tilde{n} \qquad (\tilde{n} < n).$$

Notice that if x_k is partitioned compatibly with A and B, namely as $x_k = \operatorname{col}\{x_k^{(1)}, x_k^{(2)}\}$ then $x_k^{(2)}$ and $x_k^{(2)}$ satisfy

$$x_{k+1}^{(1)} = A_{11} x_k^{(1)} + A_{12} x_k^{(2)} + B_1 u_k$$
$$x_{k+1}^{(2)} = A_{22} x_k^{(2)}.$$

This system is obviously not controllable since certain components of the state (those comprising $x_k^{(2)}$), on which the control has no effect, can be split off from the system. A very useful fact is that we can always interpret controllability as arising in this way (provided we permit a suitable transformation of the state variables).

Proposition 1.2.3

Suppose that (A, B) is not controllable. Then there exists a non-singular matrix T with the following properties: if we define

$$\tilde{A} = T^{-1}AT, \quad \tilde{B} = T^{-1}B$$

then $\tilde{A}$ and $\tilde{B}$ can be partitioned

$$\tilde{A} = \begin{bmatrix} \overbrace{\tilde{A}_{11}}^{\tilde{n}} & \tilde{A}_{12} \\ 0 & \tilde{A}_{22} \end{bmatrix} \}\tilde{n}, \qquad \tilde{B} = \begin{bmatrix} \tilde{B}_1 \\ 0 \end{bmatrix} \}\tilde{n}, \qquad (\tilde{n} < n)$$

and $(\tilde{A}_{11}, \tilde{B}_1)$ is controllable.

The matrix T of the proposition provides the required transformation, for if we introduce the new state vector z_k defined by

$$z_k = T^{-1} x_k \tag{1.2.12}$$

then substitution of (1.2.12) into (1.2.1) gives

$$z_{k+1} = T^{-1}ATz_k + T^{-1}Bu_k = \tilde{A}z_k + \tilde{B}u_k.$$

PROOF Let $W = [B \vdots AB \vdots \ldots A^{n-1}B]$ and let $\tilde{n} = \text{rank}\{W\}$. $\tilde{n}$ can be interpreted as the dimension of the space spanned by the columns of W. Since the system is not controllable, $\tilde{n} < n$. It is known that linearly independent n-vectors, $v_1, \ldots, v_n$ can be chosen such that the first $\tilde{n}$ vectors in the collection span the same space as the columns of W. Now define the non-singular matrix T as

$$T = [v_1 \vdots \ldots \vdots v_n].$$

It is convenient to partition T:

$$T = [T_1 \vdots T_2]$$

where $T_1 = [v_1 \vdots \ldots \vdots v_{\tilde{n}}]$ and $T_2 = [v_{\tilde{n}+1} \vdots \ldots \vdots v_n]$.
We shall show that T has the required properties. Notice first that

$$Av_j \text{ lies in span}\{v_1, \ldots, v_{\tilde{n}}\} \quad \text{for } j = 1, \ldots, \tilde{n}. \tag{1.2.13}$$

To see this, take v_j with $1 \leq j \leq \tilde{n}$. v_j lies in the span of the columns of W so

$$v_j = [B \vdots \ldots \vdots A^{n-1}B] \begin{bmatrix} a_1 \\ --- \\ \vdots \\ --- \\ a_n \end{bmatrix}$$

for suitable m-vectors $a_1, \ldots, a_n$ (which depend on j). Then

$$Av_j = [AB \vdots \ldots \vdots A^nB] \begin{bmatrix} a_1 \\ --- \\ \vdots \\ --- \\ a_n \end{bmatrix}.$$

But, by the Cayley–Hamilton theorem,

$$A^n = -\alpha_0 I - \cdots - \alpha_{n-1}A^{n-1},$$

where the α_i are the coefficients in the characteristic polynomial of A. It follows that

$$Av_j = [B \,\vdots\, \dots \,\vdots\, A^{n-1}B] \begin{bmatrix} 0 \\ --- \\ a_1 \\ --- \\ \vdots \\ \\ a_{n-1} \end{bmatrix} - [B \,\vdots\, \dots \,\vdots\, A^{n-1}B] \begin{bmatrix} \alpha_0 a_n \\ ----- \\ \vdots \\ ----- \\ \alpha_{n-1} a_n \end{bmatrix}.$$

We have expressed Av_j as a linear combination of the columns of W, and so have confirmed (1.2.13).

Now $\tilde{A}$ is defined by

$$AT = T\tilde{A}. \tag{1.2.14}$$

Let us partition $\tilde{A}$ as

$$\tilde{A} = \begin{bmatrix} \overbrace{\tilde{A}_{11}}^{\tilde{n}} & \tilde{A}_{12} \\ \tilde{A}_{21} & \tilde{A}_{22} \end{bmatrix} \}\tilde{n}$$

Equation (1.2.14) can be written

$$A[T_1 \,\vdots\, T_2] = [T_1 \,\vdots\, T_2] \begin{bmatrix} \tilde{A}_{11} & \tilde{A}_{12} \\ \tilde{A}_{21} & \tilde{A}_{22} \end{bmatrix}$$

or

$$[AT_1 \,\vdots\, AT_2] = [T_1\tilde{A}_{11} + T_2\tilde{A}_{21} \,\vdots\, T_1\tilde{A}_{12} + T_2\tilde{A}_{22}].$$

Equating the first blocks we obtain $AT_1 = T_1\tilde{A}_{11} + T_2\tilde{A}_{21}$. Now the columns of AT_1 lie in span$\{v_1, \dots, v_{\tilde{n}}\}$ by (1.2.13). We must therefore have that $\tilde{A}_{21} = 0$, for otherwise the v_i could not be linearly independent.

Next examine $\tilde{B}$ defined by

$$B = T\tilde{B}. \tag{1.2.15}$$

Partition B as

$$B = \begin{bmatrix} \tilde{B}_1 \\ \tilde{B}_2 \end{bmatrix} \}\tilde{n}$$

From (1.2.15) we have

$$B = [T_1 \,\vdots\, T_2] \begin{bmatrix} \tilde{B}_1 \\ \tilde{B}_2 \end{bmatrix} = T_1\tilde{B}_1 + T_2\tilde{B}_2.$$

Now the columns of B coincide with the first m columns of W, and so lie in span$\{v_1,\ldots,v_{\tilde{n}}\}$. We deduce from the linear independence of $v_1,\ldots,v_n$ that $\tilde{B}_2=0$.

We have shown that $\tilde{A}$ and $\tilde{B}$ can be partitioned

$$\tilde{A}=\begin{bmatrix}\overset{\tilde{n}}{\overbrace{\tilde{A}_{11}}} & \tilde{A}_{12}\\ 0 & \tilde{A}_{22}\end{bmatrix}\}\tilde{n}, \quad \tilde{B}=\begin{bmatrix}\tilde{B}_1\\ 0\end{bmatrix}\}\tilde{n}.$$

Finally we show that $(\tilde{A}_{11},\tilde{B}_1)$ is controllable. Since premultiplication by a non-singular matrix does not affect the rank of a matrix

$$\begin{aligned}\tilde{n}&=\operatorname{rank}[B\,\vdots\,AB\,\vdots\ldots\vdots\,A^{n-1}B]\\ &=\operatorname{rank}T^{-1}[B\,\vdots\,AB\,\vdots\ldots\vdots\,A^{n-1}B]\\ &=\operatorname{rank}[(T^{-1}B)\,\vdots\,(T^{-1}AT)(T^{-1}B)\,\vdots\ldots\vdots\,(T^{-1}AT)^{n-1}(T^{-1}B)]\\ &=\operatorname{rank}[\tilde{B}\,\vdots\,\tilde{A}\tilde{B}\,\vdots\ldots\vdots\,\tilde{A}^{n-1}\tilde{B}]\\ &=\operatorname{rank}\left[\begin{bmatrix}\tilde{B}_1\\0\end{bmatrix}\vdots\begin{bmatrix}\tilde{A}_{11}&\tilde{A}_{12}\\0&\tilde{A}_{22}\end{bmatrix}\begin{bmatrix}\tilde{B}_1\\0\end{bmatrix}\vdots\cdots\vdots\begin{bmatrix}\tilde{A}_{11}&\tilde{A}_{12}\\0&\tilde{A}_{22}\end{bmatrix}^{n-1}\begin{bmatrix}\tilde{B}_1\\0\end{bmatrix}\right]\\ &=\operatorname{rank}[\tilde{B}_1\,\vdots\,\tilde{A}_{11}\tilde{B}_1\,\vdots\ldots\vdots\,\tilde{A}_{11}^{n-1}\tilde{B}_1].\end{aligned}$$

But, in consequence of the Cayley–Hamilton theorem, the columns of $\tilde{A}_{11}^{\tilde{n}}\tilde{B}_1,\ldots,\tilde{A}_{11}^{n}\tilde{B}_1$ are expressible as linear combinations of the columns of $\tilde{B}_1,\ldots,A^{\tilde{n}-1}\tilde{B}_1$ (cf. the remarks following Proposition 1.2.2), and so if the blocks $A_{11}^{\tilde{n}}B_1,\ldots,A_{11}^{n-1}B_1$ are dropped from the matrix $[B_1\,\vdots\ldots\vdots\,A_{11}^{n}B_1]$, the rank of the matrix is unaffected. We conclude that

$$\operatorname{rank}[\tilde{B}_1\,\vdots\,\tilde{A}_{11}\tilde{B}_1\,\vdots\ldots\vdots\,\tilde{A}_{11}^{\tilde{n}-1}\tilde{B}_1]=\tilde{n}.$$

So $(\tilde{A}_{11},\tilde{B}_1)$ is controllable. This completes the proof of Proposition 1.2.3. □

An alternative to the Kalman rank condition test for controllability is due to Hautus, and is described in the following Proposition.

Proposition 1.2.4

A necessary and sufficient condition that (A, B) be controllable is

$$\operatorname{rank}[sI-A\,\vdots\,B]=n \tag{1.2.16}$$

for all eigenvalues s of A.

Testing condition (1.2.16) has the advantage that it avoids computation of powers of the matrix A, but requires knowledge of its eigenvalues. The significance of Proposition 1.2.3 in our treatment of linear systems theory is that it will illuminate the relationship between controllability and another important system theoretic property, 'stabilizability'.

Notice that we could replace the condition by the requirement that (1.2.16) hold for all complex numbers s, and not merely the eigenvalues of A since, if s is not an eigenvalue then sI-A has rank n and so (1.2.16) is automatically satisfied.

PROOF We first prove necessity of the condition. Let us assume that rank$[s_0 I - A \vdots B] < n$ for some eigenvalue s_0 of A. We must show that (A, B) is not controllable. In view of the assumption, there exists a non-zero n-vector ξ (possibly complex) such that

$$\xi^{\mathrm{T}}[s_0 I - A \vdots B] = 0.$$

This implies that $\xi^{\mathrm{T}} A = s_0 \xi^{\mathrm{T}}$ and $\xi^{\mathrm{T}} B = 0$. But then

$$\xi^{\mathrm{T}}[B \vdots AB \vdots \ldots \vdots A^{n-1}B] = [\xi^{\mathrm{T}} B \vdots s_0 \xi^{\mathrm{T}} B \vdots \ldots \vdots s_0^{n-1} \xi^{\mathrm{T}} B] = 0.$$

We suppose, of course, that A and B are real. It follows that $(\mathrm{Re}\,\xi)^{\mathrm{T}}[B \vdots AB \vdots \ldots \vdots A^{n-1}B] = (\mathrm{Im}\,\xi)^{\mathrm{T}}[B \vdots AB \vdots \ldots A^{n-1}B] = 0$. Since either $\mathrm{Re}\,\xi$ or $\mathrm{Im}\,\xi$ is non-zero, we conclude that the controllability matrix does not have linearly independent rows, i.e. (A, B) is not controllable.

And now for sufficiency. Let us assume that the system is not controllable. We must show that condition (1.2.16) fails for some s. Since (A, B) is not controllable we know (Proposition 1.2.3) that a non-singular matrix T exists such that, if we define $\tilde{A} = T^{-1}AT$ and $\tilde{B} = T^{-1}B$ then $\tilde{A}$ and $\tilde{B}$ can be partitioned

$$\tilde{A} = \begin{bmatrix} \overbrace{\tilde{A}_{11}}^{\tilde{n}} & \tilde{A}_{12} \\ 0 & \tilde{A}_{22} \end{bmatrix}\}\tilde{n}, \quad \tilde{B} = \begin{bmatrix} \tilde{B}_1 \\ 0 \end{bmatrix}\}\tilde{n} \qquad (\tilde{n} < n).$$

Let s_0 be an eigenvalue of $\tilde{A}_{22}$ and ξ^{T} be a corresponding (possibly complex) left eigenvector, that is ξ^{T} is a non-zero row vector which satisfies

$$\xi^{\mathrm{T}} \tilde{A}_{22} = s_0 \xi^{\mathrm{T}}. \tag{1.2.17}$$

We shall show that $[s_0 I - A \vdots B]$ has rank less than n. It suffices to show that the matrix

$$T^{-1}[s_0 I - A \,\vdots\, B]\begin{bmatrix} T & 0 \\ 0 & I \end{bmatrix}$$

has rank less than n since pre- and postmultiplication by square nonsingular matrices amount to performing simple row and column operations on the original matrix, and such operations leave the rank unaltered. But

$$\begin{aligned} &[0 \,\vdots\, \xi^{\mathrm{T}}]T^{-1}[s_0 I - A \,\vdots\, B]\begin{bmatrix} T & 0 \\ 0 & I \end{bmatrix} \\ &= [0 \,\vdots\, \xi^{\mathrm{T}}][s_0 I - T^{-1}AT \,\vdots\, T^{-1}B] \\ &= [0 \,\vdots\, \xi^{\mathrm{T}}]\left[\begin{bmatrix} s_0 I & 0 \\ 0 & s_0 I \end{bmatrix} - \begin{bmatrix} \tilde{A}_{11} & \tilde{A}_{12} \\ 0 & \tilde{A}_{22} \end{bmatrix} \vdots \begin{bmatrix} \tilde{B}_1 \\ 0 \end{bmatrix}\right] \\ &= [0 \,\vdots\, s_0 \xi^{\mathrm{T}} - \xi^{\mathrm{T}} \tilde{A}_{22} \,\vdots\, 0] = 0 \end{aligned}$$

by (1.2.17). This means that the rows of the matrix cannot be linearly independent and so its rank must be less than n. □

Observability

Given a linear system governed by the equations (1.2.1) and (1.2.2):

$$x_{k+1} = Ax_k + Bu_k$$
$$y_k = Hx_k$$

it is natural to ask: if we know the inputs to the system for all time, and outputs up to a certain time j, can we predict the outputs y_k for time $k > j$? Clearly the answer is yes if the initial state x_0 is known, for then we can solve the state equation for the state and obtain the output, for all time, from the output equation. The property of 'observability' concerns our ability to determine x_0 from the data. We can limit attention to the situation in which the input sequence is zero since the effect of the input is, by linearity, simply to add a known quantity to the output; this can be subtracted off and we are back to the input-free case.

Definition 1.2.5

Let $y_k(x_0)$, $k = 0, 1, \ldots$ be the solution to the system equations (1.2.1) and (1.2.2) for initial state x_0 and zero inputs. The system (1.2.1) and

(1.2.2) is *observable* when, for arbitrary x_0, there exists some time $k \geq 0$ such that x_0 can be determined as a function of $y_0(x_0)$, $y_1(x_0), \ldots, y_k(x_0)$.

Since only zero inputs are considered, the input matrix B is immaterial to the definition of observability. Observability is determined then by the nature of the matrices A and H; for this reason we often say '(H, A) is observable' in place of 'the system (1.2.1), (1.2.2) is observable'.

As with controllability, Kalman has provided a simple rank condition test for observability.

Proposition 1.2.6

(H, A) is observable if and only if

$$\operatorname{rank}\begin{bmatrix} H \\ HA \\ \vdots \\ HA^{n-1} \end{bmatrix} = n. \tag{1.2.18}$$

The matrix in (1.2.18) is called the *observability matrix*. Notice that since it is an $nm \times n$ matrix, the rank condition means that it has linearly independent columns or, in other words, the null space of the observability matrix comprises just the zero vector. In the event that the output is scalar, H is a row-vector and the observability matrix is square; here the rank condition means that the observability matrix is non-singular.

PROOF Observability of the system is equivalent to the property: if the output y_k to the system

$$x_{k+1} = Ax_k; \quad y_k = Hx_k$$

is zero for all k then x_0 must be zero. Observability obviously implies this property. On the other hand, if the system is not observable then there exist distinct initial states $x_0, \bar{x}_0$ which give rise to a sequence of states x_k, $\bar{x}_k$, $i = 1, 2, \ldots$, and an identical output sequence y_k, $k = 0, 1, \ldots$ By linearity, the output resulting from the initial state $x_0 - \bar{x}_0$ is zero; since $(x_0 - \bar{x}_0)$ is non-zero (x_0 and $\bar{x}_0$ are distinct, remember) the property above does not hold. This establishes its equivalence with observability.

If the initial condition is x_0, the corresponding output is

$$y_0 = Hx_0, \quad y_1 = HAx_0, \quad y_2 = HA^2x_0, \ldots$$

It follows that the system is observable if and only if

$$HA^k x_0 = 0 \quad \text{for all } k \text{ implies } x_0 = 0. \tag{1.2.19}$$

Now the rank condition (1.2.18) means that

$$\begin{bmatrix} H \\ HA \\ \vdots \\ HA^{n-1} \end{bmatrix} x_0 = 0 \quad \text{implies } x_0 = 0$$

or

$$Hx_0 = HAx_0 = \cdots = HA^{n-1}x_0 = 0 \quad \text{implies } x_0 = 0. \tag{1.2.20}$$

Clearly (1.2.19) implies (1.2.20). On the other hand for arbitrary k,

$$A^k = \beta_0 I + \cdots + \beta_{n-1} A^{n-1}$$

for some coefficients $\beta_0, \ldots, \beta_{n-1}$ in consequence of the Cayley–Hamilton theorem (see the remarks following Proposition 1.22). It follows that if (1.2.20) is true then

$$HA^k x_0 = \beta_0 Hx_0 + \beta_1 HAx_0 + \cdots + \beta_{n-1} HA^{n-1} x_0 = 0.$$

So (1.2.19) is true. We have shown that (H, A) is observable if and only if (1.2.18) is true. □

Actually we have proved a little more than is stated in the proposition: if (H, A) is observable then the initial state x_0 for the system $x_{k+1} = Ax_k$, $y_k = Hx_k$ can always be determined from the outputs $y_0, \ldots, y_{n-1}$ up to time $n-1$. An explicit formula for x_0 is

$$x_0 = (M^T M)^{-1} M^T \operatorname{col}\{y_0, \ldots, y_{n-1}\}$$

in which M is the observability matrix. (The matrix $(M^T M)$ is non-singular, and so the formula makes sense, since M has linearly independent columns.)

It is customary to describe two properties concerning matrices as being 'dual' when one property is equivalent to the other following transposition of matrices. The rank condition tests for controllability and observability tell us that controllability and observability are dual properties in this sense. In fact, (A, B) *is controllable if and only if*

$(B^{\mathrm{T}}, A^{\mathrm{T}})$ *is observable.* Controllability of (A, B) is equivalent to the condition

$$\operatorname{rank}[B \,\vdots\, AB \,\vdots\, \ldots \,\vdots\, A^{n-1}B] = n.$$

But rank is unaffected by transposition, so this is equivalent to

$$\operatorname{rank}\begin{bmatrix} B^{\mathrm{T}} \\ B^{\mathrm{T}}A^{\mathrm{T}} \\ \vdots \\ B^{\mathrm{T}}(A^{\mathrm{T}})^{n-1} \end{bmatrix} = n$$

which is precisely the rank condition test for observability of $(B^{\mathrm{T}}, A^{\mathrm{T}})$. Appealing to the duality of controllability and observability is a valuable, labour-saving device: it renders results on observability merely adjuncts to results on controllability, and vice versa. For example we recall (Proposition 1.2.3) that a system which is not controllable is one from which we can split off certain state components which are unaffected by the input (provided a suitable transformation of the state is first carried out); we deduce from the duality of controllability and observability that, if a system is not observable then, after the state is suitably transformed, state components can be split off which have no effect on the output, whatever the initial state. More precisely expressed we have the following.

Proposition 1.2.7

Suppose that (H, A) is not observable. Then there exists a non-singular matrix S with the following properties: if we define

$$\tilde{A} = S^{-1}AS \quad \text{and} \quad \tilde{H} = HS$$

we have that $\tilde{A}$ and $\tilde{H}$ can be partitioned

$$\tilde{A} = \begin{bmatrix} \overbrace{\tilde{A}_{11}}^{\tilde{n}} & 0 \\ \tilde{A}_{21} & \tilde{A}_{22} \end{bmatrix}\}\tilde{n}, \quad \tilde{H} = [\overbrace{\tilde{H}_{11}}^{\tilde{n}} \,\vdots\, 0] \qquad (\tilde{n} < n)$$

and $(\tilde{H}_{11}, \tilde{A}_{11})$ is observable.

PROOF If (H, A) is not observable then, by duality, $(A^{\mathrm{T}}, H^{\mathrm{T}})$ is not controllable. So there exists a matrix T with the properties described in Proposition 1.2.3. It is easy to see that the matrix $S = T^{\mathrm{T}}$ has the desired properties. □

Notice that if we transform the state:

$$z_k = S^{-1}x_k$$

then the system equations (with zero input)

$$x_{k+1} = Ax_k, \quad y_k = Hx_k$$

become

$$z_{k+1}(= S^{-1}ASz_k) = \tilde{A}z_k$$
$$y_k(= HTz_k) = \tilde{H}z_k.$$

If z_k is now partitioned as $z_k = \mathrm{col}\{z_k^{(1)}, z_k^{(2)}\}$ compatibly with the partitioning of $\tilde{A}$ and $\tilde{H}$, we have that $z_k^{(1)}$ satisfies

$$z_{k+1}^{(1)} = \tilde{A}_{11}z_k^{(1)}$$
$$y_k = \tilde{H}_{11}z_k^{(1)}.$$

We see that there exist components of the transformed state (those comprising the vector $z_k^{(2)}$) which have no effect on the output.

Likewise, the Hautus test for controllability (see Proposition 1.2.4) translates immediately into a test for observability, via duality.

Proposition 1.2.8

(H, A) is observable if and only if

$$\mathrm{rank}\begin{bmatrix} sI - A \\ \hdashline H \end{bmatrix} = n$$

for all eigenvalues s of A.

1.2.2 State feedback

Suppose that the inputs to a system with state equation

$$x_{k+1} = Ax_k + Bu_k \tag{1.2.21}$$

are chosen according to a feedback control law which specifies the input at time k as a linear function of the state at time k:

$$u_k = Kx_k. \tag{1.2.22}$$

Then the state of the system is governed by the equations obtained by substituting (1.2.22) into (1.2.21), namely

$$x_{k+1} = (A + BK)x_k.$$

Many significant qualitative properties of the state sequence $x_0, x_1, \ldots$ resulting from application of a feedback control law are expressible in terms of the eigenvalues of the 'closed-loop system matrix' $A+BK$: stability, response decay rates, frequencies of natural modes, for example. It is of interest then to know when the eigenvalues of $A+BK$ can be moved to arbitrary locations by suitable choice of the feedback matrix K. (Of course we must limit attention to eigenvalue locations which occur in complex conjugate pairs; the characteristic polynomial of $A+BK$ has real coefficients and consequently the eigenvalues must occur in complex conjugate pairs, whatever K.) Pairs of matrices (A, B) defining systems for which this is possible are called pole-assignable. The terminology reflects the fact that, in much of the control engineering literature, eigenvalues are called 'poles'.

Definition 1.2.9

(A, B) is *pole-assignable* when, given any nth-degree monic polynomial $p(s) = \alpha_0 + \alpha_1 s + \cdots + \alpha_{n-1}s^{n-1} + s^n$ (with real coefficients), there exists a (real) matrix K such that $(A+BK)$ has characteristic polynomial $p(s)$.

We can expect the presence of some relationship between the notions of pole-assignability and controllability since they are both concerned with the influence of inputs to a system on the resulting state sequence. It is one of the most striking results in linear system theory that the two notions are in fact equivalent.

Theorem 1.2.10

A necessary and sufficient condition for (A, B) to be pole-assignable is that (A, B) be controllable.

PROOF OF NECESSITY Suppose that (A, B) is not controllable. By Proposition 1.2.3, there exists a non-singular matrix T such that, taking $\tilde{A} = T^{-1}AT$, $\tilde{B} = T^{-1}B$, we have

$$\tilde{A} = \begin{bmatrix} \overset{\tilde{n}}{\tilde{A}_{11}} & \tilde{A}_{12} \\ 0 & \tilde{A}_{22} \end{bmatrix}\}\tilde{n} \qquad \tilde{B} = \begin{bmatrix} \tilde{B}_1 \\ 0 \end{bmatrix}\}\tilde{n} \qquad (\tilde{n} < n).$$

Let K be arbitrary. We partition K and T compatibly with the partitioning of A and B

$$T = \begin{bmatrix} T_{11} & T_{12} \\ T_{21} & T_{22} \end{bmatrix}, \quad K = [K_1 \mid K_2].$$

For any complex number s, we have

$$\det[sI - A - BK] = \det\{T^{-1}[sI - A - BK]T\}$$

(since $\det S_1 S_2 = \det S_1 \det S_2$ for square matrices S_1, S_2)

$$\begin{aligned}
&= \det[sI - \tilde{A} - \tilde{B}KT] \\
&= \det\begin{bmatrix} sI - \tilde{A}_{11} - \tilde{B}_1(K_1T_{11} + K_2T_{21}) & -\tilde{A}_{12} - \tilde{B}_1(K_1T_{12} + K_2T_{22}) \\ 0 & sI - \tilde{A}_{22} \end{bmatrix} \\
&= \det(sI - \tilde{A}_{11} - \tilde{B}_1(K_1T_{11} + K_2T_{21}))\det(sI - \tilde{A}_{22})
\end{aligned}$$

(by the properties of the determinants of block matrices). We see that the factor $\det(sI - \tilde{A}_{22})$ cannot be removed from the characteristic polynomial of $(A + BK)$ by choice of K, and so (A, B) is not pole-assignable. Our conclusions, otherwise expressed, are that pole-assignability implies controllability.

PROOF OF SUFFICIENCY Sufficiency of the controllability condition is rather more difficult to establish and it is convenient to break the proof down into a number of steps.

Step 1 We show that (A, B) is pole-assignable when the input is scalar valued (i.e. $B = b$, an n-vector) and when A and b take the special forms

$$A = \begin{bmatrix} 0 & 1 & \ddots & \bigcirc \\ & \bigcirc & & 1 \\ -a_0 & \cdots & & -a_{n-1} \end{bmatrix}, \quad b = \begin{bmatrix} 0 \\ \vdots \\ 0 \\ 1 \end{bmatrix}.$$

In this case, for any row vector $q^{\mathrm{T}} = [q_1 \cdots q_n]$

$$\begin{aligned}
A + bq^{\mathrm{T}} &= \begin{bmatrix} 0 & 1 & \ddots & \bigcirc \\ & \bigcirc & & 1 \\ -a_0 & \cdots & & -a_{n-1} \end{bmatrix} + \begin{bmatrix} 0 \\ \vdots \\ 0 \\ 1 \end{bmatrix}[q_1 \cdots q_n] \\
&= \begin{bmatrix} 0 & 1 & \ddots & \bigcirc \\ & \bigcirc & & 1 \\ -(a_0 - q_1) & \cdots & & -(a_{n-1} - q_n) \end{bmatrix}.
\end{aligned}$$

Now this last matrix is a matrix in 'companion form'; it is known that coefficients of its characteristic polynomial $p(s)$ can be read off the bottom line: $p(s) = s^n + (a_{n-1} - q_n)s^{n-1} + \cdots + (a_0 - q_1)$. We see that the coefficients of the characteristic polynomial of $A + bq^{\mathrm{T}}$ can be arbitrarily assigned through choice of q^{T}.

Step 2 Suppose now that A is a general $n \times n$ matrix and b an n-vector, such that (A, b) is controllable. By introducing a state transformation which brings us back to step 1, we show that in this case, also, (A, b) is pole-assignable.

Let $p(s) = \alpha_0 + \alpha_1 s + \cdots + \alpha_{n-1}s^{n-1} + s^n$ be the characteristic polynomial of A. Consider the vectors

$$b, Ab, A^2b, \ldots, A^{n-1}b.$$

These vectors are linearly independent since (A, b) is controllable. The following linear combinations of these vectors are also linearly independent:

$$\begin{aligned} e_1 &= A^{n-1}b + \alpha_{n-1}A^{n-2}b + \cdots + \alpha_1 b, \\ e_2 &= A^{n-2}b + \alpha_{n-1}A^{n-3}b + \cdots + \alpha_2 b, \\ &\vdots \\ e_n &= b. \end{aligned} \tag{1.2.23}$$

From (1.2.23) we deduce that

$$e_k = Ae_{k+1} + \alpha_k e_n$$

and so

$$Ae_{k+1} = e_k - \alpha_k e_n, \qquad k = 1, \ldots, n-1. \tag{1.2.24}$$

Also from (1.2.23), and the Cayley–Hamilton theorem,

$$Ae_1 = (A^n + \alpha_{n-1}A^{n-1} + \cdots + \alpha_1 A)e_n = -\alpha_0 e_n. \tag{1.2.25}$$

Equations (1.2.24) and (1.2.25) can be organized as follows:

$$A[e_1 \vdots \cdots \vdots e_n] = [e_1 \vdots \cdots \vdots e_n] \begin{bmatrix} 0 & 1 & \ddots & \bigcirc \\ & & \bigcirc & 1 \\ -\alpha_0 & \cdots & & -\alpha_{n-1} \end{bmatrix}.$$

From the last equation in (1.2.23) we have

$$b = [e_1 \vdots \cdots \vdots e_n] \begin{bmatrix} 0 \\ \vdots \\ 0 \\ 1 \end{bmatrix}.$$

Now take the non-singular matrix T to be

$$T = [e_1 \vdots \ldots \vdots e_n].$$

We have shown that, if $\tilde{A} = T^{-1}AT$, $\tilde{b} = T^{-1}b$, then

$$\tilde{A} = \begin{bmatrix} 0 & 1 & \ddots & \bigcirc \\ & \bigcirc & & 1 \\ -\alpha_0 & \ldots & & -\alpha_{n-1} \end{bmatrix}, \quad \tilde{b} = \begin{bmatrix} 0 \\ \vdots \\ 0 \\ 1 \end{bmatrix}. \tag{1.2.26}$$

Now consider the characteristic polynomial of the closed-loop system matrix when the feedback control law is $u_k = q^{\mathrm{T}}T^{-1}x_k$ for some n-vector q. This is

$$\begin{aligned} \det(sI - A - bq^{\mathrm{T}}T^{-1}) &= \det\{T^{-1}(sI - A - bq^{\mathrm{T}}T^{-1})T\} \\ &= \det(sI - \tilde{A} - \tilde{b}q^{\mathrm{T}}). \end{aligned}$$

The coefficients of this characteristic polynomial can be arbitrarily assigned through choice of q by (1.2.26) and by the results of step 1.

Step 3 Sufficiency of the controllability condition has been proved when the inputs are scalar-valued. We now prove a lemma which reduces the general vector inputs case to that of scalar inputs.

Lemma 1.2.11

If (A, B) is controllable, there exists an $m \times n$ matrix K and an m-vector v such that $(A + BK, Bv)$ is controllable.

First of all we prove the lemma. Suppose that (A, B) is controllable. Choose any vector v such that $Bv \neq 0$. (Such a v exists since (A, B) is controllable and therefore $B \neq 0$.) We first show that an input sequence $u_0, \ldots, u_{n-2}$ and a state sequence $x_0, \ldots, x_{n-1}$ can be chosen such that

$$\begin{aligned} x_{k+1} &= Ax_k + Bu_k \qquad \text{for } k = 0, 1, \ldots, n-2, \\ x_0 &= Bv \end{aligned} \tag{1.2.27}$$

and $x_0, \ldots, x_{n-1}$ are linearly independent. We claim that the choice can always be made as follows: x_0 ($= Bv$) is given, we choose any u_0 such that $x_1 = Ax_0 + Bu_0$ and x_1, x_0 are linearly independent, we choose any u_1 such that $x_2 = Ax_1 + Bu_1$ and x_2, x_1, x_0 are linearly independent, and so on all the way up to u_{n-2}, x_{n-1}. We argue by contradiction; if this were not possible then, for some $k < n - 1$, we would have

$$Ax_k + Bu \in \text{span}\{x_0, \ldots, x_k\} \qquad \text{for all } u. \tag{1.2.28}$$

By considering the case when $u = 0$ we see that

$$Ax_k \in \text{span}\{x_0, \ldots, x_k\}. \tag{1.2.29}$$

But then (1.2.28) implies

$$Bu \in \text{span}\{x_0, \ldots, x_k\} \qquad \text{for all } u. \tag{1.2.30}$$

Since $Ax_j = x_{j+1} - Bu_j, j = k, k-1, \ldots, 0$, we see from (1.2.29) and (1.2.30) that

$$Ax_j \in \text{span}\{x_0, \ldots, x_k\} \qquad \text{for } j = k, k-1, \ldots, 0. \tag{1.2.31}$$

Let ξ be a non-zero n-vector orthogonal to $x_0, \ldots, x_k$ (such a vector exists since $k < n - 1$). In view of (1.2.30) and (1.2.31),

$$\xi^{\mathrm{T}}[B \vdots AB \vdots \ldots \vdots A^{n-1}B] = 0.$$

This contradicts the controllability of (A, B). It follows that $x_0, \ldots, x_{n-1}, u_0, \ldots, u_{n-2}$ can be chosen to satisfy (1.2.27).

Now define K by

$$K = [u_0 \vdots u_1 \vdots \ldots \vdots u_{n-1}][x_0 \vdots x_1 \vdots \ldots \vdots x_{n-1}]^{-1}$$

in which u_{n-1} is any m-vector (the matrix inverse exists since the x_i are linearly independent). Clearly

$$K[x_0 \vdots \ldots \vdots x_{n-1}] = [u_0 \vdots \ldots \vdots u_{n-1}]$$

and so $Kx_k = u_k$ for $k = 0, \ldots, n-1$. It follows now from (1.2.27) that

$$x_{k+1} = Ax_k + BKx_k, \quad i = 0, \ldots, n-2,$$
$$x_0 = Bv.$$

Solving these equations for the x_i we obtain

$$x_k = (A + BK)^k(Bv), \quad k = 0, \ldots, n-1.$$

Since the x_k are linearly independent we conclude that the matrix

$$[Bv \vdots (A + BK)Bv \vdots \ldots \vdots (A + BK)^{n-1}Bv]$$

has rank n. But this is the controllability matrix for $(A + BK, Bv)$; we have found K and v such that $(A + BK, Bv)$ is controllable. This proves the lemma.

We can now conclude proof of the theorem. Let (A, B) be controllable. By the lemma there exists K and v such that $(A + BK, Bv)$ is controllable. Let $p(s)$ be an arbitrary monic polynomial. By the result of step 2, q can be chosen so that

$$\begin{aligned} p(s) &= \det[sI - (A + BK) - Bvq^{\mathrm{T}}] \\ &= \det[sI - (A + B[K + vq^{\mathrm{T}}])]. \end{aligned}$$

We see that the feedback control law $u_k = (K + vq^{\mathrm{T}})x_k$ yields a closed-loop system matrix with characteristic polynomial the arbitrary monic polynomial $p(s)$; (A, B) is therefore pole-assignable. □

Let us recall that, when the feedback control law

$$u_k = Kx_k,$$

specified by the matrix K, is applied to a linear system with state equation

$$x_{k+1} = Ax_k + Bu_k,$$

there results a closed-loop state equation

$$x_{k+1} = (A + BK)x_k, \tag{1.2.32}$$

and the concept of pole-assignability arises when we study whether K can be chosen so that the closed-loop state equation (1.2.32) has good characteristics. Of course, our interpretation of 'good' will depend on the application at hand. But often the components of the state represent deviations of certain variables from desired values, in which case a minimum objective in selection of the feedback control law is that the deviations diminish as time increases. This objective is achieved if the matrix $A + BK$ is stable, in the sense that its eigenvalues lie in the open unit disc†, for then the state x_i, generated by (1.2.32), decays to zero as i tends to infinity. Systems for which we can arrange that $A + BK$ is stable therefore deserve special attention: they are called 'stabilizable'.

Definition 1.2.12

(A, B) is said to be stabilizable when there exists an $m \times n$ matrix K such that the eigenvalues of $A + BK$ are contained in the open unit disc.

† The 'open unit disc' referred to here is the open subset $\{\zeta : |\zeta| < 1\}$ of the complex plane.

It is often difficult to test directly whether matrices (A, B) defining a particular linear system satisfy the conditions in the definition, since determination of a suitable matrix K is involved. Fortunately a simpler test is available. This is a variant on the Hautus controllability test (see Proposition 1.2.4). It is expressed directly in terms of the matrices A and B but does, admittedly, suffer from the disadvantage that extraction of the eigenvalues of A is involved.

Proposition 1.2.13

A necessary and sufficient condition that (A, B) be stabilizable is

$$\text{rank}\,[(sI - A) \vdots B] = n \quad \text{for all eigenvalues of } A \text{ outside the open unit disc.} \tag{1.2.33}$$

Comparison with the Hautus controllability test confirms that stabilizability is a weaker property than controllability; indeed controllability requires the rank condition in (1.2.33) to hold for *all* eigenvalues of A rather than merely those outside the open unit disc, as here.

It is clear from the definition of stabilizability that (A, B) is always stabilizable if A is a stable matrix (for then stabilization is achieved for zero input). At the other extreme, when all the eigenvectors of A lie outside the open unit disc then stabilizability and controllability are equivalent; this follows from Propositions 1.2.4 and 1.2.13.

PROOF To prove necessity, let us suppose that the rank condition (1.2.33) is violated: this means that there is some eigenvalue s_0 of A lying outside the open unit disc, and a (possibly complex) non-zero n-vector ξ such that

$$\xi^{\mathrm{T}}[s_0 I - A \vdots B] = 0.$$

This condition can be written

$$\xi^{\mathrm{T}} A = s_0 \xi^{\mathrm{T}} \quad \text{and} \quad \xi^{\mathrm{T}} B = 0.$$

It follows that, for any $m \times n$ matrix K, $A + BK$ has an eigenvalue outside the open unit disc since

$$\xi^{\mathrm{T}}[s_0 I - (A + BK)] = s_0 \xi^{\mathrm{T}} - \xi^{\mathrm{T}} A - \xi^{\mathrm{T}} BK = 0.$$

So (A, B) cannot be stabilizable, and thus the condition is necessary.

Let us now assume that the condition (1.2.33) is satisfied. Let Σ^-, Σ^+ denote the collection of eigenvalues of A lying inside and outside the open unit disc respectively. A basic result in matrix theory (see Wilkinson, 1965, p. 486) tells us that a real, non-singular matrix T exists such that

$$T^{-1}AT = \tilde{A} \tag{1.2.34}$$

where $\tilde{A}$ is a matrix which can be partitioned

$$\tilde{A} = \begin{bmatrix} \tilde{A}_1 & 0 \\ \tilde{A}_{12} & \tilde{A}_2 \end{bmatrix}.$$

Here $\tilde{A}_1$ is a square matrix, of dimension n^+ and having eigenvalues Σ^+, and $\tilde{A}_2$ is square matrix, of dimension n^- and having eigenvalues Σ^-. Condition (1.2.33) is equivalent to

$$\operatorname{rank} T^{-1}[sI - A \,\vdots\, B]\begin{bmatrix} T & 0 \\ 0 & I \end{bmatrix} = n, \qquad \text{for all } s \in \Sigma^+,$$

which can be written

$$\operatorname{rank}\left[\begin{pmatrix} sI - \tilde{A}_1 & 0 \\ -\tilde{A}_{12} & sI - \tilde{A}_2 \end{pmatrix} \vdots \begin{pmatrix} \tilde{B}_1 \\ \tilde{B}_2 \end{pmatrix}\right] = n, \qquad \text{for all } s \in \Sigma^+.$$

Here

$$\begin{bmatrix} \tilde{B}_1 \\ \tilde{B}_2 \end{bmatrix}$$

is a partitioning of $T^{-1}B$ compatible with that of $\tilde{A}$ and the I are identity matrices. The last condition is equivalent to

$$\operatorname{rank}\left[\begin{pmatrix} sI - \tilde{A}_1 & 0 \\ 0 & sI - \tilde{A}_2 \end{pmatrix} \vdots \begin{pmatrix} \tilde{B}_1 \\ 0 \end{pmatrix}\right] = n \qquad \text{for all } s \in \Sigma^+$$

since $sI - \tilde{A}_2$ is non-singular for $s \in \Sigma^+$, or

$$\operatorname{rank}[(sI_1 - \tilde{A}_1) \,\vdots\, \tilde{B}_1] + n^- = n, \qquad \text{for all } s \in \Sigma^+$$

which can be expressed as

$$\operatorname{rank}[sI_1 - \tilde{A}_1 \,\vdots\, \tilde{B}_1] = n^+ \qquad \text{for all } s \in \Sigma^+$$

since $n^- + n^+ = n$. By Hautus's criterion (Proposition 1.2.4), $(\tilde{A}_1, \tilde{B}_1)$ is controllable. We conclude from the pole-placement theorem (Theorem 1.2.10) that there exists an $m \times n^+$ matrix $\tilde{K}_1$ such that

$\tilde{A}_1 + \tilde{B}_1\tilde{K}_1$ has eigenvalues in the open unit disc. Now let

$$K = [\tilde{K}_1 \vdots 0]T^{-1},$$

and consider the eigenvalues of $A + BK$. These are the eigenvalues also of

$$T^{-1}[A+BK]T = \begin{bmatrix} \tilde{A}_1 & 0 \\ \tilde{A}_{12} & \tilde{A}_2 \end{bmatrix} + \begin{bmatrix} \tilde{B}_1 \\ \tilde{B}_2 \end{bmatrix} [\tilde{K}_1 \vdots 0]$$
$$= \begin{bmatrix} (\tilde{A}_1 + \tilde{B}_1\tilde{K}_1) & 0 \\ (\tilde{A}_{12} + \tilde{B}_2 K_1) & \tilde{A}_2 \end{bmatrix}.$$

However, due to the upper right block zero, this last matrix has eigenvalues those of $\tilde{A}_1 + \tilde{B}_1\tilde{K}_1$ together with those of $\tilde{A}_2$, all of which lie inside the open unit disc. So $A + BK$ is a stable matrix. We have shown that condition (1.2.33) is sufficient for stabilizability. □

A dual concept to that of stabilizability is detectability.

Definition 1.2.14

(H, A) is *detectable* when there exists an $n \times r$ matrix M such that all the eigenvalues of $(A + MH)$ lie in the open unit disc.

Bearing in mind that transposition does not affect the eigenvalues of a matrix, we see that (A, B) *is stabilizable if and only if* $(B^{\mathrm{T}}, A^{\mathrm{T}})$ *is detectable.* Indeed, if (A, B) is stabilizable so that there exists a matrix K such that $A + BK$ is stable, then $(B^{\mathrm{T}}, A^{\mathrm{T}})$ is detectable, since $A^{\mathrm{T}} + K^{\mathrm{T}}B^{\mathrm{T}}$ $(= (A + BK)^{\mathrm{T}})$ is stable. In a similar manner we show that detectability of $(B^{\mathrm{T}}, A^{\mathrm{T}})$ implies stabilizability of (A, B).

The duality between stabilizability and detectability enables us to deduce from Proposition 1.2.13 the following characterization of detectability.

Proposition 1.2.15

(H, A) is detectable if and only if

$$\operatorname{rank}\begin{bmatrix} sI - A \\ H \end{bmatrix} = n$$

for all eigenvalues s of A outside the open unit disc.

The terminology 'stabilizability' is a natural one, but 'detectability' requires a little explanation. This we now supply. Consider a linear system with zero input

$$x_{i+1} = Ax_i$$
$$y_i = Hx_i. \tag{1.2.35}$$

The system parameters which specify the matrices H, A are known but the initial state, x_0, is not. Suppose we wish to extract information about the state x_i at time i, from observations of present and past outputs. A simple, and widely used, procedure is the following: we take as estimate $\hat{x}_k$ of x_k the output of a replica of the system driven by an input which depends on the discrepancy between the observed output y_k and the output we would observe if x_k coincided with $\hat{x}_k$, namely $H\hat{x}_k$. More precisely stated, $\hat{x}_k$ is generated by the recursive equations

$$\hat{x}_{k+1} = A\hat{x}_k + K(y - H\hat{x}_k)$$
$$\hat{x}_0 = 0 \tag{1.2.36}$$

for some suitably chosen matrix K. The estimate of the state obtained in this way is called an *observer*. It is a deterministic analogue of the estimate supplied by the Kalman filter in a stochastic setting, which we shall study in detail in Chapter 3.

The error $\hat{e}_k$ incurred when the true state x_k is replaced by $\hat{x}_k$,

$$\hat{e}_k = x_k - \hat{x}_k,$$

is governed by the equations (obtained by subtracting (1.2.36) from (1.2.35)):

$$\hat{e}_{k+1} = (A - KH)\hat{e}_k$$
$$\hat{e}_k = x_0.$$

The error can be made to decay to zero, for arbitrary initial state, if and only if K can be chosen in such a way that $(A - KH)$ is a stable matrix. This is precisely the condition that (H, A) be detectable. Thus the detectability condition means that the state can be detected, with error which decays to zero, at the output of an appropriately designed observer.

Notes

Section 1.1.1 There are many good introductory texts on probability, for example Larson (1969). More advanced texts, such as Kingman and Taylor (1966) or Chow and Teicher (1978) involve *measure theory*, which is necessary for an adequate treatment of the more technical parts of the subject. In particular, the existence of conditional distributions for random vectors is proved in Theorem 2, Section 7.2 of Chow and Teicher (1978).

Section 1.1.2 A very readable introduction to stochastic processes and their applications in time series modelling is given by Chatfield (1979); at a more technical level, Wong (1970) can be consulted. Stationary processes and spectral analysis are covered by Cramér and Leadbetter (1967), Hannan (1970), Priestley (1982) and Wong (1970). Hannan in particular gives detailed coverage of the multivariable case.

Section 1.1.3 Convergence of sequences of real numbers is part of 'real analysis'; see for example Bartle (1964). Convergence of sequences of random variables is discussed in any book dealing with measure-theoretic probability, such as Kingman and Taylor (1966). The Borel–Cantelli lemma (Lemma 1.1.13) is given by Chow and Teicher (1978, Lemma 2, Section 2.2).

The ergodic theorem which we give as Theorem 1.1.15 is adapted from a continuous-time result of Cramér and Leadbetter (1967, Section 5.5).

Section 1.2.1 See Anderson and Moore (1971), Chen (1970), Kailath (1980) and Kwakernaak and Sivan (1972) for supplementary reading in linear systems theory. The authors of these books occupy themselves for the most part with continuous systems but, at least as far as the topics we consider are concerned, the continuous-time and discrete-time theories run in parallel. Suitable background in linear algebra can be acquired from a number of texts, for example Lang (1979). For a more advanced treatment, we refer to Gantmacher (1964).

Kalman and his co-workers (1963) provided a key early paper on controllability. The rank tests for controllability had, however, appeared independently in the literature as a technical hypothesis in optimal control theory (LaSalle, 1960). The concept of observability is due to Kalman (1960). The condition for controllability, which we refer to as the Hautus condition, first appeared in Hautus (1969).

Section 1.2.2 Early proofs of the pole assignment theorem, for multi-input systems, were provided by Popov (1964) and Wonham (1967). Wonham drew attention to the significance of stabilizability and detectability in quadratic cost control and linear filtering, and the characterization of these properties which we provide are implicit in his book (1979).

References

Probability

Bartle, R. G. (1964) *The Elements of Real Analysis*, John Wiley, New York.

Chatfield, C. (1979) *The Analysis of Time Series: Theory and Practice* (2nd edn), Chapman and Hall, London.

Chow, Y. S. and Teicher, H. (1978) *Probability Theory: Independence Interchangeability, Martingales*, Springer-Verlag, New York.

Cramér, H. and Leadbetter, M. R. (1967) *Stationary and Related Stochastic Processes*, Wiley, New York.

Davis, M. H. A. (1977) *Linear Estimation and Stochastic Control*, Chapman and Hall, London.

Hannan, E. J. (1970) *Multiple Time Series*, Wiley, New York.

Kingman, J. F. C. and Taylor, S. J. (1966) *Introduction to Measure and Probability*, Cambridge University Press.

Larson, H. J. (1969) *Introduction to Probability Theory and Statistical Inference*, Wiley, New York.

Priestley, M. B. (1982) *Spectral Analysis and Time Series*, Academic Press, London.

Wong, E. (1970) *Stochastic Processes in Information and Dynamical Systems*, McGraw-Hill, New York.

Linear system theory

Anderson, B. D. O. and Moore, J. B. (1971) *Linear Optimal Control*, Prentice-Hall, Englewood Cliffs, N.J.

Chen, C. T. (1970) *Introduction to Linear System Theory*, Holt, Rinehart and Winsten, New York.

Gantmacher, F. R. (1964) *The Theory of Matrices*, Vols 1 and 2, Chelsea, New York.

Hautus, M. L. J. (1969) Controllability and observability conditions of linear autonomous systems. *Ned. Akad. Wetenschappen*, Proc. Ser. A, **72** 443–448.

Hautus, M. L. J. (1977) A simple of proof of Heymann's lemma. *IEEE Trans. Automatic Control*, **AC-22**.

Kailath, T. (1980) *Linear Systems*, Prentice-Hall, Englewood Cliffs, N.J.

Kalman, R. E. (1960) Contributions to the theory of optimal control. *Boletín de la Sociedad Matemática Mexicana*, **5**, 102–119.

Kalman, R. E., Ho, Y. C. and Narendra, K. S. (1963) Controllability of linear dynamical systems. In *Contributions to the Theory of Differential Equations*, vol.1, Wiley-Interscience, New York.

Kwakernaak, H. and Sivan, R. (1979) *Linear Optimal Control Systems*, Wiley, New York.

Lang, S. (1979) *Linear Algebra* (2nd edn), Addison–Wesley, Reading, Mass.

La Salle, J. P. (1960) The time-optimal control problem. In *Contributions to Differential Equations*, vol. 5, Princeton University Press, Princeton, N.J.

Popov, V. M. (1964) Hyperstability and optimality of automatic systems with several control functions. *Rev. Roum. Sci-Electrotechn. et Energ.*, **9**, 629–690.

Wilkinson, J. H. (1965) *The Algebraic Eigenvalue Problem*, Clarendon Press, Oxford.

Wonham, W. M. (1967) On pole assignment in multi-input controllable linear systems. *IEEE Trans. Automatic Control*, **AC-12**, 660–665.

Wonham, W. M. (1979) *Linear Multivariable Control: A Geometric Approach* (2nd edn), Springer-Verlag, Berlin.

CHAPTER 2

Stochastic models

All science is concerned in some way with prediction, since the ultimate test of a scientific theory is its ability to predict the results of experiments which have not yet been carried out. In the context of engineering systems, a *model* is some description of a system which enables us to predict its behaviour when it is subjected to certain classes of inputs. Models may be divided into two categories, *internal* and *external*. Internal models describe the complete structure of a system, possibly including parts of it which do not contribute directly to observable outputs, whereas external models are concerned solely with describing the input/output behaviour of the system. There are two ways in which models may be arrived at: by an analysis of the components of the system using physical laws, or by a 'black box' approach whereby the contents of the 'box' are inferred from experimental data. In the former case the laws involved are those of Newtonian mechanics, electromagnetism, thermodynamics, etc. In elementary situations such as, say, describing the motion of a pendulum, Newtonian mechanics gives such good predictions that the distinction between 'model' and 'system' is almost forgotten. However, in more complicated cases – describing the motion of an aircraft, for example – it will be clear that the equations one writes down are only approximations, valid over a certain range of operating conditions. Models arrived at in this way are generally in the first instance internal ones, in that they involve the 'states' of various components comprising the system regardless of whether these states are 'observable'. An external model – which is, after all, less detailed – can often be derived from a given internal model; we study this question in Section 2.4 below. On the other hand, a model obtained by the black-box approach is necessarily external since no other information is available about the system than its input/output behaviour.

In this book we are mainly concerned with input/output models and how to obtain them by data analysis. In this chapter, however, we introduce some general classes of models, internal and external, and study their properties. We shall deal only with *linear* models, that is to say models involving a linear relationship between inputs and outputs. Of course no real system is exactly linear, but many are approximately linear at least with respect to small variations around some operating point which are often what we wish to study. Also there cannot be any satisfactory *general* theory of non-linear models – this class is simply too big to be treated as a whole – whereas for linear systems a unified theory is possible.

A system is *deterministic* if its input together with certain initial conditions and times uniquely specify the output. Otherwise it is *stochastic*. We explain the non-unique response to input signals by supposing that the system has a random 'noise' input in addition to possible control inputs. Thus denoting input, output and noise by u, y, w respectively we represent the system as in Fig. 2.1.

The actual noise in a system may well be generated internally, say by thermal noise in electronic components, but conceptually this is not different from regarding it as being injected by some external source (in either case, it is not supposed that the noise can be directly measured). An important restriction on the class of systems we consider is that the noise should be *additive*. This excludes some quite natural phenomena such as randomly varying gains but is necessary if we are to stay within the framework of linear systems. In fact if the basic input/output relationship is linear and the noise is additive then the noise can always be regarded as being added at the output, giving us a somewhat simplified model structure as depicted in Fig. 2.2.

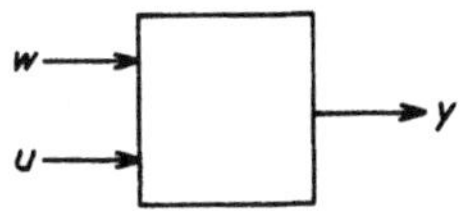

Fig. 2.1

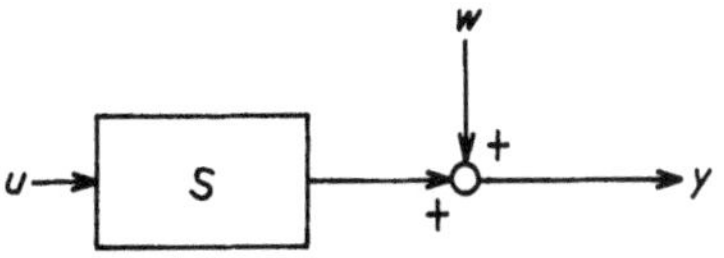

Fig. 2.2

A full description of a system model in this form must specify

(a) The input/output behaviour of system S; and,
(b) The statistical characteristics of the noise w.

So far we have said nothing as to the nature of *time*. Here we have a choice between *continuous-time* and *discrete-time* models, and a viable theory can be developed either way. In this book we only consider discrete-time models. These arise in practice either because the data for the system under study is presented in discrete form (for example, monthly or quarterly economic data) or because we wish to discretize an underlying continuous system for purposes of computer control.

If a model is to be identified from data then it is essential that it be chosen from a *finitely parametrized model set*, that is to say a set $\{M(\theta): \theta \in D\}$ indexed by a finite-dimensional parameter θ. Choice of a parameter θ from D then specifies the model $M(\theta)$ uniquely. A broad class of discrete-time linear systems with this property is that of *state-space models*, such a model being represented by the equations

$$x_{k+1} = Ax_k + Bu_k$$
$$y_k = Hx_k.$$

Here x_k is the state of the system at time k and u_k, y_k are the input and output respectively. If x_k, u_k, y_k are of dimension n, m, r then this model is specified by a parameter vector θ of dimension $n^2 + nm + nr + n$ whose components are the elements of the matrices A, B, H together with an initial state vector x_0. All of the system models we consider below can be represented in this form, though they may be parametrized in some other way.

To complete our description of the model of Fig. 2.2 we have to 'specify' the noise w. The most precise specification would be to give the finite-dimensional distributions of the process $\{w_k\}$. Generally, however, this is overambitious: estimating the distributions of random variables is not an easy task, and usually we will only be concerned with properties involving means and covariances. One can suppose that $\{w_k\}$ has zero mean (the mean is a deterministic sequence which can be modelled as part of the system S), which leaves the *covariance function* to be specified, or equivalently, if w_k is stationary, the spectral density function. The class of *all* covariance functions is, however, 'too big': again for model-fitting purposes we must have a finitely parametrized set of covariance functions. How to obtain

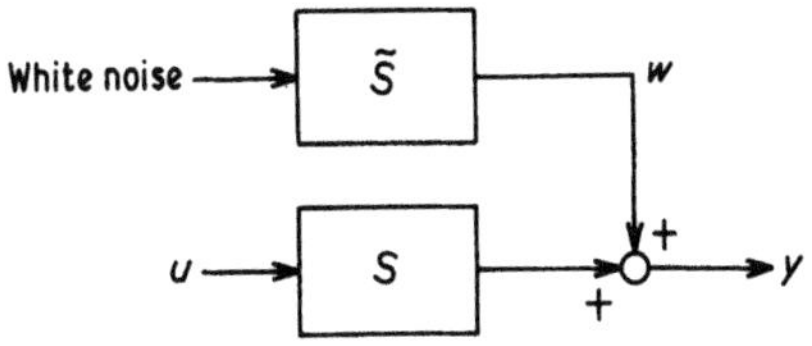

Fig. 2.3

useful classes of finitely parametrized covariance functions is in fact a major topic of this chapter. The general strategy is to start from a very simple process – white noise – and then obtain other processes by 'filtering' it, i.e. passing it through a linear system. The coefficients of this linear system then specify the noise covariance. This approach gives us the model of Fig. 2.2 in a 'symmetrized' form, as shown in Fig. 2.3. In terms of parametrization the whole model is reduced to two linear systems S and $\tilde{S}$, the coefficients of which specify the input/output relation and the noise spectral density respectively.

The chapter is organized as follows. In the first two sections we set up the mathematical machinery which will permit us to make sense of and analyse processes generated at the output of stochastic dynamical models expressed as difference equations. Considerable attention is then given to ARMA models, the important class of noise models in which the noise process is taken to be the solution of a difference equation driven by white noise. We investigate their properties and examine the nature of the assumptions on the noise process implicit in choice of such models. Finally, we incorporate the noise model in a linear system, to obtain a model of the overall stochastic dynamical model, and give some related formulations.

2.1. A general output process

Many stochastic models incorporate as a basic building block an r-vector process $\{y_k\}$ which results from passage of another, l-vector process $\{e_k\}$ through a linear system. It is necessary to attach a precise meaning to processes which arise in this way and to examine their properties. For concreteness we shall refer here to the y_k and e_k as 'outputs' and 'disturbances' respectively, although in applications of the results of this section they will often have interpretations other than the actual outputs and disturbances associated with the stochastic dynamical system in question. Initially we adopt a somewhat abstract framework for definition of the process $\{y_k\}$ because this is convenient for analysis. We shall see in the next section that our theory is consistent with an intuitive definition of the output of a system defined by stochastic difference equations.

Let $e_k, k \in \mathbb{Z}$, be a collection of l-vector random variables. Let $T(\sigma)$ be an $r \times l$ matrix of rational functions in the complex variable σ. We can represent $T(\sigma)$ in terms of a scalar polynomial $g(\sigma)$ in σ and $r \times l$ matrices $G_0, \ldots, G_n$ thus:

$$T(\sigma) = [g(\sigma)]^{-1}[G_0 + G_1\sigma + \cdots + G_n\sigma^n]. \tag{2.1.1}$$

We wish to consider the process $\{y_k\}$ obtained by passing the disturbance sequence $\{e_k\}$ through a linear system with 'transfer function' $T(z^{-1})$:

$$y_k = T(z^{-1})e_k, \quad k \in \mathbb{Z}. \tag{2.1.2}$$

Here z^{-1} is the backward shift operator defined by $z^{-1}e_k := e_{k-1}$. Powers of z^{-1} are defined recursively by $z^{-(i+1)}e_k := z^{-1}(z^{-i}e)_k$, so that $z^{-i}e_k = e_{k-i}$. When $g(\sigma) = 1$, $T(\sigma)$ is a polynomial in σ and (2.1.2) is simply an operational way of stating that y_k is given by

$$y_k = \sum_{i=0}^{n} (G_i z^{-i})e_k = \sum_{i=0}^{n} G_i e_{k-i}.$$

When $g(\sigma) \neq 1$ we express $T(\sigma)$ in the form of an infinite-degree polynomial as in (2.1.5) below and interpret (2.1.2) as the corresponding infinite sum. This approach can be made precise under the following hypotheses.

$$\text{There exists a number } c > 0 \text{ such that} \quad E\|e_k\|^d \leq c, \quad \text{for all } k \in \mathbb{Z}. \tag{2.1.3}$$

Here d is positive integer which will vary with different applications. Also, some representation (2.1.1) can be chosen for $T(z^{-1})$ such that

$$\sigma \to g(\sigma) \text{ has all zeros outside the closed unit disc.}^{\dagger} \tag{2.1.4}$$

†The 'closed unit disc' is the closed subset $\{\zeta : |\zeta| \leq 1\}$ of the complex plane.

Take $T_0, T_1, \ldots$ to be the matrix coefficients which result from formal expansion of $T(\sigma)$ about $\sigma = 0$:

$$T(\sigma) = T_0 + T_1\sigma + T_2\sigma^2 + \cdots \tag{2.1.5}$$

Such an expansion is clearly possible under hypothesis (2.1.4). The T_i are called the *Markov parameters* associated with the transfer function $T(z^{-1})$.

We now are ready to define the solution $\{y_k\}$ to the equation (2.1.2). It is

$$y_k = \lim_{N\to\infty} \sum_{i=0}^{N} T_i e_{k-i}. \tag{2.1.6}$$

The limit here is taken in the dth mean. (The positive integer d is that of hypothesis (2.1.3)). The following lemma tells us that the limit exists and that therefore the definition makes sense. It provides also the information that the y_k have uniformly bounded dth order moments.

Lemma 2.1.1

Suppose that hypotheses (2.1.3) and (2.1.4) are true. Let $T_0, T_1, \ldots$ be the Markov parameters of $T(z^{-1})$ (see (2.1.5)). Then the partial sums

$$s_k(N) = \sum_{i=0}^{N} T_i e_{k-i}, \qquad N = 1, 2, \ldots$$

converge to a limit s_k in the dth mean as $N \to \infty$, for each $k \in \mathbb{Z}$, and there exists a constant $c_1 > 0$ such that

$$E\|s_k\|^d \le c_1, \qquad \text{for all } k \in \mathbb{Z}.$$

PROOF Taking note of hypothesis (2.1.4), we deduce from Proposition D.3.3 of Appendix D that there exist numbers $c_2 > 0$ and $\lambda \in (0, 1)$ such that

$$\|T_i\| \le c_2 \lambda^i, \qquad i = 0, 1, \ldots \tag{2.1.7}$$

(The norm here is the spectral norm; see Appendix D.)

For k an integer and M, N non-negative integers such that $M \le N$, define

$$\varepsilon_k(M, N) = \sum_{i=M}^{N} T_i e_{k-i}.$$

We have

$$\|\varepsilon_k(M,N)\|^d \le \left(\sum_{i=M}^{N} \|T_i\|\,\|e_{k-i}\|\right)^d$$
$$\le c_2^d\left(\sum_{i=M}^{N} \lambda^i \|e_{k-i}\|\right)^d$$

by (2.1.7)

$$\le c_2^d\left(\sum_{i=M}^{N} \lambda^i\right)^{d-1}\left(\sum_{i=M}^{N} \lambda_i \|e_i\|^d\right)$$

by the generalized Hölder inequality (see Appendix E).

Taking expectations and noting hypothesis (2.1.3) we obtain from this inequality:

$$E\|\varepsilon_k(M,N)\|^d \le cc_2^d\left(\sum_{i=M}^{N} \lambda^i\right)^d$$

which implies

$$E\|\varepsilon_k(M,N)\|^d \le cc_2^d \lambda^{d+M}/(1-\lambda)^d. \tag{2.1.8}$$

We see that

$$\sup_{N\ge M} E\|\varepsilon_k(M,N)\|^d \to 0 \quad \text{as} \quad M\to\infty \tag{2.1.9}$$

for each $k\in\mathbb{Z}$. Noting that $\varepsilon_k(M,N)$ is related to the partial sums by

$$\varepsilon_k(M,N) = s_k(N) - s_k(M-1)$$

we deduce from (2.1.9) that $\{s_k(N)\}_N$ is a Cauchy sequence in dth mean, and therefore that the limit in dth mean $\lim_{N\to\infty} s_k(N)$ exists.

Since $s_k(N) = \varepsilon_k(0,N)$ we deduce from (2.1.8) that

$$E\|s_k(N)\|^d \le c_1, \qquad \text{for all } k\in\mathbb{Z}, N\ge 0,$$

where $c_1 = cc_2^d\lambda^d/(1-\lambda)^d$. But $s_k(N)\to s_k$ in dth mean. It follows that

$$E\|s_k\|^d \le c_1. \qquad \square$$

An important feature of the relationship between disturbances and outputs provided by the transfer function $T(z^{-1})$ is that the effect of the disturbance e_k (at time k) on subsequent outputs $(y_k, y_{k+1}, \ldots)$

falls off exponentially. In the analysis of identification algorithms transfer functions $T_\theta(z^{-1})$ need to be considered which depend on a parameter θ. Here, under suitable hypotheses, the exponential decay is uniform with respect to the parameter θ, as we now show.

Let $\{y_k(\theta)\}$ be defined by

$$y_k(\theta) = T_\theta(z^{-1})e_k, \qquad k \in \mathbb{Z} \tag{2.1.10}$$

$$T_\theta(z^{-1}) = [g_\theta(z^{-1})]^{-1} G_\theta(z^{-1})e_k \tag{2.1.11}$$

in which θ is a parameter which ranges over some compact subset $\mathscr{D}$ of $\mathbb{R}^q$. In (2.1.11), $g_\theta(z^{-1})$ is a polynomial in z^{-1}:

$$g_\theta(z^{-1}) = g_0(\theta) + g_1(\theta)z^{-1} + \cdots + g_n(\theta)z^{-n}$$

with coefficients $g_0(\theta), \ldots, g_n(\theta)$ real numbers which depend continuously on θ. $G_\theta(z^{-1})$ is a polynomial in z^{-1}:

$$G_\theta(z^{-1}) = G_0(\theta) + G_1(\theta)z^{-1} + \cdots + G_n(\theta)z^{-n},$$

with coefficients $r \times l$ matrices $G_0(\theta), \ldots, G_n(\theta)$ which depend continuously on θ. $\{e_k\}$ is a sequence of l-vector random variables. The notation $y_k(\theta)$ emphasizes that the solution depends on the parameter θ.

It will be assumed that our earlier hypotheses assuring that (2.1.10) does indeed define $\{y_k(\theta)\}$ are in force, namely there exists a number $c > 0$ such that

$$E\|e_k\|^d \leq c, \qquad \text{for all } k \in \mathbb{Z} \tag{2.1.3}$$

(d is some fixed positive integer), and

$$\text{for all } \theta \in \mathscr{D}, \sigma \to g_\theta(\sigma) \text{ has all zeros outside the closed unit disc.} \tag{2.1.12}$$

Proposition 2.1.2

Consider processes $\{y_k(\theta)\}_{k \in Z}$, $\theta \in \mathscr{D}$, defined by (2.1.10). Suppose that hypotheses (2.1.3) and (2.1.12) are true. Then there exist constants $c_3 > 0$, $\lambda \in (0, 1)$ such that

$$E\|y_k(\theta)\|^d \leq c_3 \sum_{i=0}^{\infty} \lambda^i E\|e_{k-i}\|^d, \qquad \text{for all } \theta \in \mathscr{D}.$$

PROOF Let $T_0(\theta)$, $T_1(\theta), \ldots$ be the coefficients in the formal expansion of $[g_\theta(\sigma)]^{-1}$ $G_\theta(\sigma)$ about $\sigma = 0$. By Proposition D.3.3 of

Appendix D there exist $c_1 > 0$ and $\lambda \in (0, 1)$ such that

$$\| T_i(\theta) \| \leq c_1 \lambda^i, \qquad \text{for } \theta \in \mathscr{D}, i = 0, 1, \ldots \tag{2.1.13}$$

Now define

$$s_N(\theta, k) = \sum_{i=0}^{N} T_i(\theta) e_{k-i}.$$

We have

$$\| s_N(\theta, k) \|^d \leq \left(\sum_{i=0}^{N} \| T_i(\theta) \| \cdot \| e_{k-i} \| \right)^d$$

$$\leq c_1^d \left(\sum_{i=0}^{N} \lambda^i \| e_{k-i} \| \right)^d$$

by (2.1.13)

$$\leq c_1^d \left(\sum_{i=0}^{N} \lambda^i \right)^{d-1} \sum_{i=0}^{N} \lambda^i \| e_{k-i} \|^d$$

by the generalized Hölder inequality (Appendix E).

Taking expectations, we deduce that

$$E \| s_N(\theta, k) \|^d \leq c_3 \sum_{i=0}^{N} \lambda^i E \| e_{k-i} \|^d$$

where $c_3 = c_1^d (1 - \lambda)^{-(d-1)}$. Passage to the limit $N \to \infty$ on both sides of this inequality gives

$$E \| y_k \|^d \leq c_3 \sum_{i=0}^{\infty} \lambda^i E \| e_{k-i} \|^d$$

since, by definition of $y_k(\theta)$, $s_N(\theta, k) \to y_k(\theta)$ in dth mean and since, by hypothesis, $\{E \| e_{k-i} \|^d\}_{i=0}^{\infty}$ is a sequence of uniformly bounded numbers. □

Notice that Proposition 2.1.2 provides bounds for *changes* in the output $\{y_k\}$ resulting from *changes* in the disturbances $\{e_k\}$ since the system of equations (2.1.10) is linear; it will be in this guise that Proposition 2.1.2 will prove most useful.

2.2 Stochastic difference equations

In this section we examine systems in which the output sequence $\{y_k\}$ is related to the input sequence by difference equations:

$$A_0 y_k + A_1 y_{k-1} + \cdots + A_n y_{k-n} = B_0 e_k + B_1 e_{k-1} + \cdots + B_n e_{k-n}$$

where $A_0, \ldots, A_n$ are $r \times r$ matrices and $B_0, \ldots, B_n$ and $r \times l$ matrices. It is assumed that A_0 is non-singular. We shall show that initial data can be supplied in the form of the finite past of both $\{y_k\}$ and $\{e_k\}$ (here the definition of $\{y_k\}$ in terms of $\{e_k\}$ is elementary) or alternatively in the form of the infinite past of just $\{e_k\}$ (in this case we must draw on the theory of Section 2.1). The output processes corresponding to these two different frameworks will then be related.

Recall that we interpret z^{-1} as the backward shift operator: more precisely if $\{u_k\}$ is any sequence then $z^{-1}\{u_k\}$ denotes the same sequence shifted backwards by one time unit, i.e. the ith element of $z^{-1}\{u_k\} = u_{i-1}$, for all i. Powers of z^{-1} are defined in the following manner: the ith entry of $z^{-n}\{u_k\} = u_{i-n}$, for all i.

The difference equations can be expressed in terms of the shift operator as

$$A(z^{-1})y_k = B(z^{-1})e_k \tag{2.2.1}$$

in which $A(z^{-1})$, $B(z^{-1})$ are polynomials in z^{-1} with matrix coefficients:

$$A(z^{-1}) = A_0 + A_1 z^{-1} + \cdots + A_n z^{-n},$$
$$B(z^{-1}) = B_0 + B_1 z^{-1} + \cdots + B_n z^{-n}.$$

Eqn. (2.2.1) should be regarded as shorthand for the equation for sequences

$$\sum_{i=0}^{n} A_i z^{-i}\{y_k\} = \sum_{i=0}^{n} B_i z^{-i}\{u_k\}.$$

Here $A_0\{y_k\}$ denotes $\{A_0 y_k\}$ etc., and sums of sequences are defined in an obvious manner.

Notice that our formulation does not restrict us to consideration of situations in which the output and disturbances are subject to the same number of delays, since we can take certain of the A_i or B_i to be zero.

Initial data must be supplied if the outputs $y_k, k = 0, 1, \ldots$ are to be well-defined, given e_k, $k = 0, 1, \ldots$. Two forms of initial data are commonly considered. On the one hand, we can take (2.2.1) at face value as a difference equation which we can rewrite

$$y_k = A_0^{-1}[-A_1 y_{k-1} - \cdots - A_n y_{k-n} + B_0 e_k + \cdots + B_n e_{k-n}],$$
$$k = 0, 1, \ldots. \tag{2.2.2}$$

Initial data y_k, $k = -1, -2, \ldots, -n$, e_k, $k = -1, -2, \ldots, -n$ which may possibly be random, is appropriate here. Knowledge of this initial data, together with the disturbances e_k, $k=0,1,\dots$ clearly permits us to generate the outputs y_k, $k=0,1,\dots$ by recursive solution of equations (2.2.2). On the other hand, we can take as initial data the 'infinite past' of the disturbance process at time 0, namely $e_{-1}, e_{-2},\dots$. In this setting restrictions must be placed on the equation parameters, the disturbances and the initial data, in order that they well-define the output $\{y_k\}$.

In order to interpret (2.2.1) when the initial data is the infinite past of $\{e_k\}$ we rewrite (2.2.1) as

$$y_k = T(z^{-1})e_k, \tag{2.2.3}$$

with

$$T(z^{-1}) = [g(z^{-1})]^{-1}G(z^{-1})$$

in which $g(z^{-1}) = \det[A(z^{-1})]$ and $G(z^{-1}) = (\operatorname{Adj}[A(z^{-1})])B(z^{-1})$. Then y_k is defined by the limit (2.1.6). This is possible, in view of the theory of Section 2.1, provided there exists a number $c>0$ such that

$$E\,\|e_k\|^d \le c, \quad k\in\mathbb{Z} \quad (\text{for some } d) \tag{2.2.4}$$

and

the zeros of $\sigma \to \det A(\sigma)$ lie outside the closed unit disc. (2.2.5)

Now suppose that the initial data is the infinite past of $\{e_k\}$. Assuming of course that (2.2.4), (2.2.5) hold, we shall show that in this case y_k satisfies

$$y_k = A_0^{-1}[-A_1y_{k-1} - \cdots - A_ny_{k-n} + B_0e_k + \cdots + B_ne_{k-n}],$$
$$k=0,1,\dots. \tag{2.2.2$'$}$$

The last equation can be interpreted as a recursive equation for $y_0, y_1,\dots$ with (random) starting values $e_{-1},\dots,e_{-n}, y_{-1},\dots,y_{-n}$ (The random variables $y_{-1},\dots,y_{-n}$ are obtained from (2.2.3)). In other words, the process $y_0, y_1,\dots$ defined by (2.2.3) (and (2.1.5)) coincides with the solutions to the recursive equation

$$A(z^{-1})y_k = B(z^{-1})e_k, \quad k=1,2,\dots$$

provided the initial data on the variable y_k are chosen appropriately. Of course the converse is not necessarily true. Definition (2.2.2) makes sense if merely $\det[A(0)] \neq 0$ and arbitrary initial data $y_{-1},\dots,y_{-n}$, $e_{-1},\dots,e_{-n}$ are given, whereas the definition provided by (2.2.3) can be used only when $\sigma \to \det[A(\sigma)]$ has zeros outside the closed unit disc and when the initial output data $y_{-1},\dots,y_{-n}$ is compatible with past disturbances $e_{-1}, e_{-2}, \ldots$, in the sense that

$$y_k = T(z^{-1})e_k, \quad k = -1, \ldots, -n.$$

One instance when they are compatible is when $e_{-1} = 0, e_{-2} = 0, \ldots$ and $y_{-1} = 0$, $y_{-2} = 0, \ldots, y_{-n} = 0$; the difference equations

$$y_k = A_0^{-1}[A_1 y_{k-1} + \cdots + A_n y_{k-n} + B_0 e_k + \cdots + B_n e_{k-n}], \quad k = 0, 1, \ldots$$
$$y_k = 0, \qquad e_k = 0, \qquad k = -1, \ldots, -n$$

generate the same process y_k, $k = 0, 1, \ldots$ as

$$y_k = T(z^{-1})e_k, \qquad k = 0, 1, \ldots$$

provided $e_k = 0$, $k = -1, -2, \ldots$, and the zeros of $\sigma \rightarrow \det A(0)$ lie outside the closed unit disc. This example is important since it is natural to formulate models for identification as difference equations with initial data specified over a finite time interval, yet the analysis is often simpler if we treat the output as a function of the infinite past of the disturbances. The example tells us this can be done when the initial data on the difference equation is zero and the system is stable.

Let us confirm that y_k defined by (2.2.3) satisfies equation (2.2.2). The Markov parameters $T_0, T_1, \ldots$ of $T(z^{-1})$ satisfy

$$(A_0 + A_1\sigma + \cdots + A_n\sigma^n)(T_0 + T_1\sigma + \cdots) = B_0 + B_1\sigma + \cdots + B_n\sigma^n.$$

We deduce (on equating powers of σ)

$$\sum_{i=0}^{j} A_i T_{j-i} = B_j, \qquad j = 0, 1, \ldots, n$$

$$\sum_{i=0}^{\min\{j,n\}} A_i T_{j-i} = 0, \qquad j > n. \tag{2.2.6}$$

Define for $N > n$ and $k = 0, 1, \ldots$

$$y_k(N) = \sum_{j=0}^{k+N} T_j e_{k-j}.$$

We have, by definition of y_k,

$$y_k(N) \rightarrow y_k \text{ in } d\text{th mean as } N \rightarrow \infty. \tag{2.2.7}$$

Now

$$A(z^{-1})y_k(N) = \sum_{i=0}^{n} \sum_{j=0}^{k-i+N} A_i T_j e_{k-i-j}.$$

The double summation on the right-hand side can be rearranged and written

$$\sum_{j=0}^{k+N}\left(\sum_{i=0}^{\min\{j,n\}} A_i T_{j-i}\right)e_{k-j}.$$

In view of (2.2.6), this is precisely $\sum_{j=0}^{n} B_j e_{k-j}$, since $N > n$, $k \geq 0$. It follows that

$$y_k(N) = A_0^{-1}[-A_1 y_{k-1}(N) - \cdots - A_n y_{k-n}(N) + B_0 e_k + \cdots + B_n e_{k-n}].$$

Bearing in mind (2.2.7) and taking the limit $N \to \infty$, we deduce that y_k, $k = 0, 1, \ldots$ satisfies the difference equations (2.2.2) as claimed.

2.3 ARMA noise models

A widely employed and versatile noise model takes a noise process $\{n_k\}$ as the output of a linear system, defined by difference equations driven by white noise. According to this model $\{n_k\}$ is given by

$$A(z^{-1})n_k = B(z^{-1})e_k, \qquad k \in \mathbb{Z} \tag{2.3.1}$$

in which

$$A(\sigma) = A_0 + A_1\sigma + \cdots + A_{d_1}\sigma^{d_1}$$

and

$$B(\sigma) = B_0 + B_1\sigma + \cdots + B_{d_2}\sigma^{d_2}.$$

The matrices $A_0, \ldots, A_{d_1}$ are $r \times r$ and the matrices $B_0, \ldots, B_{d_2}$ are $r \times l$. (It is convenient to emphasize here that $A(\sigma)$, $B(\sigma)$ might be of different degrees.) It is assumed that the roots of $\sigma \to \det A(\sigma)$ lie outside the closed unit disc. We take $\{e_k\}_{k \in Z}$ to be a sequence of zero-mean, uncorrelated second-order vector random variables with common covariance matrix W.

One important special case occurs when $A_1, \ldots, A_{d_1}$ take value zero. The equations can then be organized

$$n_k = \tilde{B}_0 e_k + \tilde{B}_1 e_{k-1} + \cdots + \tilde{B}_{d_2} e_{k-d_2} \tag{2.3.2}$$

in which $\tilde{B}_0 = A_0^{-1}B_0$, $\tilde{B}_1 = A_0^{-1}B_1, \ldots$. Models of the form (2.3.2) are called *moving average* models (MA for short) since the noise variable is expressed as a weighted average of present and past values of the white noise.

The situation when $B_1,\ldots,B_{d_2}$ take value zero is another important special case. In this case the equations can be organized

$$n_k+\tilde{A}_1 n_{k-1}+\cdots+\tilde{A}_{d_1} n_{k-d_1}=\tilde{e}_k$$

where $\tilde{A}_1=A_0^{-1}A_1$, $\tilde{A}_2=A_0^{-1}A_2,\ldots$ and $\{\tilde{e}_k\}$ is the process $\{A_0^{-1}B_0e_k\}$. Models of this kind are called *autoregressive* (AR for short) models since the equations for the current value of the process $\{n_k\}$ involve a linear combination of (or in classical statistical parlance 'regression on') past values of the same process.

The model (2.3.1) in its full generality, which contains moving average and autoregressive terms, is referred to as an *autoregressive moving average* model (ARMA for short).

Notice that, by the theory of Section 2.1, $\{n_k\}$ is a well-defined second-order process. We have

$$n_k=\sum_{i=0}^{\infty} T_i e_{k-i} \qquad k\in\mathbb{Z}$$

where $T_0, T_1,\ldots$ are the coefficients in the formal expansion of $\sigma\rightarrow A^{-1}(\sigma)B(\sigma)$ about 0:

$$A^{-1}(\sigma)B(\sigma)=T_0+T_1\sigma+\cdots$$

and the infinite summation indicates a limit in mean square. (The formal expansion is possible since, by assumption, $\det A_0$ $(=\det A(0))\neq 0)$. Actually $\{n_k\}$ is a wide-sense stationary process with spectral density function which is related in a simple way to the transfer function $A(\sigma)^{-1}B(\sigma)$ as follows.

Proposition 2.3.1

The process $\{n_k\}$ defined by equations (2.3.1) has zero mean and is wide-sense stationary. The process has covariance function

$$R(l)=\begin{cases}\sum_{j=0}^{\infty} T_{j+l}WT_j^{\mathrm{T}} & l\geq 0\\ \sum_{j=0}^{\infty} T_jWT_{j-l}^{\mathrm{T}} & l<0\end{cases}$$

and spectral density function

$$\Phi(\omega)=A^{-1}(e^{-i\omega})B(e^{-i\omega})W[A^{-1}(e^{-i\omega})B(e^{-i\omega})]^*$$

(the star * indicates complex conjugate transpose).

PROOF By definition

$$n_k = \lim_{N \to \infty} s_k(N)$$

where $s_k(N) = \sum_{i=0}^{N} T_i e_{k-i}$. The limit is taken in mean square.

For each N, $s_k(N)$ obviously has zero mean. The random variable n_k therefore has zero mean since it is the mean square limit of a sequence of zero-mean random variables.

We now calculate the covariance function $R(l)$.

$$\begin{aligned} R(l) &= E\{n_k n_{k-l}^{\mathrm{T}}\} \\ &= E\{(\lim_{N \to \infty} s_k(N))(\lim_{N \to \infty} s_{k-l}(N))^{\mathrm{T}}\}. \end{aligned}$$

We claim that we can bring the limiting operations outside the expectation operator and write

$$E\{n_k n_{k-l}^{\mathrm{T}}\} = \lim_{N \to \infty} E\{s_k(N) s_{k-l}^{\mathrm{T}}(N)\} \tag{2.3.3}$$

To see this observe that, for arbitrary N, n_k can be written

$$n_k = s_k(N) + \varepsilon_k(N)$$

where

$$\varepsilon_k = \sum_{j=N+1}^{\infty} T_j e_{k-j}.$$

The infinite summation indicates a limit in mean square. Now $E\{n_k n_{k-l}^{\mathrm{T}}\}$ can be written

$$\begin{aligned} E\{n_k n_{k-l}^{\mathrm{T}}\} &= E\{[s_k(N) + \varepsilon_k(N)][s_{k-l}^{\mathrm{T}}(N) + \varepsilon_{k-l}^{\mathrm{T}}(N)]\} \\ &= E\{s_k(N) s_{k-l}^{\mathrm{T}}(N)\} + q(N) \end{aligned} \tag{2.3.4}$$

where

$$q(N) = E\varepsilon_k(N) s_{k-l}^{\mathrm{T}}(N) + E s_k(N) \varepsilon_{k-l}^{\mathrm{T}}(N) + E\varepsilon_k(N) \varepsilon_{k-l}^{\mathrm{T}}(N).$$

By taking spectral norms across this last equation, and by appealing to the Schwarz inequality (Proposition 1.1.2) and the properties of the spectral norm (see Appendix D) we deduce

$$\begin{aligned} \|q(N)\| \leq\ & (E\|\varepsilon_k(N)\|^2)^{1/2} (E\|s_{k-l}(N)\|^2)^{1/2} \\ &+ (E\|s_k(N)\|^2)^{1/2} (E\|\varepsilon_{k-l}(N)\|^2)^{1/2} \\ &+ (E\|\varepsilon_k(N)\|^2)^{1/2} (E\|\varepsilon_{k-l}(N)\|^2)^{1/2}. \end{aligned}$$

It now follows from the facts that, for fixed $k, l, E\|s_k(N)\|^2$ and $E\|s_{k-l}(N)\|^2$, $N = 0, 1, \ldots$ are uniformly bounded and $\varepsilon_{k+l}(N) \to 0$,

$\varepsilon_k(N) \to 0$ as $N \to \infty$ (see Proposition 2.1.2), that

$$q(N) \to 0 \qquad \text{as } N \to \infty. \tag{2.3.5}$$

Since (2.3.4) is true for arbitrary N,

$$\begin{aligned} E\{n_k n_{k-l}^{\mathrm{T}}\} &= \lim_{N \to \infty} [E\{s_k(N) s_{k-l}^{\mathrm{T}}(N)\} + q(N)] \\ &= \lim_{N \to \infty} E\{s_k(N) s_{k-l}^{\mathrm{T}}(N)\} \end{aligned}$$

by (2.3.5). We have verified (2.3.3).

For each N,

$$\begin{aligned} E\{s_k(N) s_{k-l}^{\mathrm{T}}(N)\} &= \sum_{j=0}^{N} \sum_{p=0}^{N} T_j E\{e_{k-j} e_{k-l-p}^{\mathrm{T}}\} T_p^{\mathrm{T}} \\ &= \sum_{j=0}^{N} \sum_{p=l}^{N+l} T_j E\{e_{k-j} e_{k-p}^{\mathrm{T}}\} T_{p-l}^{\mathrm{T}} \\ &= \sum_{j=\max\{0,l\}}^{\min\{N,N+l\}} T_j W T_{j-l}^{\mathrm{T}} \\ &= \begin{cases} \displaystyle\sum_{j=0}^{N-l} T_{j+l} W T_j^{\mathrm{T}} & l \geq 0 \\ \displaystyle\sum_{j=0}^{N+l} T_j W T_{j-l}^{\mathrm{T}} & l < 0. \end{cases} \end{aligned}$$

It follows from these expressions and (2.3.3) that

$$E\{n_k n_{k-l}^{\mathrm{T}}\} = \begin{cases} \displaystyle\sum_{j=0}^{\infty} T_{j+l} W T_j^{\mathrm{T}} & l \geq 0 \\ \displaystyle\sum_{j=0}^{\infty} T_j W T_{j-l}^{\mathrm{T}} & l < 0. \end{cases}$$

We see that $\{u_k\}$ is a wide-sense stationary process and the covariance function is as claimed.

Notice next that, by Proposition D.3.3 of Appendix D, there exist numbers $c > 0$, $\lambda \in (0, 1)$ such that

$$\|T_j\| \leq c\lambda^j, \qquad j = 0, 1, \ldots \tag{2.3.6}$$

From these estimates and properties of the spectral norm it is not difficult to deduce that the covariance function $R(\cdot)$ satisfies

$$\|R(l)\| \leq \left\| \sum_{j=\max\{0,l\}}^{\infty} T_j W T_{j-l}^{\mathrm{T}} \right\|$$

$$\leq \sum_{j=\max\{0,l\}}^{\infty} \|T_j\| \, \|W\| \, \|T_{j-l}\|$$

$$\leq c_1 \lambda^{|l|}, \qquad \text{for } l = 0, \pm 1, \ldots$$

for some constant c_1. So certainly $\sum_{l=-\infty}^{\infty} \|R(l)\| < \infty$ and therefore $\{n_k\}$ had a spectral density function $\Phi(\omega)$ which is given by

$$\Phi(\omega) = \sum_{l=-\infty}^{\infty} R(l)e^{-il\omega}, \qquad \omega \in [-\pi, +\pi]. \tag{2.3.7}$$

For the purposes of calculating $\Phi(\omega)$ it is convenient to introduce the convention $T_k = 0$ for $j < 0$. Then $R(l)$ can be expressed

$$R(l) = \sum_{j=-\infty}^{+\infty} T_j W T_{j-l}^{\mathrm{T}} \qquad l = 0, \pm 1, \ldots$$

Substitution into (2.3.7) gives

$$\Phi(\omega) = \sum_{l=-\infty}^{\infty} \sum_{j=-\infty}^{\infty} T_j W T_{j-l}^{\mathrm{T}} e^{-il\omega}.$$

However, we easily deduce from (2.3.6) that

$$\sum_{l=-\infty}^{+\infty} \sum_{j=-\infty}^{+\infty} \|T_j W T_{j-l}^{\mathrm{T}}\| < \infty.$$

Under these circumstances we are justified in changing the order of summation and writing

$$\Phi(\omega) = \sum_{j=-\infty}^{+\infty} T_j W \sum_{l=-\infty}^{\infty} T_{j-l}^{\mathrm{T}} e^{-il\omega}$$

$$= \sum_{j=-\infty}^{+\infty} T_j W \sum_{m=-\infty}^{+\infty} T_m^{\mathrm{T}} e^{-i(j-m)\omega}$$

$$= \left(\sum_{j=-\infty}^{+\infty} T_j e^{-ij\omega} \right) W \left(\sum_{m=-\infty}^{+\infty} T_m^{T} e^{im\omega} \right)$$

$$= A^{-1}(e^{-i\omega})B(e^{-i\omega})W[A^{-1}(e^{-i\omega})B(e^{-i\omega})]^*.$$

This completes the proof of Proposition 2.3.1. □

We have expressed the covariance function $R(\cdot)$ of the process $\{n_k\}$ as an infinite sum. Sometimes we require expressions for values of the covariance function in closed form. These can be obtained from the *Yule–Walker* equations:

$$\sum_{i=0}^{d_1} A_i R(i-l) = \begin{cases} \sum_{i=\max\{0,l\}}^{d_2} B_i W T_{i-l}^{\mathrm{T}} & l \le d_2 \\ 0 & l > d_2. \end{cases}$$

Here the T_i are, as usual, the Markov parameters. To check these equations we post-multiply (2.3.1) by n_{k-l}^{T} and take expectations:

$$\sum_{i=0}^{d_1} A_i E\{n_{k-i} n_{k-l}^{\mathrm{T}}\} = \sum_{i=0}^{d_2} B_i E\{e_{k-i} n_{k-l}^{\mathrm{T}}\}. \tag{2.3.8}$$

The left-hand side is simply

$$\sum_{i=0}^{d_1} A_i R(l-i).$$

To evaluate the right-hand side recall that

$$n_k = \sum_{j=0}^{\infty} T_j e_{k-j},$$

and so

$$E\{e_{k-i} n_{k-l}^{\mathrm{T}}\} = \begin{cases} W T_{i-l}^{\mathrm{T}} & i-l \ge 0 \\ 0 & i-l < 0 \end{cases}. \tag{2.3.9}$$

Substitution of (2.3.9) into (2.3.8) yields the Yule–Walker equations.

These equations can be solved for $R(0)$, $R(\pm 1)$, $R(\pm 2), \ldots,$ under suitable non-degeneracy conditions. Let us see, first of all, how we can obtain $R(0)$, $R(\pm 1), \ldots, R(\pm d_1)$.Consideration of the values $l = 0, 1, \ldots, d_1$ yields $d_1 + 1$ linear $r \times r$ matrix equations for the $2d_1 + 1$ unknown $r \times r$ matrices $R(-d_1)$, $R(-d_1+1), \ldots, R(d_1)$. However,

$$R(-i)^{\mathrm{T}} = R(i), \qquad \text{all } i,$$

so the linear equations really involve just $d_1 + 1$ unknown $r \times r$ matrices, say $R(0), \ldots, R(d_1)$. If the Jacobian matrix in question is non-singular the equations can be solved for $R(0), \ldots, R(d_1)$ (and hence $R(-d_1), \ldots, R(d_1)$). We may now regard the Yule–Walker equations as recursive relations which yield the remaining $R(k)$ given the starting values $R(-d_1), \ldots, R(d_1)$.

In practice it is often more convenient, instead of using the Yule–Walker equations themselves, to use the idea behind their derivation. That is to say we obtain relationships between the $R(j)$ as a result of multiplying across the ARMA model equations by outputs or disturbances at different times and taking expectations. We shall illustrate this shortly.

An alternative approach is to evaluate contour integrals. Here we note that the $R(j)$ are the coefficients in the Fourier expansion of the function $\Phi(\omega)$ on $[-\pi, \pi]$:

$$\Phi(\omega) = \sum_{j=-\infty}^{+\infty} R(j)e^{-ij\omega}.$$

Consequently the $R(j)$ can be recovered from $\Phi(\omega)$ by use of the 'inverse' formulae:

$$R(j) = \frac{1}{2\pi}\int_{-\pi}^{\pi} \Phi(\omega)e^{ij\omega}\,\mathrm{d}\omega, \qquad j = 0, \pm 1, \ldots$$

Now

$$\int_{-\pi}^{\pi} \Phi(\omega)e^{ij\omega}\,\mathrm{d}\omega = \int_{-\pi}^{\pi} \Lambda(e^{-i\omega})e^{ij\omega}\,\mathrm{d}\omega$$

(where $\Lambda(\sigma)$ is taken to be $[A(\sigma)]^{-1}B(\sigma)W[A^{-1}(\sigma^{-1})B(\sigma^{-1})]^{\mathrm{T}}$)

$$= \int_{-\pi}^{\pi} \Lambda(e^{-i\omega})(e^{i\omega})^{j-1}e^{i\omega}\,\mathrm{d}\omega$$

$$= i^{-1}\int_{\Gamma} \Lambda(\xi^{-1})\xi^{j-1}\,\mathrm{d}\xi.$$

The last integral is a contour integral in the complex plane around the unit circle, denoted Γ. It follows that

$$R(j) = \frac{1}{2\pi i}\int_{\Gamma} \Lambda(\xi^{-1})\xi^{j-1}\,\mathrm{d}\xi, \qquad j = 0, \pm 1, \ldots \tag{2.3.10}$$

The right-hand side will be recognized as the sum of the residues of poles of $\xi \to \Lambda(\xi^{-1})\xi^{j-1}$ which lie inside the open unit disc. (No difficulties arise with poles on the contours because of our hypotheses on $A(z^{-1})$). The problem of calculating the covariance function then reduces to one in residue calculus. In the event that $\xi \to \Lambda(\xi^{-1})\xi^{j-1}$ has only simple poles (let us write them $\xi_1, \ldots, \xi_p$) inside the open unit disc, we have

$$R(j) = \sum_i (\text{residue of the pole at } z_i)$$

$$= \sum_i \left[\lim_{\xi \to \xi_i} (\xi - \xi_i)\Lambda(\xi^{-1})\xi^{j-1}\right].$$

***Example* 2.3.2**

Consider a scalar autoregressive process $\{y_k\}$ described by the equations

$$y_k - ay_{k-1} = e_k, \qquad k \in \mathbb{Z}, \tag{2.3.11}$$

Here $\{e_k\}$ is a sequence of zero mean, second order, uncorrelated random variables, each of unit variance. We assume that $|a| < 1$. The Markov parameters associated with the transfer function $(1 - az^{-1})^{-1}$ are $1, a, a^2, \ldots$. Consequently the Yule–Walker equations take the form:

$$R(-l) - aR(1-l) = \begin{cases} 0 & l > 0 \\ a^{|l|} & l \le 0 \end{cases}. \tag{2.3.12}$$

Setting $l = 0$ and 1, as prescribed above, yields equations for starting values $R(0)$, $R(-1)$ $(= R(+1))$:

$$R(0) - aR(-1) = 1$$
$$R(-1) - aR(0) = 0.$$

These simultaneous equations for $R(-1)$, $R(0)$ have solution

$$R(0) = (1 - a^2)^{-1}, \quad R(-1) = a(1 - a^2)^{-1}.$$

Knowing $R(0)$ we can solve (2.3.12) recursively for $R(-1), R(-2), \ldots$:

$$R(l) = a^{|l|}(1 - a^2)^{-1} \qquad \text{for} \qquad l = -1, -2, \ldots.$$

A more direct approach is to multiply across (2.3.11) by y_{k-l} and take expectations:

$$E\{y_k y_{k-l}\} - aE\{y_{k-1} y_{k-l}\} = E\{e_k y_{k-l}\}. \tag{2.3.13}$$

Now y_{k-l} and e_k have mean zero and are uncorrelated for $l \ge 1$. Consequently

$$R(l) - aR(l-1) = 0, \qquad l \ge 1. \tag{2.3.14}$$

Setting $l = 0$ in (2.3.13) gives

$$R(0) - aR(1) = E\{e_k y_k\}. \tag{2.3.15}$$

In order to evaluate $E\{e_k y_k\}$, we multiply across (2.3.11) by e_k and take expectations:

$$E\{e_k y_k\} - aE\{e_k y_{k-1}\} = E\{e_k^2\}.$$

The second term on the left is zero and we deduce

$$E\{e_k y_k\} = 1.$$

From (2.3.15) then

$$R(0) - aR(1) = 1. \tag{2.3.16}$$

It follows from (2.3.14) and (2.3.16) that

$$R(l) = a^l(1-a^2)^{-1} \qquad l = 0, 1, 2, \ldots,$$

in agreement with our earlier calculation.

Finally we illustrate computation of the covariance function by contour integration. According to Proposition 2.3.1, the process $\{y_k\}$ has spectral density

$$\Phi(\omega) = (1 - ae^{-i\omega})^{-1}(1 - ae^{i\omega})^{-1}.$$

Formula (2.3.10) for the covariance function gives

$$R(j) = \frac{1}{2\pi i}\int_{\Gamma}(1 - a\xi^{-1})^{-1}(1 - a\xi)^{-1}\xi^{j-1}d\xi.$$

The integrand can be expressed

$$\frac{\xi^j}{(\xi - a)(1 - a\xi)}.$$

For $j = 0, 1, \ldots$ this function has just one first-order pole in the unit disc; it is at $\xi = a$. The residue at $\xi = a$ is

$$\lim_{\xi \to a}\frac{(\xi - a)\xi^j}{(\xi - a)(1 - a\xi)} = \frac{a^j}{(1 - a^2)}.$$

It follows that

$$R(j) = a^j(1-a^2)^{-1}, \qquad j = 0, 1, \ldots$$

whence

$$R(j) = a^{|j|}(1-a^2)^{-1}, \qquad j = 0, \pm 1, \ldots$$

as before.

It is natural to enquire into the nature of the assumptions implicit in consideration of ARMA noise models. As we have seen, the ARMA noise model (2.3.1), under the stated hypotheses, has spectral density function $\Phi(\omega)$:

$$\Phi(\omega) = A^{-1}(e^{-i\omega})B(e^{-i\omega})\, W[A^{-1}(e^{i\omega})B(e^{i\omega})]^{\mathrm{T}}.$$

By multiplying numerators and denominators of entries of this matrix by a sufficiently high power of $e^{-i\omega}$ we can arrange that Φ has the property that its entries are rational functions of $e^{-i\omega}$. It is essentially this property which characterizes a process whose spectral density function coincides with that of some ARMA noise model. In other words, consideration of an ARMA model amounts to assuming that the spectral density function is rational in $e^{-i\omega}$. To be more precise, we have the following theorem.

Theorem 2.3.3

Let $\Phi(\omega)$ be a matrix spectral density function. The following is a necessary and sufficient condition that $\Phi(\omega)$ be the matrix spectral density function of some ARMA noise model†:

$$\Phi(\omega) = \Lambda(e^{-i\omega}) \text{ a.e. } \omega \in [-\pi, +\pi]$$

for some matrix $\Lambda(\sigma)$ of rational functions with real coefficients which is such that

$$\Lambda(\sigma) = \Lambda^{\mathrm{T}}(\sigma^{-1}) \text{ a.e. } \sigma \in \mathbb{C}. \tag{2.3.17}$$

Furthermore, any $r \times r$ matrix spectral density function which satisfies this condition is the matrix spectral density function of some ARMA noise model in which the noise vectors have dimension r and common covariance matrix I_r (the $r \times r$ identity matrix).

Observe that, for any $\Lambda(\sigma)$ related to $\Phi(\omega)$ by $\Phi(\omega) = \Lambda(e^{-i\omega})$, (2.3.17) is automatically true for almost every σ on the unit circle by properties of the spectral density function. Condition (2.3.17) requires the relation to hold almost everywhere on the complex plane, not just on the unit circle.

Necessity of the condition given in the proposition is a simple consequence of the representation of the spectral density function of an ARMAX noise model provided by Proposition 2.3.1. To prove the rest of the theorem we need to show that if $\Phi(\omega)$ is an $r \times r$ matrix

† 'Almost every' (a.e. for short) means 'for all except a finite number of values of the variable in question'.

spectral density function expressible in terms of Λ as described in the proposition, then Λ can be written

$$\Lambda(\sigma)=[A^{-1}(\sigma)B(\sigma)][A^{-1}(\sigma^{-1})B(\sigma^{-1})]^{T}$$

where A, B are polynomials with $r\times r$ matrix coefficients and such the $\sigma\to\det A(\sigma)$ has no zeros in the closed unit disc; these polynomials will then serve to define an ARMA noise model (2.3.1) with spectral density function $\Phi(\omega)$ when we take $\operatorname{cov}\{e_k\}=I_r$.

We limit ourselves now to consideration to the scalar case. The matrix case, which is complicated, is treated by Hannan (1970), p. 128.

COMPLETION OF PROOF OF THEOREM 2.3.1 (SCALAR CASE) Let us suppose that $\Lambda(\sigma)$ is not identically zero since otherwise $\Phi(\omega)$ is obviously associated with some ARMA noise model.

We show first of all that none of the poles of $\Lambda(\sigma)$ lie on the unit circle. Suppose this were not the case. Then there is some $\theta\in[-\pi,+\pi]$ such that $e^{-i\theta}$ is a pole of $\Lambda(\sigma)$, of multiplicity v. By means of partial fraction expansion we can express $\Lambda(\sigma)$:

$$\Lambda(\sigma)=\frac{r(\sigma)}{(e^{-i\theta}-\sigma)^{v}}+s(\sigma)$$

where $r(\sigma)$ is a polynomial such that $r(e^{-i\theta})\neq 0$ and $s(\sigma)$ is a rational function of which $e^{-i\theta}$ is not a pole. It is not difficult to see that, as $\omega\to\theta$, $\Lambda(e^{-i\omega})$ deviates by an arbitrarily small amount from $\psi(\omega)$:

$$\psi(\omega)=i^{v}r(e^{-i\theta})e^{vi\theta}(\omega-\theta)^{-v}+s(e^{-i\omega}).$$

But because of the $\omega\to(\omega-\theta)^{-v}$ singularity $\psi(\omega)$ is not an integrable function. It follows that neither is $\Lambda(e^{-i\omega})$. This contradicts our assumption that $\Lambda(e^{-i\omega})$ is a spectral density function. So $\Lambda(\sigma)$ can have no poles on the unit circle.

Next note that, in the scalar case which we consider here, (2.3.17) can be expressed

$$\Lambda(\sigma)/\Lambda(\sigma^{-1})=1,\qquad \text{a.e. } \sigma\in\mathbb{C}. \tag{2.3.18}$$

Suppose now that b ($b\neq 0$) is a zero of $\Lambda(\sigma)$ of multiplicity μ. It follows from (2.3.18) and the fact that the coefficients of $\Lambda(\sigma)$ are real that b^{-1}, b^* and $(b^*)^{-1}$ are also zeros of multiplicity μ. The zero b occurs then as a member of a certain configuration of 4, 2 or 1 distinct zeros, each of multiplicity μ, depending on how many distinct complex numbers are generated by the operations of inversion, complex conjugation and inversion of the complex conjugate. Let us state these conclusions more precisely. An arbitrary complex number c, if it is not 0, 1 or -1, is of one of the following three types:

Type 1: $\mathrm{Im}\{c\} \neq 0$ and $|c| \neq 1$
Type 2: c is real and $c \neq 0$, 1 or -1
Type 3: $\mathrm{Im}\{c\} \neq 0$ and $|c| = 1$.

If b is a zero of type 1 and with multiplicity μ then b occurs in a configuration of four distinct zeros, each of multiplicity μ and of the same type. This is true also for both type 2 and type 3 zeros, except that in these two cases the configurations are of just two distinct zeros.

The poles of $\Lambda(\sigma)$ have analogous properties. Note however that, in view of earlier remarks, poles of type 3 (which lie on the unit circle) cannot arise.

If then b is a zero of multiplicity μ and of type 1, $\Lambda(\sigma)$ can be factored:

$$\Lambda(\sigma) = D(\sigma)h(\sigma)h(\sigma^{-1})$$

in which $h(\sigma)$ is the polynomial with real coefficients

$$h(\sigma) = (\sigma - b)^{\mu}(\sigma - b^*)^{\mu}$$

and $D(\sigma)$ has neither poles nor zeros at $b, b^*, b^{-1}, (b^*)^{-1}$. Notice that we can always arrange that the roots of the polynomial $h(\sigma)$ lie outside the closed unit disc by modifying $h(\sigma)$ and $D(\sigma)$ in the factorization if necessary. Indeed

$$h(\sigma)h(\sigma^{-1}) = (bb^*)^{2\mu}\, \tilde{h}(\sigma)\tilde{h}(\sigma^{-1})$$

where

$$\tilde{h}(\sigma) = (\sigma - b^{-1})^{\mu}(\sigma - (b^*)^{-1})^{\mu}).$$

If b lies inside the open unit disc then b^{-1} and $(b^*)^{-1}$ lie outside the closed unit disc, so the desired factorization can be achieved if we replace $h(\sigma)$ by $\tilde{h}(\sigma)$ and multiply $D(\sigma)$ by $(bb^*)^{2\mu}$.

Such reasoning applied also to the poles of type 1 and to the poles and zeros of type 2 leads to the conclusion that $\Lambda(\sigma)$ can be factored

$$\Lambda(\sigma) = k\sigma^{p}(\sigma - 1)^{q}(\sigma + 1)^{r}P(\sigma)\tilde{G}(\sigma)\tilde{G}(\sigma^{-1}). \qquad (2.3.19)$$

Here k is a (possibly complex) number and p, q and r are integers. $\tilde{G}(\sigma)$, constructed from poles and zeros of types 1 and 2, is a rational polynomial which has real coefficients and which has all its poles and zeros outside the closed unit disc. $P(\sigma)$, which arises from zeros of type 3, is identically 1 or has roots on the unit circle and is of the form

$$P(\sigma) = \prod_k (\sigma - 2\cos\theta_k + \sigma^{-1})^{\mu_k}. \tag{2.3.20}$$

In this last expression the θ_k's are distinct real numbers lying in the set $(-\pi, 0) \cup (0, \pi)$ and the μ_k's are positive integers. We claim that the μ_k's must be even. Otherwise we can arrange by reordering that μ_1 is odd. Then $\Lambda(\sigma)$ can be factored

$$\Lambda(\sigma) = (\sigma - 2\cos\theta_1 + \sigma^{-1})^{\mu_1} F(\sigma)$$

in which $F(\sigma)$ is a rational function which does not vanish at $\sigma = e^{-i\theta_k}$. Consider now

$$\Lambda(e^{-i(\theta_1+\delta)}) = 2^{\mu_1}(\cos(\theta_1+\delta) - \cos\theta_1)^{\mu_1} F(e^{-i(\theta_i+\delta)})$$

as a function of δ on some neighbourhood of 0. Since μ_1 is assumed odd, this function changes sign as δ passes through 0. This contradicts the non-negativity of the spectral density function; we conclude that the μ_k's must be even.

Since the μ_k's in (2.3.20) are even, we can factor $P(\sigma)$, if it is present in (2.3.19), as

$$P(\sigma) = Q(\sigma)Q(\sigma^{-1})$$

in which $Q(\sigma) = Q(\sigma^{-1})$. Writing $G(\sigma)$ for $Q(\sigma)\tilde{G}(\sigma)$ we obtain the representation

$$\Lambda(\sigma) = k\sigma^p(\sigma - 1)^q(\sigma + 1)^r G(\sigma)G(\sigma^{-1}) \tag{2.3.21}$$

in which $G(\sigma)$ is a rational function with real coefficients which has no zeros inside the open unit disc and no poles inside the closed unit disc.

Because $\Lambda(\sigma)$ cannot have poles on the unit circle we deduce from (2.3.21) that $q, r \geq 0$. From (2.3.18)

$$\sigma^{(2p+q+r)} \cdot (-1)^q = 1 \qquad \text{a.e. } \sigma \in \mathbb{C}.$$

This identity can be satisfied only if q is even and $2p + q + r = 0$. But then r must also be even and

$$\sigma^p(\sigma-1)^q(\sigma+1)^r = (\sigma-1)^{q/2}(\sigma+1)^{r/2}(\sigma^{-1}-1)^{q/2}(\sigma^{-1}+1)^{r/2}(-1)^{q/2}.$$

We have shown that

$$\Lambda(\sigma) = \bar{k}H(\sigma)H(\sigma^{-1}) \qquad \text{a.e. } \sigma \in \mathbb{C} \tag{2.3.22}$$

where $H(\sigma)$ is the rational function with real coefficients $H(\sigma) = (\sigma + 1)^{r/2}(\sigma - 1)^{q/2} G(\sigma)$ and $\bar{k} = k(-1)^{q/2}$. The facts that both $\Lambda(\sigma)$ and $H(\sigma)$ have real coefficients and are not identifically zero lead us to the conclusion that $\bar{k}$ is real.

Let us now note that $\bar{k}$ must be positive. Since $\Lambda(e^{-i\omega})$ is a (scalar) spectral density function it assumes real, non-negative values. By assumption it is not identically zero however and so $\Lambda(e^{-i\theta}) > 0$ for some $\theta \in [-\pi, +\pi]$. Then

$$\bar{k}|H(e^{-i\theta})|^2 = \bar{k}H(e^{-i\theta})H(e^{i\theta}) = \Lambda(e^{-i\theta}) > 0.$$

This inequality implies $\bar{k} > 0$.

Since $\bar{k}$ is positive, we may remove it from (2.3.22) by absorbing $\bar{k}^{1/2}$ into $H(\sigma)$. There results a representation of $\Lambda(\sigma)$ of the form (2.3.18) in which $A(\sigma)$ has no zeros in the closed unit disc. The proof is complete in the scalar case. □

Scrutiny of the proof will reveal that we have actually established (in the scalar case) rather more than is claimed in the theorem, namely: a scalar spectral density function which satisfies the condition of the theorem can be realized by an ARMA noise model (2.3.1) in which the polynomial $B(\sigma)$ has no zeros in the open unit disc.

2.4 Stochastic dynamical models

We now describe a number of important stochastic dynamical models. They all conform to the description of Fig. 2.3, namely the output supplied by the model can be interpreted as the output to a deterministic system S driven by the input, to which has been added a noise process $\{w_k\}$ expressible as the output of a second linear system $\tilde{S}$ driven by white noise. It will turn out that all the models which we describe in this section are essentially different forms of the same model. There is a point none the less in separating them out, since different forms of the model suggest different controller design and identification procedures.

2.4.1 General stochastic dynamical models

According to this model the sequence of r-vector outputs $\{y_k\}$ and m-vector inputs $\{u_k\}$ are related by

$$y_k = P(z^{-1})u_k + Q(z^{-1})e_k. \tag{2.4.1}$$

Here $P(\sigma)$, $Q(\sigma)$ are $r \times m$, $r \times r$ matrices of rational functions in σ expressible as

$$P(\sigma) = p^{-1}(\sigma)\tilde{P}(\sigma), \quad Q(\sigma) = q^{-1}(\sigma)\tilde{Q}(\sigma).$$

In these expressions $p(\sigma)$, $q(\sigma)$ are polynomials in σ such that $p(0) \neq 0$, $q(0) \neq 0$. $\tilde{P}(\sigma)$, $\tilde{Q}(\sigma)$ are polynomials in σ with coefficients $r \times m$, $r \times r$ matrices respectively. The driving noise process, $\{e_k\}$, is a collection of zero mean, uncorrelated r-vector random variables. We make few restrictions on the nature of $\{e_k\}$ at this stage, but will impose additional conditions on $\{e_k\}$ in the future as the need arises (e.g. the e_k's have common covariance matrix, are independent, etc.).

In accordance with our earlier remarks $\{y_k\}$ is expressible as the sum of the output $\{y_k^*\}$ from a deterministic linear system given by the input:

$$y_k^* = P(z^{-1})u_k$$

and a noise process which is the output $\{w_k\}$ of a linear system driven by white noise $\{e_k\}$:

$$w_k = Q(z^{-1})e_k.$$

Notice that we have taken the e_k's to be of the same dimension as the outputs. This is not unreasonable since if the noise at the output, $\{w_k\}$, is a wide sense stationary ARMA process then we can assume without loss of generality, so far as second order statistics of $\{w_k\}$ are concerned, that the driving noise has the same dimension as $\{w_k\}$ (see Theorem 2.3.3).

2.4.2 ARMAX models

ARMAX models are obtained by appending a moving average of the input to the ARMA noise model of Section 2.3. The 'X' in the label ARMAX attached to these models refers to the terminology 'exogeneous variable' used in the econometrics literature to mean 'external inputs to the system'. The r-vector outputs y_k and m-vector inputs u_k are related then by

$$A(z^{-1})y_k = B(z^{-1})u_k + C(z^{-1})e_k. \tag{2.4.2}$$

In this equation, $A(\sigma)$, $B(\sigma)$, $C(\sigma)$ are polynomials in σ with coefficients $r \times r$, $r \times m$, $r \times r$ matrices, and $A(\sigma)$ satisfies $\det A(0) \neq 0$. $\{e_k\}$ is a collection of zero-mean, uncorrelated r-vector random variables.

2.4.3 Stochastic state-space models

Stochastic state-space models result from adding noise processes to the state and observation equations of the linear system model studied in Section 1.2. Thus the r-vector outputs $\{y_k\}$ are related to the n-vector states $\{x_k\}$ by the equations

$$\begin{aligned} x_{k+1} &= Ax_k + Bu_k + Ce_k \\ y_k &= Hx_k + Ge_k. \end{aligned} \tag{2.4.3}$$

Here A, B, C, H, G are $n \times n$, $n \times m$, $n \times l$, $r \times n$, $r \times l$ matrices respectively. $\{e_k\}$ is a sequence of zero-mean uncorrelated l-vector random variables, and the initial state x_0 is uncorrelated with $\{e_k\}$. Note that by superposition x_k, y_k can be written as

$$\begin{aligned} x_k &= \bar{x}_k + x_k^*, \\ y_k &= \bar{y}_k + y_k^* \end{aligned}$$

where

$$\begin{aligned} x_{k+1}^* &= Ax_k^* + Bu_k,\ x_0^* = Ex_0 \\ y_k^* &= Hx^* \end{aligned} \tag{2.4.4}$$

and

$$\begin{aligned} \bar{x}_{k+1} &= A\bar{x}_k + Ce_k, \qquad \bar{x}_0 = x_0 - Ex_0 \\ \bar{y}_k &= H\bar{x}_k + Ge_k. \end{aligned} \tag{2.4.5}$$

Referring back to Fig. 2.3 at the beginning of this chapter, we see that (2.4.4) and (2.4.5) represent the 'system' and 'noise' models S and $\tilde{S}$ respectively, both in state space form.

Some covariance calculations for state space models

Consider the stochastic state space model when the input is zero:

$$\begin{aligned} x_{k+1} &= Ax_k + Ce_k \\ y_k &= Hx_k + Ge_k. \end{aligned} \tag{2.4.6}$$

We assume that the e_k's are uncorrelated, have zero mean and $\operatorname{cov}\{e_k\} = I$ for all k. In future chapters we require detailed information about the covariance matrices of the state and output processes. We collect together the necessary results in the following proposition.

Proposition 2.4.1

Suppose in (2.4.6) that the time set is $\mathbb{Z}^+$, and the initial state x_0 has mean m_0 and covariance P_0. Then $P(k) := \operatorname{cov}\{x_k\}$ satisfies

$$P(k+1) = AP(k)A^{\mathrm{T}} + CC^{\mathrm{T}}, P(0) = P_0 \tag{2.4.7}$$

and

$$\operatorname{cov}\{y_k, y_{k-j}\} = \begin{cases} HP(k)H^{\mathrm{T}} + GG^{\mathrm{T}}, & j=0 \\ HA^jP(k-j)H^{\mathrm{T}} + HA^{j-1}CG^{\mathrm{T}}, & j = 1, 2, \dots, k. \end{cases} \tag{2.4.8}$$

If A is stable then $P(k) \to P$ as $k \to \infty$ where P is the unique solution of the *Lyapunov equation*

$$P = APA^{\mathrm{T}} + CC^{\mathrm{T}}.$$

Furthermore if $P_0 = P$ then $P(k) = P$ for all $k \geq 0$. The mean $m(k) := E\{x_k\}$ satisfies

$$\begin{aligned} m(k+1) &= Am(k) \\ m(0) &= m_0. \end{aligned} \tag{2.4.9}$$

Now suppose that the time set is $\mathbb{Z}$ and that A is stable. In this case $\{x_k\}$ and $\{y_k\}$ are widesense stationary processes,

$$E\{x_k\} = 0, \quad \operatorname{cov}\{x_k\} = P$$

and

$$\operatorname{cov}\{y_k, y_{k-j}\} = \begin{cases} HPH^{\mathrm{T}} + GG^{\mathrm{T}}, & j=0 \\ HA^jPH^{\mathrm{T}} + HA^{j-1}CG^{\mathrm{T}}, & j>0. \end{cases}$$

Here P is once again the solution to the Lyapunov equation.

PROOF Suppose the time set is $\mathbb{Z}^+$. The equations (2.4.9) follow from taking expectations across the state equation. Defining $\bar{x}_k := x_k - m(k)$, $\bar{y}_k := y_k - E\{y_k\}$ we deduce from (2.4.9) and (2.4.6) that $\{\bar{x}_k\}$, $\{\bar{y}_k\}$ satisfy (2.4.5). By (2.4.5)

$$\bar{x}_k = A^j\bar{x}_{k-j} + A^{j-1}Ce_{k-j} + \cdots + Ce_{k-1}$$

for $1 \leq j \leq k$. Since $\bar{x}_{k-j}$ is uncorrelated with $e_{k-j}, \dots, e_{k-1}$ we deduce from this equation that

$$E\{\bar{x}_k\bar{x}_{k-j}^{\mathrm{T}}\} = A^jP(k-j)$$

and

$$E\{\bar{x}_k e_{k-j}^{\mathrm{T}}\} = A^{j-1}C.$$

It follows that, for $1 \le j \le k$,

$$\begin{aligned}\operatorname{cov}\{y_k, y_{k-j}\} &= E\{\bar{y}_k \bar{y}_{k-j}^{\mathrm{T}}\} = E\{(H\bar{x}_k + Ge_k)(H\bar{x}_{k-j} + Ge_{k-j})^{\mathrm{T}}\}\\ &= HA^jP(k-j)H^{\mathrm{T}} + HA^{j-1}CG^{\mathrm{T}}.\end{aligned}$$

By (2.4.5) and since $\bar{x}_k$ and e_k are uncorrelated, $\{P(k)\}$ satisfies (2.4.7) and

$$\operatorname{cov}\{y_k\} = HP(k)H^{\mathrm{T}} + GG^{\mathrm{T}}.$$

we have proved (2.4.7) and (2.4.8).

Now suppose that A is a stable matrix, let $\mathscr{M}$ denote the set of $n \times n$ matrices and define for any $D \in \mathscr{M}$

$$\tilde{P}(D) := \sum_{k=0}^{\infty} A^k D (A^{\mathrm{T}})^k. \tag{2.4.10}$$

$\tilde{P}(D)$ is well-defined since for any $x \in \mathbb{R}^n$

$$|x^{\mathrm{T}} A^k D (A^{\mathrm{T}})^k x| \le c\lambda^k \|x\|^2$$

for some $c > 0$, $\lambda \in (0, 1)$ under the stability condition (see Appendix D). It is easy to see that $\tilde{P}(D)$ satisfies

$$\tilde{P}(D) = A\tilde{P}(D)A^{\mathrm{T}} + D.$$

Let L be the map from $\mathscr{M}$ to $\mathscr{M}$ defined by

$$L(P) = P - APA^{\mathrm{T}}.$$

L can be thought of as a map from $\mathbb{R}^{n^2}$ to $\mathbb{R}^{n^2}$ since each matrix can be identified with the point in $\mathbb{R}^{n^2}$ whose coordinates are its n^2 entries. L is linear and its range is all of $\mathbb{R}^{n^2}$ since for any matrix D there is a P such that $L(P) = D$, namely $P = \tilde{P}(D)$. But it is a standard result in linear algebra that if the range of L is full then its null space (i.e. the set of P such that $L(P) = 0$) consists of only the zero element. Taking $D = CC^{\mathrm{T}}$ this shows that the Lyapunov equation $P = APA^{\mathrm{T}} + CC^{\mathrm{T}}$ has unique solution $P = \tilde{P}(CC^{\mathrm{T}})$ and P is non-negative in view of (2.4.10). The nth partial sum of the right hand side of (2.4.10) coincides with $P(k)$ given by (2.4.7) with $P_0 = 0$. If $P_0 \ne 0$ then there is an additional term $A^{n+1}P_0(A^{\mathrm{T}})^{n+1}$ and this converges to 0 as $n \to \infty$. Thus $P(k) \to P$ as $k \to \infty$ regardless of the initial condition P_0.

Finally consider the case when the time set is $\mathbb{Z}$. Since A is stable,

$\{x_k\}$ and $\{y_k\}$ can be expressed as outputs of ARMA models (see below). By Proposition 2.3.1 then $\{x_k\}$ and $\{y_k\}$ are widesense stationary processes. It follows from the state equation in (2.4.6) that $\operatorname{cov}\{x_k\}$ satisfies the Lyapunov equation and therefore, by uniqueness, $\operatorname{cov}\{x_k\} = P$. We show much as before that, for $j \geq 1$,

$$E\{x_k x_{k-j}^{\mathrm{T}}\} = A^j P.$$

We now deduce the formulae for the covariance function of $\{y_k\}$ from the output equation in (2.4.6).

2.4.4 Initial conditions

For each of the preceding models we can take the underlying time set to be either $\mathbb{Z}$ or $\mathbb{Z}^+$.

Consider first the situation in which the time set is $\mathbb{Z}$ (which can be viewed as the case when initial data comes in the form of the infinite past of $\{u_k\}$ and $\{e_k\}$). The output $\{y_k\}$ of the general stochastic dynamical model (2.4.1) is defined by

$$y_k = T(z^{-1})\tilde{e}_k$$

where

$$T = [P \,\vdots\, Q]], \quad \tilde{e}_k = \begin{bmatrix} u_k \\ e_k \end{bmatrix}, \quad P = p^{-1}\tilde{P}, \quad Q = q^{-1}\tilde{Q},$$

according to the theory of Section 2.1. This is possible under the additional hypotheses that $\{e_k\}$, $\{u_k\}$ have uniformly bounded moments of an appropriate order and that the zeros of $p(\sigma)$ and $q(\sigma)$ lie outside the closed unit disc. The output $\{y_k\}$ of the ARMAX model (2.4.2):

$$A(z^{-1})y_k = B(z^{-1})u_k + C(z^{-1})e_k$$

is taken in this setting to be the output of the general stochastic dynamical model

$$y_k = A^{-1}(z^{-1})B(z^{-1})u_k + A^{-1}(z^{-1})C(z^{-1})e_k$$

in the sense just described. The output is well defined if $\sigma \to \det A(\sigma)$ has all zeros outside the unit disc and if the moments of $\{e_k\}$, $\{u_k\}$ are suitably bounded. As for the space model (2.4.3):

$$\begin{aligned} x_{k+1} &= Ax_k + Bu_k + Ce_k \\ y_k &= Hx_k + Ge_k, \end{aligned}$$

when the time set is $\mathbb{Z}$, the output y_k is taken to be that of the general stochastic dynamical model (2.4.1):

$$y_k = P(z^{-1})u_k + Q(z^{-1})e_k$$

with

$$P(\sigma) = \sigma H[I - \sigma A]^{-1}B, \quad Q(\sigma) = \sigma H[I - \sigma A]^{-1}C + G.$$

It is not difficult to show that the hypotheses under which (2.4.1) defines $\{y_k\}$ are satisfied if A has all its eigenvalues in the open unit disc, and the moments of $\{u_k\}$, $\{e_k\}$ are suitably bounded.

The other case to be considered is that when the time set is $\mathbb{Z}^+$. Here the ARMAX model equation (2.4.2) can be solved recursively to yield y_k, $k = 0, 1, \ldots$ (as a function of the inputs $u_0, u_1, \ldots$ and noise $e_0, e_1, \ldots$, provided initial data $y_{-1}, \ldots, y_{-n}$, $u_{-1}, \ldots, u_{-n}$, $e_{-1}, \ldots, e_{-n}$ is supplied). There is no difficulty either with defining the output to the state-space model (2.4.3) when the time set is $\mathbb{Z}^+$. The state-space equations can be solved recursively given x_0 as initial data. It remains to consider general stochastic dynamical model (2.4.1). By extracting the least common multiple $g(\sigma)$ of the polynomials comprising the denominators of entries of $P(\sigma)$ and $Q(\sigma)$, we can always express the general stochastic dynamical model equations as:

$$y_k = g^{-1}(z^{-1})\tilde{P}(z^{-1})u_k + g^{-1}(z^{-1})\tilde{Q}(z^{-1})e_k$$

for some polynomials $P(\sigma)$, $Q(\sigma)$ in σ with matrix coefficients. For the purposes of computing the output to (2.4.1) when the time set is $\mathbb{Z}^+$, the model is treated as the ARMAX model (2.4.2) with

$$A(\sigma) = g(\sigma)I, B(\sigma) = P(\sigma), \; C(\sigma) = Q(\sigma).$$

Initial data in the form of $y_{-1}, \ldots, y_{-\tilde{n}}, u_{-1}, \ldots, u_{-\tilde{n}}$ and $e_{-1}, \ldots, e_{-\tilde{n}}$ is required, where $\tilde{n} = \max$ degree $\{g(\sigma), \tilde{P}(\sigma), \tilde{Q}(\sigma)\}$.

The arguments of Section 2.1 applied to the equations

$$A(z^{-1})y_k = [B(z^{-1}) \vdots C(z^{-1})]\begin{bmatrix} u_k \\ e_k \end{bmatrix}$$

establish that our two notions of solutions to the ARMAX model equations for the output y_k are compatible to the extent that, if y_k is a solution on $\mathbb{Z}$ then $y_0, y_1, \ldots,$ is a solution to the recursive equations specified by

$$A(z^{-1})y_k = B(z^{-1})u_k + C(z^{-1})e_k, \qquad k \in \mathbb{Z}^+,$$

when initial data on the y_k variable is suitably chosen. It is not difficult to show that the two notions of solutions are compatible (in an analogous sense) for general stochastic dynamical models and state-space models also.

2.4.5 Interchangeability of models

We limit attention to models for which the time set is $\mathbb{Z}$. As a by-product of our discussion in the preceding subsection, we see that ARMAX models (2.4.2) and stochastic state-space models (2.4.3) can always be reformulated as general stochastic dynamical models (2.4.1) (under appropriate stability and boundedness assumptions, of course). It is clear, too, that a general stochastic dynamical model can always be rewritten as an ARMAX model:

$$g(z^{-1})y_k = \tilde{P}(z^{-1})u_k + \tilde{Q}(z^{-1})e_k$$

where the polynomial $g(\sigma)$ is the least common multiple of the denominators of $P(\sigma)$ and $Q(\sigma)$, $\tilde{P}(\sigma)$, $\tilde{Q}(\sigma)$ are suitable polynomials with matrix coefficients.

In fact it is true that we can pass freely between all the models considered (subject to mild qualifications). To confirm this, it remains to show that an ARMAX model can be reformulated as a stochastic state-space model. This final step is supplied by the next proposition.

Proposition 2.4.2

Suppose that the processes $\{y_k\}$, $\{u_k\}$ and $\{e_k\}$ are related by the ARMAX model equations

$$y_k + A_1 y_{k-1} + \cdots + A_n y_{k-n} = B_1 u_{k-1} + \cdots + B_n u_{k-n} + C_0 e_k + C_1 e_{k-1} + \cdots + C_n e_{k-n}, \qquad k \in \mathbb{Z},$$

where $A_1, \ldots, A_n$ are $r \times r$ matrices, $B_1, \ldots, B_n$ are $r \times m$ matrices, $C_0, \ldots, C_n$ are $r \times r$ matrices. Then there exists some 'state' process $\{x_k\}$ such that the state-space equations

$$x_{k+1} = \begin{bmatrix} 0 & & -A_n \\ & & --- \\ I & & \vdots \\ & \ddots & ---- \\ & I & -A_1 \end{bmatrix} x_k + \begin{bmatrix} B_n \\ ---- \\ \vdots \\ --- \\ B_1 \end{bmatrix} u_k + \begin{bmatrix} C_n - A_n C_0 \\ ----- \\ \vdots \\ C_1 - A_1 C_0 \end{bmatrix} e_k,$$

$$y_k = [0 \mid \ldots \mid I] x_k + e_k, \quad k \in \mathbb{Z},$$

are satisfied by $\{u_k, y_k, e_k\}$.

Here I is the $r \times r$ identity matrix.

PROOF The ARMAX model equations can be organized as a 'nested' sum:

$$\begin{aligned} y_k = (&-A_1 y_{k-1} + B_1 u_{k-1} + C_1 e_{k-1} \\ &+(-A_2 y_{k-2} + B_2 u_{k-2} + C_2 e_{k-2} \\ &+(\ldots(-A_n y_{k-n} + B_n u_{k-n} + C_n e_{k-n})\ldots) + C_0 e_k. \end{aligned} \tag{2.4.11}$$

Let us now introduce auxiliary variables defined by the recursive relations:

$$\left.\begin{aligned} x_k^1 &= -A_n y_{k-1} + B_n u_{k-1} + C_n e_{k-1} \\ x_k^2 &= -A_{n-1} y_{k-1} + B_{n-1} u_{k-1} + C_{n-1} e_{k-1} + x_{k-1}^1 \\ &\vdots \\ x_k^{n-1} &= -A_2 y_{k-1} + B_2 u_{k-1} + C_2 e_{k-1} + x_{k-1}^{n-2} \end{aligned}\right\} \tag{2.4.12}$$

and

$$x_k^n = y_k - C_0 e_k. \tag{2.4.13}$$

Elimination of $x_k^1, \ldots, x_k^{n-2}$ from (2.4.12) gives

$$\begin{aligned} x_{k-1}^{n-1} = (&-A_2 y_{k-2} + B_2 u_{k-2} + C_2 e_{k-2} \\ &+(\ldots(-A_n y_{k-n} + B_n u_{k-n} + C_n e_{k-n})\ldots). \end{aligned} \tag{2.4.14}$$

Comparing this equation with (2.4.11) we see that

$$y_k = -A_1 y_{k-1} + B_1 u_{k-1} + C_1 e_{k-1} + x_{k-1}^{n-1} + C_0 e_k. \tag{2.4.15}$$

Substitution of (2.4.13) into (2.4.12) gives

$$\begin{aligned} x_k^1 &= -A_n x_{k-1}^n + B_n u_{k-1} + (C_n - A_n C_0) e_{k-1} \\ x_k^2 &= (-A_{n-1} x_{k-1}^n + x_{k-1}^1) + B_{n-1} u_{k-1} + (C_{n-1} - A_{n-1} C_0) e_{k-1} \\ &\vdots \\ x_k^{n-1} &= (-A_2 x_{k-1}^n + x_{k-1}^{n-2}) + B_2 u_{k-1} + (C_2 - A_2 C_0) e_{k-1}, \end{aligned}$$

Finally, from (2.4.13) and (2.4.15) we have

$$x_k^n = -A_1 x_{k-1}^n + x_{k-1}^{n-1} + B_1 u_{k-1} + (C_1 - A_1 C_0) e_{k-1}.$$

The last n equations, together with (2.4.13), can be organized as the state-space equations of the proposition. □

The principal restriction on an ARMAX model

$$A(z^{-1}) y_k = B(z^{-1}) u_k + C(z^{-1}) e_k$$

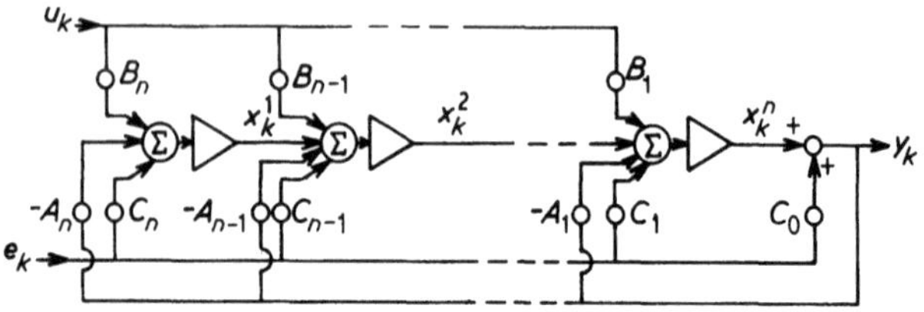

Fig. 2.4

in order that it can be reformulated as a stochastic state space model is that $B(0) = 0$. In other words, there must be a pure delay between application of a control and its effect on the output. In addition, we require that $A(0) = I$. Since of course it is assumed that $\det A(0) \neq 0$, this last requirement can always be achieved by transformations.

Readers may find it helpful to note that the choice of state components (for the state-space representation of an ARMAX model) which we have adopted in the proof of Proposition 2.4.2 is summarized by the 'analogue circuit' diagram for the ARMAX model shown in Fig. 2.4, in which small circles represent multiplication by the given constant factor, the symbol $\sum$ denotes summation, and the triangle a unit delay.

2.5 Innovations representations

Suppose an output process $\{y_k\}$ is generated by ARMAX model equations (2.4.2):

$$y_k + A_1 y_{k-1} + \cdots + A_n y_{k-n} = B_0 u_k + B_1 u_{k-1} + \cdots + B_n u_{k-n} \\ + e_k + C_1 e_{k-1} + \cdots + C_n e_{k-n}.$$

Here $e_0, e_1, \ldots$ is a sequence of zero-mean, independent random variables. The inputs $u_0, u_1, \ldots$ are taken to be deterministic. Values of $y_{-1}, \ldots, y_{-n}, u_{-1}, \ldots, u_{-n}, e_{-1}, \ldots, e_{-n}$ are supplied as initial data. In these circumstances it can be shown that the conditional expectation of y_k given $y_0, \ldots, y_{k-1}$, written $\hat{y}_{k|k-1}$, is

$$\hat{y}_{k|k-1} = -A_1 y_{k-1} - \cdots - A_n y_{k-n} + B_0 u_k + \cdots + B_n u_{k-n} \\ + C_1 e_{k-1} + \cdots + C_n e_{k-n}.$$

It follows that

$$e_k = y_k - \hat{y}_{k|k-1}. \tag{2.5.1}$$

This is the 'innovations process', a sequence of zero-mean, independent random variables which plays a central role in the theory of filtering and stochastic control, as we shall see in future chapters.

Because we are permitted to interpret the noise in the system equations (2.5.1) as the innovations process associated with $\{y_k\}$, the representation of $\{y_k\}$ provided by (2.4.2) is called the *innovations representation* of $\{y_k\}$. The terminology 'innovations representation' is often loosely attached to ARMAX model equations even in the absence of assumptions on initial conditions and inputs and on independence of the e_k's which make the interpretation of $\{e_k\}$ as an innovations process valid. The closely related general stochastic dynamical model equations (2.4.1) are also so named.

Consider next a stochastic state-space model description (2.4.3) of the process $\{y_k\}$,

$$x_{k+1} = Ax_k + Bu_k + Ce_k$$
$$y_k = Hx_k + Ge_k.$$

This system is said to be *in innovations form* if

$$G \text{ is a square non-singular matrix} \tag{2.5.2}$$

The most obvious consequence of this property is that if the initial state x_0 is given, then the state x_k can be reconstructed exactly from observed inputs and outputs, since

$$x_{k+1} = Ax_k + Bu_k + CG^{-1}(y_k - Hx_k) \qquad k = 0, 1, \ldots$$

Thus, regarded as a 'black box', the only 'uncertainty' in an innovations-form model is the value of the initial state. The noise process e_k is closely related to the so-called 'innovations process' of Kalman filtering theory, discussed in Chapter 3, and this is the reason for saying that the model is 'in innovations form' if (2.5.2) is satisfied.

An important consequence of the filtering theory of Chapter 3 is that, given a stochastic state-space model description of a process $\{y_k\}$, there is essentially no loss of generality in assuming that it provides an innovations representation for $\{y_k\}$. This observation coupled with the assertion of Proposition 2.4.2 (note that, given an ARMAX description, Proposition 2.4.2 provides us with a state-space description having output equation $y_k = Hx_k + Ge_k$ with $G = I$) leads to the conclusion that ARMAX models and stochastic state-space models are interchangeable even if we stipulate that the stochastic state-space model provides an innovations representation of the output.

Let us be more precise. Suppose an output process $\{y_k\}$ is generated by the equations of a stochastic state-space model

$$\tilde{x}_{k+1} = A\tilde{x}_k + Bu_k + \tilde{C}\tilde{e}_k$$
$$y_k = H\tilde{x}_k + \tilde{G}\tilde{e}_k$$

in which A is a stable matrix. Here $u_0, u_1, \dots$ are assumed to be deterministic, x_0 and $\tilde{e}_0, \tilde{e}_1, \dots$ are zero-mean independent random variables, and $\tilde{e}_0, \tilde{e}_1, \dots$ have the same covariance matrix. (No assumptions are made here concerning non-singularity of $\tilde{G}$.) Then, for the covariance matrix $\operatorname{cov}\{x_0\}$ of x_0 appropriately chosen, $\{y_k\}$ is generated also by stochastic state-space model equations

$$x_{k+1} = Ax_k + Bu_k + Ce_k$$
$$y_k = Hx_k + e_k, \qquad k \geq 0 \tag{2.5.3}$$

in which $\{e_k\}$ is the innovations process associated with $\{y_k\}$, i.e. by equations which provide an innovations representation. Even if the matrix $\operatorname{cov}\{x_0\}$ is arbitrary, (2.5.3) will still describe y_k to a very good approximation, for large k.

These considerations lie behind the fact that, when stochastic state-space models are adopted in the field of identification (where we are interested in external models), attention is usually limited to those which provide an innovations representation (2.5.3). An economy in the number of parameters specifying the models can usually thereby be achieved and no loss of generality is involved. On the other hand, stochastic state-space models (general form) (2.4.3) are important too, since they arise from internal modelling of systems for which there are certain natural choices of state components and interconnections.

2.6 Predictor models

We describe now a rather general class of models, models whose main value will prove to be their suitability for the formulation of identification procedures and analysis of their convergence properties. The models are called predictor models†. This class of models subsumes in essential respects the stochastic dynamical models of Section 2.4. We must assume however that the driving noise is a sequence of independent random variables. Rather than present and analyse identification algorithms associated with, say, ARMAX

† The name 'prediction error model' is often used in the literature.

models or state-space models individually, we shall for the most part work with predictor models and specialize down to individual cases for detailed description of results. We will thereby emphasize common themes and avoid needless duplication of effort.

The r-vector output $\{y_k\}$ of a prediction error model at time k is related to past outputs and past m-vector inputs $u_{k-1}, u_{k-2}, \ldots$ according to

$$y_k = f_k(y^{k-1}, u^{k-1}) + e_k, \qquad k = 0, 1, \ldots \tag{2.6.1}$$

In these equations y^{k-1} and u^{k-1} denote col $[y_{k-1}, y_{k-2}, \ldots, y_0]$ and col $[u_{k-1}, u_{k-2}, \ldots, u_0]$ respectively; $f_k: \mathbb{R}^{kr} \times \mathbb{R}^{kf} \to \mathbb{R}^p$, $k = 0, 1, \ldots,$ are given deterministic functions of past inputs and outputs; $\{e_k\}_{k \in \mathbb{Z}^+}$ is a sequence of independent, zero-mean r-vector random variables.

It is clear that $f_k(y^{k-1}, u^{k-1})$ is the expected value of y_k given y^{k-1}, u^{k-1}, and is therefore the best 'one-step-ahead predictor' in the mean-square sense (see Proposition 1.1.5). Thus a predictor is built explicitly into equations (2.6.1). This accounts for our calling the models 'predictor models'.

Let us now suppose that the noise vectors e_k entering into the stochastic models of Section 2.4 are independent. Under these conditions we substantiate our claim that the class of predictor models essentially subsumes these models. This amounts to solving, in each case, the one-step-ahead prediction problem.

Consider the general stochastic dynamical model of Section 2.4 (we limit attention to systems in which there is a unit delay in the implementation of the input and for which the initial inputs and outputs are taken to be zero):

$$y_k = P(z^{-1})u_{k-1} + Q(z^{-1})e_k \qquad k \geq 0. \tag{2.6.2}$$

with initial conditions

$$u_k = 0, y_k = 0, e_k = 0, \quad k < 0.$$

We suppose that, in (2.6.2), the polynomial $Q(\sigma)$ has coefficients $r \times r$ matrices and $Q(0) = I$. This can always be arranged by providing, if necessary, fictitious additional noise components of zero mean and variance, and by application of linear transformations to the disturbances.

Rearrangement of the system equations gives

$$y_k = [I - Q^{-1}(z^{-1})]y_k + Q^{-1}(z^{-1})P(z^{-1})u_{k-1} + e_k \qquad k \geq 0,$$
$$u_k = 0, y_k = 0, \qquad k < 0.$$

Notice that $[I - Q^{-1}(0)] = 0$ since $Q(0) = I$, and we can therefore express $\{y_k\}$ as the solution to the predictor model equations

$$y_k = f_k(y^{k-1}, u^{k-1}) + e_k$$

in which

$$f_k(y^{k-1}, u^{k-1}) = [I - Q^{-1}(z^{-1})]y_k + Q^{-1}(z^{-1})P(z^{-1})u_{k-1}.$$

The right-hand side of this equation defines a function of y^{k-1} and u^{k-1} in view of the initial conditions.

Consider next an ARMAX model (Section 2.4)

$$A(z^{-1})y_k = B(z^{-1})u_{k-1} + C(z^{-1})e_k \qquad k \geq 0$$

when we take as initial conditions

$$y_k = 0, u_k = 0, e_k = 0, k < 0.$$

If it is assumed that the e_k are independent and have zero mean, $A(0) = I$ and $C(0) = I$ then the model can be expressed as a predictor model (2.6.1) in which $f(y^{k-1}, u^{k-1})$ is the function $\hat{y}_k$ calculated from the recursive equations

$$C(z^{-1})\hat{y}_i = [C(z^{-1}) - A(z^{-1})]y_i + B(z^{-1})u_{i-1} \qquad i = 0, 1, \ldots$$

with initial condition

$$\hat{y}_i = 0, y_i = 0, u_i = 0, \quad \text{for } i < 0.$$

Consider finally a stochastic state-space model provided with an innovations representation (Section 2.5)

$$\begin{aligned} x_{k+1} &= Ax_k + Bu_k + Ke_k \\ y_k &= Hx_k + Ge_k, \qquad k \geq 0 \end{aligned}$$

with initial condition $x_0 = 0$. Here G is a non-singular square matrix. This model, too, can be expressed as a predictor model (2.6.1) provided the e_k are independent and have zero mean. Now we take the function $f_k(y^{k-1}, u^{k-1})$ to be $\hat{y}_k$, where $\hat{y}_k$ is obtained by solving the equations

$$x_{i+1} = Ax_i + Bu_i + KG^{-1}(y_i - Hx_i) \qquad i \geq 0$$

with initial condition $x_0 = 0$, and by setting

$$\hat{y}_k = H\hat{x}_k.$$

Notes

Sections 2.1–2.3. Processes defined through stochastic difference equations are studied in a number of books; for example, Åström (1970), Hannan (1970) and Whittle (1963). Detailed information about various specific ARMA models can be found in Box and Jenkins (1976). In our treatment we have emphasized stability aspects in preparation for the convergence analysis of identification algorithms, which is given in Chapter 5. The spectral factorization theorem, Theorem 2.3.2, is proved here only in the scalar case. A proof of the theorem when the process considered is vector valued can be found in Hannan's book (1970, p. 129).

Sections 2.4–2.5 We follow Ljung (1974) in interpreting standard stochastic dynamical models as special cases of predictor models. For material on the detailed structure of stochastic dynamical models suitable for identification we refer to some of the literature on 'canonical forms': Denham (1974), Dickinson (1974), Glover and Willems (1974) and Mayne (1972).

References

Åström, K. J. (1970) *Introduction to Stochastic Control Theory*, Academic Press, New York.

Box, G. E. P. and Jenkins, G. M. (1976) *Time Series Analysis, Forecasting and Control*, 2nd Edition, Holden-Day, San Francisco.

Denham, M. J. (1974) Canonical forms for the identification of multivariable linear systems. *IEEE Trans. Automatic Control*, **AC-19**(5) 646–655.

Dickinson, B. W., Kailath, T. and Morf, M. (1974) Canonical matrix fraction and state space descriptions for deterministic and stochastic linear systems. *IEEE Trans. Automatic Control*, **AC-19**(5), 656–666.

Glover, K. and Willems, J. C. (1974) Parametrizations of linear dynamical systems: canonical forms and identifiability. *IEEE Trans. Automatic Control*, **AC-19**(5), 640–645.

Hannan, E. J. (1970) *Multiple Time Series*, Wiley, New York.

Ljung, L. (1974) On consistency for prediction error identification methods. *Div. Automat. Contr.*, Lund Inst. of Technology, Lund, Sweden, Rep. 7405.

Mayne, D. Q. (1972) A canonical form for identification of multivariable linear systems. *IEEE Trans. Automatic Control*, **AC-17**, 728–729.

Whittle, P. (1963) *Prediction and Regulation*, The English Universities Press, London. (Reprinted 1984 by Basil Blackwell, Oxford.)

CHAPTER 3

Filtering theory

The stochastic state space model introduced in Section 2.4 is an internal model: its states x_k are not observed directly but do contribute to the observed outputs y_k as specified by the observation equation in (2.4.3). It is natural then to consider the problem of forming 'best estimates' of the state x_k give the available data $(y_0, y_1, \ldots, y_k)$. This procedure is known as *filtering*. There are at least three situations in which filtering is required. Firstly, it may be an end in itself: this is the case when, as often happens, the state variables x_k^i represent important physical quantities in a system which we need to know as accurately as possible even though they cannot be measured directly. Secondly, if we wish to control systems described by state space models then the natural class of controls to consider is that of state feedback controls where the control variable u_k takes the form $u_k = u(k, x_k)$. If x_k is not 'known' then in some circumstances it can be replaced by a best estimate $\hat{x}_k$ produced by filtering; this topic is described at length in Chapter 6. Finally, filtering is relevant when we wish to replace the state space model by an 'equivalent' external model; see section 3.4 below.

Initially we will consider the filtering or estimation problem in a more general setting than that described above, specializing to state space models later. The general problem may be described as follows: one observes the values of random variables $Y_1, \ldots, Y_n$ and wishes to 'estimate' the value of another random variable Y_0. Here $\bar{Y}^T := (Y_0, Y_1, \ldots, Y_n)$ is a vector random variable with a given joint distribution function F. An *estimator* is any function $g(Y)$ of the observed vector $Y^T := (Y_1, \ldots, Y_n)$ and this is to be chosen so as to minimize the *mean square error*

$$\mathscr{E} = E[Y_0 - g(Y)]^2. \tag{3.0.1}$$

We have already seen in Section 1.1 that the function g which minimizes the mean square error is the conditional expectation

$$E[Y_0|Y] = \int_{-\infty}^{\infty} y_0 \mathrm{d}F_{Y_0|Y}(y_0|Y).$$

However, this may be hard to compute and in any case we may only know certain parameters of the joint distribution of $\bar{Y}$ rather than the function F itself. For these reasons and others which will emerge later, we are led to study the *linear estimation problem* where the choice of estimators g is limited to linear functions, i.e. those of the form

$$g(Y) = \alpha_1 Y_1 + \alpha_2 Y_2 + \cdots + \alpha_n Y_n. \tag{3.0.2}$$

This is much simpler since we are now just searching for the n-vector $\alpha^{\mathrm{T}} = (\alpha_1, \ldots, \alpha_n)$ which mininizes (3.0.1) with g given by (3.0.2). Notice that in this case

$$E[Y_0 - g(Y)]^2 = E \sum_{i,j=0}^{n} \alpha_i \alpha_j Y_i Y_j$$

$$= \sum_{i,j}^{n} \alpha_i \alpha_j E Y_i Y_j$$

where for notational convenience we have defined $\alpha_0 = -1$. Suppose that all the random variables have zero mean. Then EY_iY_j is just the (i,j)th entry of the covariance matrix $\mathrm{cov}(\bar{Y})$, and this shows that in order to solve the linear estimation problem we only need to know the means ($=0$) and covariances of the random variables. This is much more reasonable than requiring that the whole joint distribution function be known. (Of course, the theory only applies when all the random variables have finite variance, but this is hardly a restriction in practice.)

The solution of the linear estimation problem in principle is quite straightforward and in fact a formula for α is given in Theorem 3.1.1 below. The key idea is that the best linear estimate can be thought of geometrically as the 'orthogonal projection' of Y_0 onto the observations Y. Section 3.1 is devoted to explaining this idea and its relation to the conditional expectation mentioned above. What remains is to develop effective ways of calculating this projection. The main application we have in mind is estimating the state vector of the state-space model of Section 2.4 from the output. This problem has a dynamic structure in that the output values $y_0, y_1, \ldots$ are measured at successive instants of time and we wish to 'keep track of' the state vector x_k as it evolves. Thus a *recursive algorithm* is required which will take the estimate at time k and, using the new observation y_{k+1}, update it to give the estimate at time $k+1$. Such recursive estimators, or *filters*, are discussed in general terms in Section 3.2. We then derive in Section 3.3 the *Kalman filter* equations which provide a recursive estimator for the state-space model. Kalman filtering theory is applied in Section 3.4 to derive the *innovations representation* of the state-space model mentioned at the end of Chapter 2.

3.1 The geometry of linear estimation

To introduce the geometric picture of linear estimation let us consider the problem introduced above with $n=1$. Thus (Y_0, Y_1) are jointly distributed zero-mean random variables, and we wish to find the number α which minimizes

$$\mathscr{E} = E[Y_0 - \alpha Y_1]^2 = E(Y_0^2) - 2\alpha E(Y_0 Y_1) + \alpha^2 E(Y_1^2).$$

Elementary calculus shows that the right choice is

$$\alpha = \frac{E(Y_0 Y_1)}{E(Y_1^2)} \tag{3.1.1}$$

(provided that $E(Y_1^2) \neq 0$) resulting in a minimum error of

$$\mathscr{E} = E(Y_0^2) - \frac{1}{E(Y_1^2)}(E(Y_1 Y_0))^2.$$

Let σ_0, σ_1, ρ be the standard deviations and correlation coefficient of Y_0, Y_1 (see Section 1.1.1). Then the best estimator is

$$\hat{Y}_0 = \alpha Y_1 = \rho \frac{\sigma_0}{\sigma_1} Y_1 \tag{3.1.2}$$

and the error is

$$\tilde{Y}_0 = Y_0 - \alpha Y_1 = \sigma_0 \left(\frac{Y_0}{\sigma_0} - \rho \frac{Y_1}{\sigma_1} \right) \tag{3.1.3}$$

with variance

$$\mathscr{E} = \sigma_0^2 (1 - \rho^2).$$

Now note the crucial fact that *the error* $\tilde{Y}$ *is uncorrelated with the observed random variable* Y_1, i.e.

$$E(\tilde{Y}_0 Y_1) = 0.$$

This is easily seen from (3.1.3). It is also easily seen that the value of α

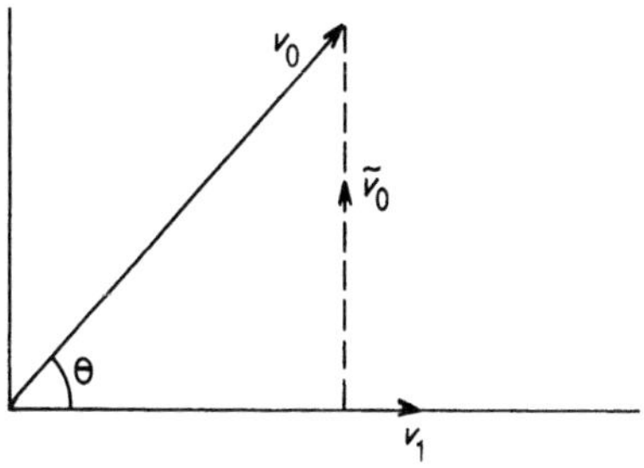

Fig. 3.1

given by (3.1.1) is the only one such that $(Y_0 - \alpha Y_1)$ and Y_1 are uncorrelated.

The geometric picture that goes along with this is as follows: Suppose $\mathbf{v}_0$, $\mathbf{v}_1$ are vectors in the plane which have lengths σ_0, σ_1 respectively and intersect at an angle θ such that $\cos\theta = \rho$ (see Fig. 3.1). The vector $\mathbf{v}_0$ can be expressed as the vector sum of its projection $\hat{\mathbf{v}}_0$ on to $\mathbf{v}_1$ and the difference $\tilde{\mathbf{v}}_0 = \mathbf{v}_0 - \hat{\mathbf{v}}_0$ which is orthogonal to $\mathbf{v}_1$. The projection $\hat{\mathbf{v}}_0$ is given by

$$\hat{\mathbf{v}}_0 = \sigma_0 \cos\theta\left(\frac{1}{\sigma_1}\mathbf{v}_1\right) = \rho\frac{\sigma_0}{\sigma_1}\mathbf{v}_1 \tag{3.1.4}$$

Comparing (3.1.2) and (3.1.4) we see that if the random variables Y_0, Y_1 are identified with the vectors $\mathbf{v}_0$, $\mathbf{v}_1$ respectively then the best linear estimate $\hat{Y}_0$ corresponds to the projection $\hat{\mathbf{v}}_0$ of $\mathbf{v}_0$ onto $\mathbf{v}_1$. The *inner* (or *dot*) *product* of the vectors $\mathbf{v}_0$ and $\mathbf{v}_1$ is

$$\mathbf{v}_0 \cdot \mathbf{v}_1 = \sigma_0\sigma_1 \cos\theta = \sigma_0\sigma_1\rho = EY_0Y_1 = \operatorname{cov}(Y_0, Y_1).$$

Thus the vectors representing the random variables have lengths equal to the standard deviations of the random variables and inner product equal to the covariance. Notice in particular that if $\theta = 0$ or $\theta = \pi$ then the vectors are colinear and $\hat{\mathbf{v}}_0 = \pm\mathbf{v}_0 = \pm(\sigma_0/\sigma_1)\mathbf{v}_1$. Since $\rho = \cos\theta$ the equivalent condition on ρ is that $\rho = \pm 1$. But we already saw in Chapter 1 that if Y_0, Y_1 have correlation coefficient ± 1 then they are linearly related: $Y_0 = \pm(\sigma_0/\sigma_1)Y_1$. Thus 'linear estimation' can be done with zero error, as the geometric picture indicates.

In order to formalize the above discussion and generalize it to higher dimensions we need to review the geometrical properties of $\mathbb{R}^d$ considered as a vector space. Elements or *vectors* $\mathbf{x}$ of $\mathbb{R}^d$ are n-tuples of real numbers $\mathbf{x} = (x_1, x_2, \ldots, x_d)$. Addition and scalar multiplication are defined componentwise: $\mathbf{x}+\mathbf{y}=(x_1+y_1,\ldots,x_d+y_d)$ and $\alpha\mathbf{x}=(\alpha x_1,\ldots,\alpha x_d)$ for $\alpha\in\mathbb{R}$. The *inner* or *dot* product of two vectors $\mathbf{x}$, $\mathbf{y}$ is defined by

$$\mathbf{x}\cdot\mathbf{y}=\sum_{i=1}^{d} x_i y_i.$$

The vectors $\mathbf{x}$ and $\mathbf{y}$ are *orthogonal* ($\mathbf{x}\perp\mathbf{y}$) if $\mathbf{x}\cdot\mathbf{y}=0$. The *norm* of a vector is

$$\|\mathbf{x}\|=\sqrt{(\mathbf{x}\cdot\mathbf{x})}.$$

For $i=1, 2,\ldots,d$ define

$$\mathbf{z}:=(0,\ldots,0,\ 1,\ 0,\ldots,0)\ (1 \text{ in the } i\text{th position}).$$

These are the *coordinate vectors*. They have the following properties:

(a) They are normalized and mutually orthogonal:

$$\mathbf{z}_i\cdot\mathbf{z}_j=\begin{cases}1 & i=j\\ 0 & i\neq j\end{cases}$$

(b) They form a *basis* for $\mathbb{R}^d$: any $\mathbf{x}\in\mathbb{R}^d$ can be expressed as $\mathbf{x}=\sum_1^d a_i\mathbf{z}_i$ for some coefficients a_i.

It is clear from the definitions that the coefficients a_i in (b) are given by $a_i=\mathbf{x}\cdot\mathbf{z}_i$, so that each $\mathbf{x}\in\mathbb{R}^d$ has the representation:

$$\mathbf{x}=\sum_{i=1}^{d}(\mathbf{x}\cdot\mathbf{z}_i)\mathbf{z}_i.$$

Any set of vectors $\mathbf{z}_i$ satisfying (a) and (b) is called an *orthonormal basis* of $\mathbb{R}^d$. A *subspace* $\mathscr{L}$ of $\mathbb{R}^d$ is a subset with the property that if $\mathbf{x}, \mathbf{y}\in\mathscr{L}$ then $\alpha\mathbf{x}+\beta\mathbf{y}\in\mathscr{L}$ for any $\alpha, \beta\in\mathbb{R}$. The subspace *generated* or *spanned* by any collection of vectors $\mathbf{u}_1,\ldots,\mathbf{u}_m$ is denoted by $\mathscr{L}(\mathbf{u}_1,\ldots,\mathbf{u}_m)$ and is the smallest subspace containing the generating vectors. It is easy to see that

$$\mathscr{L}(\mathbf{u}_1,\ldots,\mathbf{u}_m)=\left\{\sum_{i=1}^{m} a_i\mathbf{u}_i : \mathbf{a}=(a_1,\ldots,a_m)\in\mathbb{R}^m\right\}.$$

It is always possible to construct an orthonormal basis $\mathbf{x}_1,\ldots,\mathbf{x}_d$ of $\mathbb{R}^d$ such that $\mathscr{L}(\mathbf{u}_1,\ldots,\mathbf{u}_m)=\mathscr{L}(\mathbf{x}_1,\ldots,\mathbf{x}_k)$ for some $k\leq\min\{d,m\}$. This can be done by using the *Gram–Schmidt orthogonalization procedure*, which we describe next. Supppose, to avoid triviality, that $\|\mathbf{u}_i\| > 0$ for some i (otherwise $\mathscr{L}(\mathbf{u}_1, \ldots, \mathbf{u}_m) = \{0\}$); we can then assume, permuting indices if necessary, that $\|\mathbf{u}_1\| > 0$. Define

$$\mathbf{x}_1 = \frac{1}{\|\mathbf{u}_1\|}\mathbf{u}_1.$$

Now suppose that orthogonal vectors $\mathbf{x}_1, \ldots, \mathbf{x}_{k(l)}$ have been found for some number $k(l) \leq \min\{d, m\}$ such that $\mathscr{L}(\mathbf{u}_1, \ldots, \mathbf{u}_l) = \mathscr{L}(\mathbf{x}_1, \ldots, \mathbf{x}_{k(l)})$. Define

$$\mathbf{v} := \mathbf{u}_{l+1} - \sum_{i=1}^{k(l)} (\mathbf{u}_{l+1} \cdot \mathbf{x}_i)\mathbf{x}_i.$$

Then $\mathbf{v} \perp \mathbf{x}_i$ for $i = 1, \ldots, k(l)$. If $\|\mathbf{v}\| = 0$, set $k(l+1) := k(l)$; otherwise, set $k(l+1) := k(l) + 1$ and $\mathbf{x}_{k(l+1)} = \mathbf{v}/\|\mathbf{v}\|$. Then $\mathbf{x}_1, \ldots, \mathbf{x}_{k(l+1)}$ are orthonormal and $\mathscr{L}(\mathbf{u}_1, \ldots, \mathbf{u}_{l+1}) = \mathscr{L}(\mathbf{x}_1, \ldots, \mathbf{x}_{k(l+1)})$. Since clearly $\mathscr{L}(\mathbf{u}_1) = \mathscr{L}(\mathbf{x}_1)$ we conclude by induction that $\mathscr{L}(\mathbf{u}_1, \ldots, \mathbf{u}_m) = \mathscr{L}(\mathbf{x}_1, \ldots, \mathbf{x}_{k(m)})$. By construction $k := k\,(m) \leq m$, and $k \leq d$ since d is the maximum number of linearly independent vectors in $\mathbb{R}^d$. If $k < d$ then orthonormal vectors $\mathbf{x}_{k+1}, \ldots, \mathbf{x}_d$ can be constructed in a similar way so that $\mathbf{x}_1, \ldots, \mathbf{x}_d$ form a basis of $\mathbb{R}^d$. We leave it to the reader to supply the details.

The *orthogonal projection* $\hat{\mathbf{v}}$ of $\mathbf{v} \in \mathbb{R}^d$ onto a subspace $\mathscr{U} := \mathscr{L}(\mathbf{u}_1, \ldots, \mathbf{u}_m)$ is defined by

$$\hat{\mathbf{v}} = \sum_{i=1}^{k} (\mathbf{v} \cdot \mathbf{x}_i)\mathbf{x}_i$$

where $\mathbf{x}_1, \ldots, \mathbf{x}_d$ is an orthonormal basis such that $\mathscr{U} = \mathscr{L}(\mathbf{x}_1, \ldots, \mathbf{x}_k)$. $\hat{\mathbf{v}}$ can be characterized in the following two equivalent ways.

(a) $\hat{\mathbf{v}}$ is the unique vector satisfying

$$\hat{\mathbf{v}} \in \mathscr{U}$$
$$\mathbf{v} - \hat{\mathbf{v}} \perp \mathscr{U}$$

(b) $\hat{\mathbf{v}}$ is the closest point in $\mathscr{U}$ to $\mathbf{v}$, i.e.

$$\|\mathbf{v} - \hat{\mathbf{v}}\| = \min_{\mathbf{u} \in \mathscr{L}} \|\mathbf{v} - \mathbf{u}\|.$$

In (a), $\mathbf{v} - \hat{\mathbf{v}} \perp \mathscr{U}$ means that $(\mathbf{v} - \hat{\mathbf{v}}) \perp \mathbf{u}$ for all $\mathbf{u} \in \mathscr{L}$.

Both (a) and (b) are very easily established using the basis $\mathbf{x}_1, \ldots, \mathbf{x}_d$, but note that the statements themselves do not involve any particular choice of basis. For part (b) we have

$$\mathbf{v} - \mathbf{u} = (v_1 - u_1)\mathbf{x}_1 + \cdots + (v_k - u_k)\mathbf{x}_k + v_{k+1}\mathbf{x}_{k+1} + \cdots + v_d\mathbf{x}_d$$

where $v_i = \mathbf{v}\cdot\mathbf{x}_i$, $u_i = \mathbf{u}\cdot\mathbf{x}_i$. Thus

$$\|\mathbf{v} - \mathbf{u}\|^2 = \sum_{i=1}^{k} (v_i - u_i)^2 + \sum_{i=k+1}^{d} v_i^2$$

and this is clearly minimized by taking $u_i = v_i$, $i \leq k$.

Let us denote $\hat{\mathbf{v}} = \mathscr{P}\mathbf{v}$. Then $\mathscr{P}$ is a *projection operator* which maps each vector in $\mathbb{R}^d$ to its projection onto the subspace $\mathscr{U}$. We note the following properties of the projection operator:

(a) $\mathscr{P}$ is linear: $\mathscr{P}(\alpha\mathbf{v}_1 + \beta\mathbf{v}_2) = \alpha\mathscr{P}\mathbf{v}_1 + \beta\mathscr{P}\mathbf{v}_2$, $\alpha, \beta \in \mathbb{R}$

(b) $\mathscr{P}^2 = \mathscr{P}$ (Here $\mathscr{P}^2\mathbf{v} := \mathscr{P}(\mathscr{P}\mathbf{v})$)

(c) If $\mathscr{U}'$ is a subspace such that $\mathscr{U}' \supset \mathscr{U}$ and $\mathscr{P}'$ is the projection onto $\mathscr{U}'$ then for any $\mathbf{v} \in \mathbb{R}^d$

$$\mathscr{P}\mathbf{v} = \mathscr{P}(\mathscr{P}'\mathbf{v}).$$

The first two of these are evident. For (c), suppose that $\mathscr{U}' = \mathscr{L}(\mathbf{u}_1, \ldots, \mathbf{u}_{m'})$ for some $m' > m$. By means of the Gram–Schmidt procedure we can construct a basis $\mathbf{x}_1, \ldots, \mathbf{x}_d$ and numbers k, k' with $k \leq k' \leq d$ such that $\mathscr{U} = \mathscr{L}(\mathbf{x}_1, \ldots, \mathbf{x}_k)$ and $\mathscr{U}' = \mathscr{L}(\mathbf{x}_1, \ldots, \mathbf{x}_{k'})$. For $\mathbf{v} \in \mathbb{R}^d$, denote $v_i = \mathbf{v}\cdot\mathbf{x}_i$. Then $\mathscr{P}'\mathbf{v} = \sum_1^{k'} v_i\mathbf{x}_i$ so that $\mathscr{P}(\mathscr{P}'\mathbf{v}) = \sum_1^k v_i\mathbf{x}_i = \mathscr{P}\mathbf{v}$.

Now back to random variables. Suppose as before that $\bar{Y} := (Y_0, Y_1, \ldots, Y_n)^{\mathrm{T}}$ is a random $(n+1)$-vector such that for each i

$$EY_i = 0, \operatorname{var}(Y_i) < \infty,$$

and denote $Q := \operatorname{cov}(\bar{Y})$. We wish to associate these random variables with vectors $\mathbf{v}_0, \ldots, \mathbf{v}_n$ in such a way that

$$\mathbf{v}_i\cdot\mathbf{v}_j = \operatorname{cov}(Y_i, Y_j) = EY_iY_j$$

More precisely, let $\mathscr{H}$ denote the set of all linear combinations of the random variables $Y_0, \ldots, Y_n$, i.e.

$$\mathscr{H} = \left\{ \sum_{i=0}^{n} \alpha_i Y_i : \alpha = (\alpha_0, \ldots, \alpha_n) \in \mathbb{R}^{n+1} \right\}.$$

We take the function $U, V \to EUV$ as an 'inner product' for $U, V \in \mathscr{H}$. Note that EUV is entirely determined by the covariance matrix Q

since if $U, V \in \mathscr{H}$ then $U = \mathbf{a}^T \bar{Y}$ and $V = \mathbf{b}^T \bar{Y}$ for some $\mathbf{a}, \mathbf{b} \in \mathbb{R}^{n+1}$ and then $EUV = \mathbf{a}^T Q \mathbf{b}$. We wish to construct a function $\boldsymbol{\varphi}: \mathscr{H} \to \mathbb{R}^d$ for some integer d with the following properties

(a) $\boldsymbol{\varphi}$ is linear, one-to-one and onto

(b) $\boldsymbol{\varphi}$ is inner product preserving:

$$\boldsymbol{\varphi}(U) \cdot \boldsymbol{\varphi}(V) = EUV. \tag{3.1.5}$$

Such a function $\boldsymbol{\varphi}$ always exists. Recall from Proposition 1.1.3 that by factoring Q in the form $Q = AA^T$ we can express Y in the form

$$\bar{Y} = AZ$$

where $Z^T = (Z_1, \ldots, Z_d)$ is a vector of unit variance uncorrelated random variables and $d \le n + 1$. Now define

$$\boldsymbol{\varphi}(Z_i) := \mathbf{z}_i$$

where $\mathbf{z}_1, \ldots, \mathbf{z}_d$ is the coordinate basis of $\mathbb{R}^d$, and

$$\boldsymbol{\varphi}(\mathbf{a}^T Z) := \sum_{i=1}^{d} a_i \mathbf{z}_i \quad \text{for } \mathbf{a} \in \mathbb{R}^d.$$

Since $\mathscr{H} = \{\mathbf{a}^T Z : \mathbf{a} \in \mathbb{R}^d\}$ this defines $\boldsymbol{\varphi}(U)$ for all $U \in \mathscr{H}$. By construction, $\boldsymbol{\varphi}$ is linear and onto, and an immediate calculation shows that (3.1.5) holds. In particular if we define $\mathbf{v}_i := \boldsymbol{\varphi}(Y_i) = \boldsymbol{\varphi}(\sum_k a_{ik} Z_k)$ then we see that

$$\mathbf{v}_i \cdot \mathbf{v}_j = EY_i \cdot Y_j.$$

To check that $\boldsymbol{\varphi}$ is one-to-one, suppose that $\boldsymbol{\varphi}(U) = \boldsymbol{\varphi}(V)$; then $\boldsymbol{\varphi}(U - V) = \boldsymbol{\varphi}(U) - \boldsymbol{\varphi}(V) = 0$ so that $E(U - V)^2 = \boldsymbol{\varphi}(U - V) \cdot \boldsymbol{\varphi}(U - V) = 0$. Recall by the way that $E(U - V)^2 = 0$ if and only if $P[U = V] = 1$. Thus this theory does not distinguish between equivalent random variables, i.e. if $P[U = V] = 1$ then $\boldsymbol{\varphi}(U) = \boldsymbol{\varphi}(V)$.

The existence of the map $\boldsymbol{\varphi}$ means that the geometrical properties of the space $\mathscr{H}$ with 'inner product' EUV and 'distance' $[E(U - V)^2]^{1/2}$ are identical to those of Euclidean space $\mathbb{R}^d$. To illustrate the utility of this, let $\mathscr{V}$ be the subspace spanned by $\mathbf{v}_1, \ldots, \mathbf{v}_n$ where $\mathbf{v}_i = \boldsymbol{\varphi}(Y_i)$ and let $\hat{\mathbf{v}}_0$ be the projection of $\mathbf{v}_0$ on to V. Then

$$\hat{\mathbf{v}}_0 = \alpha_1 \mathbf{v}_1 + \cdots + \alpha_n \mathbf{v}_n$$

for some constants $\alpha_1, \ldots, \alpha_n$, and the corresponding element of $\mathscr{H}$ is

$$\hat{Y}_0 := \boldsymbol{\varphi}^{-1}(\hat{\mathbf{v}}_0) = \alpha_1 Y_1 + \cdots + \alpha_n Y_n.$$

Now recall that $\hat{\mathbf{v}}_0$ is the closest point in $\mathscr{V}$ to $\mathbf{v}_0$:

$$\| \hat{\mathbf{v}}_0 - \mathbf{v}_0 \| = \min_{\mathbf{u} \in V} \| \mathbf{v}_0 - \mathbf{u} \|.$$

It follows from (3.1.5) that $E(U - V)^2 = \| \boldsymbol{\varphi}(U) - \boldsymbol{\varphi}(V) \|^2$; thus $\hat{Y}_0$ satisfies

$$E(Y_0 - \hat{Y}_0)^2 = \min_U E(Y_0 - U)^2$$

where the minimum is taken over all linear combinations $U = \sum_1^n \alpha_i Y_i$. But this means that $\hat{Y}_0$ solves the linear estimation problem. $\hat{Y}_0$ has the property that $E[(Y_0 - \hat{Y}_0)Y_i] = 0$, $i = 1, 2, \ldots, n$, i.e. the error $Y_0 - \hat{Y}_0$ is uncorrelated with the observed random variables $Y_1, \ldots, Y_n$ just as in the scalar case.

We can dispense with explicit mention of the map $\boldsymbol{\varphi}$ and Euclidean space $\mathbb{R}^d$. Just think of the random variables as 'vectors' with lengths equal to their standard deviations and 'inner product' given by the covariance. Thus two random variables are 'orthogonal' (and we write $U \perp V$) if they are uncorrelated, and the best linear estimator $\hat{Y}_0$ is the 'projection' of Y_0 onto the 'subspace' $\mathscr{L}(Y_1, \ldots, Y_n)$ spanned by $Y_1, \ldots, Y_n$.

Let us summarize the results we have obtained. At the same time we generalize to the vector case, replacing Y_0 by a p-vector X.

Theorem 3.1.1

Let X and Y be random p- and n-vectors respectively, all components having zero mean and finite variance. (Here, $Y^{\mathrm{T}} = (Y_1, \ldots, Y_n)$.) Then for each $j = 1, \ldots, p$ there is a unique (up to equivalence) random variable $\hat{X}_j$ such that:

(a) $\hat{X}_j \in \mathscr{L}(Y)$
(b) $X_j - \hat{X}_j \perp \mathscr{L}(Y)$.

$\hat{X}^{\mathrm{T}} := (\hat{X}_1, \ldots, \hat{X}_p)$ is the minimum mean-square error estimate of X given Y, i.e. for any $\beta \in \mathbb{R}^p$

$$E[(\beta^{\mathrm{T}}(X - \hat{X})]^2 = \min_{U \in \mathscr{L}(Y)} E[\alpha^{\mathrm{T}} X - U]^2.$$

If $\operatorname{cov}(Y)$ is non-singular then $\hat{X}$ is given by

$$\hat{X} = E[XY^{\mathrm{T}}][\operatorname{cov}(Y)]^{-1} Y. \tag{3.1.6}$$

REMARK By a slight abuse of terminology, $\hat{X}$ is referred to as the 'projection of X onto $\mathscr{L}(Y)$'.

PROOF Only the last part remains to be established. By definition, $\hat{X} = AY$ for some $p \times n$ matrix A. Using the orthogonality relation (b) we see that for any $\beta \in \mathbb{R}^p$, $\gamma \in \mathbb{R}^n$

$$E[\beta^{\mathrm{T}}(X - AY)(\gamma^{\mathrm{T}} Y)] = 0$$

i.e.

$$\beta^{\mathrm{T}}[E(XY^{\mathrm{T}} - AYY^{\mathrm{T}})]\gamma = 0.$$

This implies that

$$E[XY^{\mathrm{T}}] - AE[YY^{\mathrm{T}}] = 0$$

and hence that $A = E[XY^{\mathrm{T}}][\mathrm{cov}(Y)]^{-1}$ if $\mathrm{cov}(Y)$ is non-singular. If $\mathrm{cov}(Y)$ is singular then some components of Y are linearly related and it may be possible to express $\hat{X}$ in several different but equivalent ways.

Random variables with non-zero mean

Let us consider the same problem as above (with scalar Y_0) but supposing now that the random variables have possibly non-zero means

$$EY_i = m_i \qquad i = 0, 1, \ldots, m.$$

This situation easily reduces to the zero-mean case. Rather than a linear estimator, it is preferable now to use an *affine* (linear + constant) estimator:

$$\hat{Y}_0 = \alpha_1 Y_1 + \cdots + \alpha_n Y_n + \beta.$$

We have to choose $\alpha_1, \ldots, \alpha_n$, β to minimize $E[Y_0 - \hat{Y}_0]^2$. Minimization can be carried out over these coefficients in any order, so let us fix $\alpha_1, \ldots, \alpha_n$ and minimize first over β. Define

$$U = Y_0 - \alpha_1 Y_1 - \cdots - \alpha_n Y_n.$$

Then

$$E[Y_0 - \hat{Y}_0]^2 = E[U - \beta]^2$$

It was shown in Proposition 1.1.1 that this is minimized by taking

$$\beta = EU = m_0 - \alpha_1 m_1 - \cdots - \alpha_n m_n.$$

Incidentally, this justifies our previous implicit choice $\beta = 0$ for the zero-mean case. With the above choice of β we see that

$$E[Y_0 - \hat{Y}_0]^2 = E[Y_0^c - (\alpha_1 Y_1^c + \cdots + \alpha_n Y_n^c)]^2 \qquad (3.1.7)$$

where Y_i^c is the 'centered' random variable $Y_i^c = Y_i - m_i$. We now have to choose $\alpha_1, \ldots, \alpha_n$ to minimize (3.1.7), but this is the zero-mean problem that was solved before. Let P be the covariance matrix of Y, now given by

$$P_{ij} = E[(Y_i - m_i)(Y_j - m_j)].$$

If P is non-singular, then from Theorem 3.1.1

$$\hat{Y}_0 = (Y - m)^T P^{-1} E[(Y - m)(Y_0 - m_0)] + m_0 \qquad (3.1.8)$$

where $m^T = (m_1, \ldots, m_n)$. Notice that *the error* $Y_0 - \hat{Y}_0$ *always has zero mean.*

To get the geometric picture for this case we adopt the rather artificial, but convenient, stratagem of adjoining to the observations another random variable denoted $\mathbb{1}$ which takes on the value 1 with probability one (thus no new 'information' has been added). $\hat{Y}_0$ can then be regarded as a *linear* (no longer affine) combination of the observations:

$$\hat{Y}_0 = \beta \mathbb{1} + \alpha_1 Y_1 + \cdots + \alpha_n Y_n$$

As before, random variables U, V are regarded as vectors with inner product EUV, but note that this is *not* now the covariance, which is $E(U - EU)(V - EV)$. Now $U \perp \mathbb{1}$ if $E(\mathbb{1}U) = EU = 0$ and thus if we express U as

$$U = (EU)\mathbb{1} + U^c$$

then the first term on the right is the projection of U onto the one-dimensional subspace spanned by the random-variable $\mathbb{1}$. Thus the random variables $\mathbb{1}, Y_0, \ldots, Y_n$ form a vector space of dimension $k \leq n + 2$ consisting of a $(k - 1)$-dimensional subspace of zero-mean random variables (spanned, in fact, by $Y_0^c, \ldots, Y_n^c$) and a 1-dimensional subspace spanned by $\mathbb{1}$. The best estimate of Y_0 is the sum of its projection into $\mathscr{L}(Y_1^c, \ldots, Y_n^c)$ and its projection onto $\mathscr{L}(\mathbb{1})$ and these projections are the two terms on the right of (3.1.8), respectively.

The normal case

As pointed out at the beginning of this chapter, only means and covariances are required to calculate best linear estimators. If we suppose that the random variables involved are jointly normally distributed then we get the following result strengthening Theorem 3.1.1.

Theorem 3.1.2

Let X and Y be as in Theorem 3.1.1 but with possibly non-zero means and suppose that X and Y are jointly normally distributed. Then the best affine estimate of X given Y coincides with the conditional expectation $E[X|Y]$.

PROOF Consider first the zero mean case. Since $\hat{X} = AY$ for some matrix A, the random variables $(X, \hat{X}, Y)$ are jointly normally distributed, and $(X_i - \hat{X}_i)$ is uncorrelated with and hence independent of Y_j for each i, j. Using the properties of conditional expectation given in Proposition 1.1.4 we see that, with $\tilde{X} = X - \hat{X}$,

$$\begin{aligned} E[X|Y] &= E[\hat{X} + \tilde{X}|Y] \\ &= \hat{X} + E[\tilde{X}|Y] \\ &= \hat{X} + E\tilde{X} = \hat{X}. \end{aligned}$$

If X, Y have non-zero means m_X, m_Y, write $X^c = X - m_X$, $Y^c = Y - m_Y$. Then

$$E[X|Y] = E[X^c + m_X|Y] = m_X + E[X^c|Y].$$

It follows from Proposition 1.1.7 that $E[X^c|Y] = E[X^c|Y^c]$ and the latter expression coincides with the best linear estimator. This completes the proof. □

This result shows that in the normal case $\hat{X}$ is the best estimate of X not only in the class of affine functions $AY + b$ but also in the class of all finite-variance functions $g(Y)$. It also shows that the *conditional distribution* of X given Y is normal with mean $\hat{X}$ and covariance $\text{cov}(X - \hat{X})$. This follows from the fact that $X = \hat{X} + \tilde{X}$ where $\hat{X}$ is a function of Y and $\tilde{X}$ is independent of Y. We have thus, somewhat belatedly, completed the proof of Proposition 1.1.7(e) of Chapter 1.

3.2 Recursive estimation

The idea of recursive estimation arises when random variables $Y_1, Y_2, \ldots$ are observed sequentially and we wish to process them in real time to form successive best estimates of an unobserved random variable Y_0. At time n we can form the best linear estimate $\hat{Y}_{0,n}$ of Y_0 given $Y_1, \ldots, Y_n$ by using formula (3.1.9) (supposing that all means are zero and that the covariance matrix $P_n = \text{cov}(Y_1, \ldots, Y_n)$ is non-singular). Note that this involves inverting the $n \times n$ matrix P_n. At the next time instant we have one more observation, Y_{n+1}. How are we to compute $\hat{Y}_{0,n+1}$? The most obvious way would be to apply the same formula again. However, if we do this successively for $n = 1, 2, 3, \ldots$, then:

(a) It is necessary to store the entire observation record as this becomes available; and,
(b) At each time n, an $n \times n$ matrix must be inverted.

Obviously, the computational effort required to do this becomes massive even for moderate n. Is it really necessary, at each stage, to throw away the results of all previous calculations, or is there some method by which $\hat{Y}_{0,n}$ can be updated using the new observation Y_{n+1} to give $\hat{Y}_{0,n+1}$? The simplest form such an updating could take is as follows:

$$\hat{Y}_{0,n+1} = a_n \hat{Y}_{0,n} + b_n Y_{n+1} \tag{3.2.1}$$

i.e. the next estimate is a linear combination of the current estimate and the next observation. Only in special cases will a formula such as (3.2.1) be possible, but these include important applications such as the Kalman filter discussed in Section 3.3.

In this section we discuss the general relation between successive estimates. In view of later applications it is convenient to deal from the outset with the vector case. Thus suppose x is an n-vector random variable and $y_1, y_2, \ldots$ are r-vectors of observed random variables.† All random variables will be taken to have zero mean and finite variance, and to avoid difficulties with non-uniqueness it will be supposed that the covariance matrix of the rk-vector $y^k = \text{col}\{y_1, y_2, .. y_k\}$ is non-singular for each k.

† In accordance with the established notational conventions of Kalman filtering theory these are denoted by lower-case letters.

Denote by $\mathscr{L}(y^k)$ the linear subspace spanned by the observations up to time k, and by $\hat{x}_k$ the best linear estimate of x given y^k, i.e. the projection of x onto $\mathscr{L}(y^k)$. (Recall the notational conventions for projection of vector r.v.s introduced in Section 3.1).

$\mathscr{L}(y^{k-1})$ is a subspace of $\mathscr{L}(y^k)$. Let $\hat{y}_{k|k-1}$ be the projection of y_k onto $\mathscr{L}(y^{k-1})$ and $\tilde{y}_{k|k-1}$ the error: $\tilde{y}_{k|k-1} = y_k - \hat{y}_{k|k-1}$. The random variables $\{\tilde{y}^i_{k|k-1}, i = 1, 2, \ldots, r\}$ span the orthogonal complement of $\mathscr{L}(y^{k-1})$ in $\mathscr{L}(y^k)$, so that any r.v. Z in $\mathscr{L}(y^k)$ has a unique orthogonal decomposition

$$Z = Z_1 + Z_2$$

where $Z_1 \in \mathscr{L}(y^{k-1})$ and Z_2 is a linear combination of $\{\tilde{y}^i_{k|k-1}, i = 1, \ldots, r\}$. Take in particular $Z = \hat{x}^i_k$; then we claim that $Z_1 = \hat{x}^i_{k-1}$. Indeed, let $\tilde{x}^i_k = x^i - \hat{x}^i_k$ be the estimation error at time k. Then

$$x^i = \hat{x}^i_k + \tilde{x}^i_k = Z_1 + (Z_2 + \tilde{x}^i_k)$$

where $Z_1 \in \mathscr{L}(y^{k-1})$ and $(Z_2 + \tilde{x}^i_k) \perp \mathscr{L}(y^{k-1})$. But we also have

$$x^i = \hat{x}^i_{k-1} + \tilde{x}^i_{k-1}$$

and again $\hat{x}^i_{k-1} \in \mathscr{L}(y^{k-1})$, $\tilde{x}^i_{k-1} \perp \mathscr{L}(y^{k-1})$. Since such orthogonal decompositions are unique, it must be the case that $Z_1 = \hat{x}^i_{k-1}$, as claimed. As to Z_2, this is the projection of $\hat{x}^i_k$ onto $\mathscr{L}(\tilde{y}_{k|k-1})$ and this is the same as the projection of x^i onto $\mathscr{L}(\tilde{y}_{k|k-1})$ since $\mathscr{L}(\tilde{y}_{k|k-1}) \subset \mathscr{L}(y^k)$, But this projection can be calculated using formula (3.1.9) again. Collecting the above results we see that $\hat{x}_k$ can be written in the form

$$\hat{x}_k = \hat{x}_{k-1} + E[x\tilde{y}^{\mathrm{T}}_{k|k-1}](E[\tilde{y}_{k|k-1}\tilde{y}^{\mathrm{T}}_{k|k-1}])^{-1}(y_k - \hat{y}_{k|k-1}). \qquad (3.2.2)$$

In general this is *not* a recursive formula for $\hat{x}_k$, since $\hat{y}_{\mu|k-1}$ depends on $y_1, \ldots, y_{k-1}$. It *is* a recursive formula precisely when this dependence factors through $\hat{x}_{k-1}$. Let us examine an important example where this occurs.

Example 3.2.1

Suppose

$$y_k = Hx + z_k$$

where H is an $r \times n$ matrix and $z_1, z_2, \ldots$ is a sequence of mutually uncorrelated random variables with zero mean and common covariance $\operatorname{cov}(z_k) = N > 0$. We also suppose x and z_k are uncorrelated for each k. Thus y_k represents a sequence of 'measurements' of x with uncorrelated measurement errors z_k. Let P be the covariance matrix of x.

In this example $\hat{y}_{k|k-1}$ is the projection of $y_k = Hx + z_k$ onto $\mathscr{L}(y^{k-1})$ and this is the same as the projection of Hx onto $\mathscr{L}(y^{k-1})$ since $z_k \perp \mathscr{L}(y^{k-1})$. Thus

$$\hat{y}_{k|k-1} = H\hat{x}_{k-1}$$

and (3.2.2) becomes

$$\hat{x}_k = \hat{x}_{k-1} + K(k)(y_k - H\hat{x}_{k-1}) \tag{3.2.3}$$

where $K(k)$ denotes the matrix coefficient in (3.2.2). This is a recursive formula for $\hat{x}_k$ and it only remains to calculate $K(k)$. We will do this in two ways: the 'slick' way specifically adapted to this problem, and by use of a general technique which will be useful in connection with the Kalman filter in the next section.

The slick way is to notice that the y_k are *interchangeable*, in that if

$$\hat{x}_k = A_1 y_1 + \cdots + A_k y_k$$

then all the A_i must be the same, since the correlation structure of the random variables would be completely unchanged if any two observations y_i and y_j were permuted. Denote by $\bar{y}_k$ the sample mean

$$\bar{y}_k = \frac{1}{k}\sum_{i=1}^{k} y_i = Hx + \frac{1}{k}\sum_{i=1}^{k} z_i = Hx - \bar{z}_k.$$

The noise sample mean $\bar{z}_k$ has covariance N/k and our contention is that

$$\hat{x}_k = A\bar{y}_k$$

for some $n \times r$ matrix A. The orthogonality condition is

$$x - \hat{x}_k = (I - AH)x - A\bar{z}_k \perp y_i = Hx + z_i \qquad i = 1, \ldots, k.$$

Since x is uncorrelated with z_i and $\bar{z}_k$, this is equivalent to requiring that

$$(I - AH)E[xx^{\mathrm{T}}]H^{\mathrm{T}} - AE[\bar{z}_k z_i^{\mathrm{T}}] = 0.$$

Now $E[xx^{\mathrm{T}}] = \operatorname{cov}(x) = P$ and $E[\bar{z}_k z_i^{\mathrm{T}}] = N/k$ since the z_j are mutually uncorrelated. The fact that this expression is independent of i

confirms the 'interchangeability' argument. Thus the orthogonality requirement is:

$$(I - AH)PH^{\mathrm{T}} = \frac{1}{k}AN$$

and hence A is given by

$$A = PH^{\mathrm{T}}\left[HPH^{\mathrm{T}} + \frac{1}{k}N\right]^{-1}.$$

Notice that $(HPH^{\mathrm{T}} + (1/k)N)$ is non-singular, since by assumption $N > 0$. Thus

$$\hat{x}_k = A\bar{y}_k = \frac{1}{k}PH^{\mathrm{T}}\left[HPH^{\mathrm{T}} + \frac{1}{k}N\right]^{-1}\left(\sum_{i=1}^{k} y_k\right). \qquad (3.2.4)$$

Comparing this with (3.2.3), we see that the coefficient of y_k is $K(k)$ and hence

$$K(k) = \frac{1}{k}PH^{\mathrm{T}}\left[HPH^{\mathrm{T}} + \frac{1}{k}N\right]^{-1}. \qquad (3.2.5)$$

The more general method of obtaining this result is to calculate $K(k)$ from the expression for it in (3.2.2). Now

$$\begin{aligned}\tilde{y}_{k|k-1} = y_k - \hat{y}_{k|k-1} &= (Hx + z_k) - H\hat{x}_{k-1}\\ &= H\tilde{x}_{k-1} + z_k \end{aligned} \qquad (3.2.6)$$

where $\tilde{x}_{k-1} = x - \hat{x}_{k-1}$ is the error at time $k-1$. Thus

$$\begin{aligned}E[x\tilde{y}^{\mathrm{T}}_{k|k-1}] &= E[x\tilde{x}^{\mathrm{T}}_{k-1}]H^{\mathrm{T}}\\ &= E[\tilde{x}_{k-1}\tilde{x}^{\mathrm{T}}_{k-1}]H^{\mathrm{T}}\end{aligned}$$

since $x = \hat{x}_{k-1} + \tilde{x}_{k-1}$ and $\hat{x}_{k-1} \perp \tilde{x}_{k-1}$. We denote $P(k-1) = \mathrm{cov}(\tilde{x}_{k-1})$ (the error covariance at time $k-1$). Similarly,

$$\begin{aligned}E[\tilde{y}_{k|k-1}\tilde{y}^{\mathrm{T}}_{k|k-1}] &= E[(H\tilde{x}_{k-1} + z_k)(H\tilde{x}_{k-1} + z_k)^{\mathrm{T}}]\\ &= HP(k-1)H^{\mathrm{T}} + N.\end{aligned}$$

This is non-singular since $N > 0$, and hence

$$K(k) = P(k-1)H^{\mathrm{T}}[HP(k-1)H^{\mathrm{T}} + N]^{-1}$$

It remains to calculate $P(k-1)$. Subtracting x from both sides of

(3.2.2) and using (3.2.6) gives

$$\begin{aligned}\tilde{x}_k &= \tilde{x}_{k-1} - K(k)(H\tilde{x}_{k-1} + z_k)\\ &= (I - K(k)H)\tilde{x}_{k-1} + K(k)z_k.\end{aligned}$$

The two terms in this expression are orthogonal since $\tilde{x}_{k-1} \in \mathscr{L}(x, z_1, \ldots, z_{k-1})$. Thus

$$P(k) = E[\tilde{x}_k\tilde{x}_k^{\mathrm{T}}] = (I - K(k)H)P(k-1)(I - K(k)H)^{\mathrm{T}} + K(k)NK(k). \tag{3.2.7}$$

Now substitute for $K(k)$ from (3.2.5). After a little algebra one finds that (3.2.7) becomes simply

$$P(k) = P(k-1) - P(k-1)H^{\mathrm{T}}[HP(k-1)H^{\mathrm{T}} + N]^{-1}HP(k-1). \tag{3.2.8}$$

Together with the initial condition $P(0) = P = \mathrm{cov}(x)$ this provides a recursive algorithm for generating $P(1), P(2), \ldots$ and hence $K(k)$ from (3.2.5). In this example one can in fact obtain a closed-form expression for $P(k)$ from (3.2.4). Indeed, subtracting x from both sides of (3.2.4) and using the fact that $\bar{y}_k = Hx + \bar{z}_k$ we see that

$$\begin{aligned}\tilde{x}_k = &\left(I - PH^{\mathrm{T}}\left[HPH^{\mathrm{T}} + \frac{1}{k}N\right]^{-1}H\right)x\\ &+ PH^{\mathrm{T}}\left[HPH^{\mathrm{T}} + \frac{1}{k}N\right]^{-1}\bar{z}_k.\end{aligned}$$

Again, the two terms on the right-hand side are orthogonal, and calculating the sum of their covariances we find that

$$P(k) = P - PH^{\mathrm{T}}\left[HPH^{\mathrm{T}} + \frac{1}{k}N\right]^{-1}HP.$$

Some laborious algebra confirms that this indeed satisfies (3.2.8).

In this example the recursive estimator (3.2.3) offers no advantages over the non-recursive form (3.2.4): in either case the main computational task at each stage is to invert an $r \times r$ matrix, so the general problem of having to invert matrices of growing dimensions has been avoided. The storage requirements are also similar: in (3.2.4) one requires the sample mean $\bar{y}_k$ at each stage and this can be updated as follows:

$$\bar{y}_k = \left(\frac{k-1}{k}\right)\bar{y}_{k-1} + \frac{1}{k}y_k.$$

Thus in neither case is it necessary to store the complete observation record. In more general problems such as the Kalman filter considered below, it is usually not possible to obtain simple closed-form expressions for the estimator, but the recursive solution may still be viable. From the implementation point of view this is perfectly satisfactory. The coefficient matrices $K(k)$ can be computed in advance and then the 'data processing' consists of on-line implementation of the very simple algorithm (3.2.3).

3.3 The Kalman filter

The Kalman filter is a recursive algorithm for estimating the state x_k of a state-space model given the values of the observed outputs $y^{k-1}(= y_0, y_1, y_2, \ldots, y_{k-1})$. The equations describing the model are

$$x_{k+1} = A(k)x_k + B(k)u_k + C(k)w_k \tag{3.3.1}$$

$$y_k = H(k)x_k + G(k)w_k. \tag{3.3.2}$$

Here, $\{w_k\}$ is an l-vector white-noise process with unit covariance $(Ew_k w_k^{\mathrm{T}} = I_l)$ and the initial random variable x_0 is uncorrelated with $\{w_k\}$, with known mean and covariance m_0, P_0 respectively. The coefficient matrices $A(k)$, etc., may be time-varying, as indicated by their dependence on k in (3.3.1), (3.3.2). The model is, in this respect, more general than that of Section 2.4. We assume that

$$G(k)G^{\mathrm{T}}(k) > 0 \tag{3.3.3}$$

(in particular this implies that $l \geq r$, r being the dimension of y_k). If this were not the case then there would exist vectors λ such that $\lambda^{\mathrm{T}}G(k) = 0$, so that, from (3.3.2),

$$\lambda^{\mathrm{T}} y_k = \lambda^{\mathrm{T}} H(k) x_k$$

i.e. certain linear combinations of components of x_k could be measured *exactly*. Thus (3.3.3) says essentially that all observations and linear combinations of observations are 'noisy'.

The sequence u_k is the m-vector control input. In this section we suppose that this is a *deterministic sequence*. In future sections we shall wish to consider feedback controls, where u_k depends on the observed outputs y^k, but this presents a more delicate situation, consideration of which we defer to Section 6.3 below.

The example considered in the preceding section is a special case of the model (3.3.1), (3.3.2): take $A(k) = I_n$, $B(k) = C(k) = 0$, $H(k) = H$ and $G(k) = N^{1/2}$ (so that $r = m$). We saw there that the estimators $\hat{x}_k$ could be computed recursively, and the same is true for the general state-space model considered here. The situation is complicated somewhat by the fact that the signal being estimated is not constant but is itself a stochastic process, and by the possible correlation between signal and observation noise. Nonetheless, the derivation of the Kalman filter equations follows exactly the same approach as used in the example.

We denote by $\hat{x}_{i|j}$ the best linear (or affine) estimator of x_i given $y^j = (y_0, y_1, \ldots, y_j)$, i.e. the projection of x_i onto $\mathscr{L}(y^j)$, and by $\tilde{x}_{i|j}$ the error $(x_i - \hat{x}_{i|j})$, with similar notation for other random variables. It turns out that the most useful form of estimator is the 'one-step-ahead' estimator $\hat{x}_{k|k-1}$.

Theorem 3.3.1 (Kalman filter)

For the system (3.3.1), (3.3.2) with the above assumptions, the estimator $\hat{x}_{k|k-1}$ satisfies the recursive equation

$$\hat{x}_{k+1|k} = A(k)\hat{x}_{k|k-1} + B(k)u_k + K(k)[y_k - H(k)\hat{x}_{k|k-1}] \quad k = 0, 1, \ldots \tag{3.3.4}$$

$$\hat{x}_{0|-1} = m_0.$$

The $n \times r$ gain matrix $K(k)$ is given by

$$K(k) = [A(k)P(k)H^{\mathrm{T}}(k) + C(k)G^{\mathrm{T}}(k)][H(k)P(k)H^{\mathrm{T}}(k) + G(k)G^{\mathrm{T}}(k)]^{-1} \tag{3.3.5}$$

where $P(k)$ is the error covariance

$$P(k) = E[(x_k - \hat{x}_{k|k-1})(x_k - \hat{x}_{k|k-1})^{\mathrm{T}}]$$

$P(k)$ satisfies the recursive *Riccati equation*:

$$\begin{aligned} P(k+1) = A(k)P(k)A^{\mathrm{T}}(k) + C(k)C^{\mathrm{T}}(k) - [A(k)P(k)H^{\mathrm{T}}(k) \\ + C(k)G^{\mathrm{T}}(k)][H(k)P(k)H^{\mathrm{T}}(k) + G(k)G^{\mathrm{T}}(k)]^{-1} \\ \cdot[A(k)P(k)H^{\mathrm{T}}(k) + C(k)G^{\mathrm{T}}(k)]^{\mathrm{T}} \end{aligned} \tag{3.3.6}$$

$$P(0) = P_0.$$

The innovations process

$$\nu_k := y_k - H(k)\hat{x}_{k|k-1}$$

is a wide-sense white-noise process with covariance function

$$E[\nu_k \nu_j^{\mathrm{T}}] = [H(k)P(k)H^{\mathrm{T}}(k) + G(k)G^{\mathrm{T}}(k)]\delta_{kj}.$$

If in addition to the above assumptions $(x_0, w_0, w_1, \ldots)$ are jointly normally distributed, so that in particular $\{w_k\}$ is a gaussian white-noise process, then

$$\hat{x}_{k|k-1} = E[x_k | y^{k-1}].$$

REMARKS To implement the Kalman filter, the sequence $P(k)$ is computed from the Riccati equation and the corresponding sequence of $n \times r$ 'gain' matrices $K(k)$ is computed using (3.3.5). All of this can be done off-line, i.e. before any observations are taken. Calculation of $\hat{x}_{k|k-1}$ can now be done recursively using (3.3.4) as successive observations become available. The fact that all the coefficients in (3.3.4) can be pre-computed (are not data-dependent) means that the amount of on-line signal processing required is very modest, and this is important in applications where computing power is at a premium. Note, however, that it is assumed that all coefficients appearing in the problem – i.e. the matrices A, B, C, H, G as well as the initial state mean and covariance m_0, P_0 – are exactly known.

PROOF Suppose to start with that $m_0 = 0$ and $u_k = 0$ for all k. Then $Ex_k = 0$ for all k and hence all random variables in system (3.3.1), (3.3.2) have zero mean. From (3.3.2) we see that†

$$\hat{y}_{k|k-1} = H\hat{x}_{k|k-1}$$

since $w_k \perp \mathscr{L}(y^{k-1})$, and hence the basic recursive formula (3.2.2) with $x = x_k$ gives

$$\hat{x}_{k|k} = \hat{x}_{k|k-1} + E[x_k \tilde{y}^{\mathrm{T}}_{k|k-1}](E[\tilde{y}_{k|k-1}\tilde{y}^{\mathrm{T}}_{k|k-1}])^{-1}(y_k - H\hat{x}_{k|k-1}). \quad (3.3.7)$$

(It will be verified below that $\text{cov}(\tilde{y}_{k|k-1})$ is non-singular). Now

$$\tilde{y}_{k|k-1} = y_k - \hat{y}_{k|k-1} = H\tilde{x}_{k|k-1} + Gw_k \quad (3.3.8)$$

(this coincides with the innovations process v_k of the theorem statement) and hence

$$\begin{aligned} E[x_k \tilde{y}^{\mathrm{T}}_{k|k-1}] &= E[x_k(\tilde{x}^{\mathrm{T}}_{k|k-1}H^{\mathrm{T}} + w_k^{\mathrm{T}}G^{\mathrm{T}})] \\ &= P(k)H^{\mathrm{T}} \end{aligned}$$

where $P(k) := \text{cov}(\tilde{x}_{k|k-1})$. The last equality follows by noting that $x_k \perp w_k$ and that x_k has the orthogonal decomposition $x_k = \tilde{x}_{k|k-1} +$

† For notational simplicity we write H for $H(k)$, etc., throughout the following argument.

$\hat{x}_k$. Similarly,

$$E[\tilde{y}_{k|k-1}\tilde{y}_{k|k-1}^{\mathrm{T}}] = HP(k)H^{\mathrm{T}} + GG^{\mathrm{T}}.$$

This is strictly positive definite, and hence non-singular, since $HP(k)H^{\mathrm{T}} \geq 0$ and $GG^{\mathrm{T}} > 0$. We now have to relate $\hat{x}_{k+1|k}$ to $\hat{x}_{k|k}$. From (3.3.1) we see that (since $u_k = 0$)

$$\hat{x}_{k+1|k} = A\hat{x}_{k|k} + C\hat{w}_{k|k}.$$

Now $w_k \perp \mathscr{L}(y^{k-1})$ and hence the best estimate $\hat{w}_{k|k}$ of w_k given y^k is equal to the best estimate given $\tilde{y}_{k|k-1}$ which, according to (3.1.6), is

$$\hat{w}_{k|k} = E[w_k\tilde{y}_{k|k-1}^{\mathrm{T}}]E[\tilde{y}_{k|k-1}\tilde{y}_{k|k-1}^{\mathrm{T}}]^{-1}\tilde{y}_{k|k-1}. \tag{3.3.9}$$

Using (3.3.8) we obtain

$$E[w_k\tilde{y}_{k|k-1}^{\mathrm{T}}] = E[w_k w_k^{\mathrm{T}} G^{\mathrm{T}}] = G^{\mathrm{T}}. \tag{3.3.10}$$

Combining (3.3.7)–(3.3.10) gives

$$\begin{aligned}\hat{x}_{k+1|1} = A[\hat{x}_{k|k-1} &+ PH^{\mathrm{T}}(HPH^{\mathrm{T}} + GG^{\mathrm{T}})^{-1}\tilde{y}_{k|k-1}] \\ &+ CG^{\mathrm{T}}(HPH^{\mathrm{T}} + GG^{\mathrm{T}})^{-1}\tilde{y}_{k|k-1}\end{aligned} \tag{3.3.11}$$

which is equivalent to (3.3.4)–(3.3.5). The best estimate of x_0 with *no* observations is 0 since $Ex_0 = 0$ and hence the initial condition for (3.3.11) is $x_{0|-1} = 0$. To compute the conditional covariance we use the same technique as in the example of the preceding section. Subtracting (3.3.4) from (3.3.1) and using (3.3.8) shows that the error $\tilde{x}_{k|k-1}$ satisfies the recursive equation

$$\tilde{x}_{k+1|k} = (A - KH)\tilde{x}_{k|k-1} + (C - KG)w_k. \tag{3.3.12}$$

We can therefore compute the covariance by using the general results given for the state-space model in Proposition 2.4.1†. Indeed, replacing A and C in (2.4.7) by $(A - KH)$ and $(C - KG)$ respectively, we see from (2.4.7) that $P(k) = \mathrm{cov}(\tilde{x}_{k|k-1})$ satisfies

$$P(k+1) = (A - KH)P(k)(A^{\mathrm{T}} - H^{\mathrm{T}}K^{\mathrm{T}}) + (C - KG)(G^{\mathrm{T}} - G^{\mathrm{T}}K^{\mathrm{T}}). \tag{3.3.13}$$

Substituting from (3.3.5) the expression for K in terms of $P(k)$, one obtains, after a little algebra, the variance equation (3.3.6).

Finally, suppose $Ex_0 = m_0 \neq 0$ and that u_k is also non-zero. Then $m(k) = Ex_k$ satisfies

$$\begin{aligned}m(k+1) &= Am(k) + Bu(k) \\ m(0) &= m_0\end{aligned} \tag{3.3.14}$$

† Equation (2.4.7) is valid for time varying models with $A = A(k)$ etc.

and

$$Ey_k = Hm(k).$$

As shown in Section 3.1, the best affine estimator of x_k given y^{k-1} is now

$$\hat{x}_{k|k-1} = \hat{x}^c_{k|k-1} + m(k)$$

where $\hat{x}^c_{k|k-1}$ is the projection of $x^c_k = x_k - m(k)$ onto $\mathscr{L}((y^c)^{k-1})$. But x^c_k, y^c_k satisfy the equations

$$\begin{aligned} x^c_{k+1} &= Ax^c_k + Cw_k \\ x^c_0 &= x_0 - m_0 \\ y^c_k &= Hx^c_k + Gw_k \end{aligned}$$

so that the computation of $\hat{x}^c_{k|k-1}$ is the zero-mean estimation problem we have just solved, i.e. $\hat{x}^c_{k|k-1}$ satisfies

$$\begin{aligned} \hat{x}^c_{k+1|k} &= Ax^c_{k|k-1} + K(k)(y^c_k - Hx^c_{k|k-1}) \\ \hat{x}^c_{0|-1} &= 0. \end{aligned} \tag{3.3.15}$$

Note that

$$\begin{aligned} y^c_k - H\hat{x}^c_{k|k-1} &= (y_k - Hm(k)) - H(\hat{x}_{k|k-1} - m(k)) \\ &= y_k - H\hat{x}_{k|k-1}. \end{aligned}$$

Thus, adding (3.3.14) and (3.3.15), we obtain (3.3.4). $P(k)$ given by (3.3.6) is still the error covariance, $\operatorname{cov}(x_k - \hat{x}_{k|k-1})$, since covariances are unaffected by a shift of mean.

Finally, suppose $x_0, w_0, w_1, \ldots$ are jointly normal. Then (x_k, y_k) is a normal process, since (3.3.1) (3.3.2) are linear equations, and it follows from Theorem 3.1.2 that $\hat{x}_{k|k-1} = E[x_k | y^{k-1}]$. □

Example 3.3.2

The example considered in Section 3.2 above is a Kalman filtering problem but a somewhat special one in that there are no 'system dynamics'. As the simplest example involving dynamics, let us consider estimating the autoregression

$$x_{k+1} = ax_k + v_k \tag{3.3.16}$$

given noisy observations

$$y_k = x_k + w_k.$$

Here all quantities are scalars and we assume that v_k, w_k are uncorrelated unit variance white-noise processes. The initial random variable x_0 is supposed to have mean and variance m_0 and P_0 respectively. The Kalman filtering equations (3.3.4)–(3.3.6) become

$$\hat{x}_{k+1|k} = a\hat{x}_{k|k-1} + \frac{aP(k)}{1+P(k)}(y_k - \hat{x}_{k|k-1}) \tag{3.3.17}$$

$$\hat{x}_{0|-1} = m_0$$

$$P(k+1) = a^2P(k) + 1 - \frac{a^2P^2(k)}{1+P(k)}$$

$$= \frac{(1+a^2)P(k)+1}{P(k)+1}.$$

It is interesting to note the behaviour of $P(k)$. Figure 3.2 shows the evolution of $P(k)$ starting from $P_0 = 2$ for $a = 1/2, 2$. It converges very rapidly towards a steady-state value, which in fact is the positive solution $P^* = P^*(a)$ of the *algebraic Riccati equation*

$$P^* = \frac{(1+a^2)P^*+1}{P^*+1}. \tag{3.3.18}$$

This solution is given by

$$P^* = \tfrac{1}{2}(a^2 + \sqrt{(a^4+4)})$$

(the other solution of (3.3.18) is negative). If $P_0 = P^*(a)$ then $P(k) = P^*(a)$ for all k and the Kalman filter (3.3.17) is time invariant (has

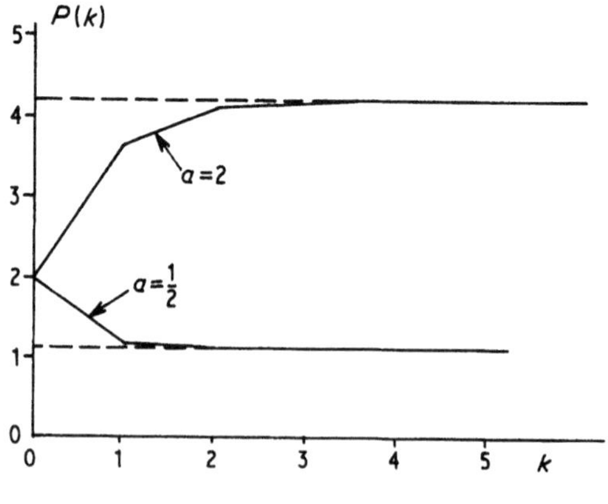

Fig. 3.2

constant coefficients). Otherwise the filter is asymptotically time-invariant, the gain $K(k)$ converging to the steady-state value $K^* = aP^*(a)/(1 + P^*(a))$. This is true even if the state equation (3.3.16) is unstable ($a > 1$). In this case $\mathrm{var}(x_k) \to \infty$ as $k \to \infty$ but the *conditional* variance $P(k)$ remains bounded. One of the curves in Fig. 3.2 shows this for $a = 2$. Now $P^*(2) = 4.2361$ so that the steady-state Kalman gain is $K^* = 1.618$. From (3.3.12) the error $\tilde{x}_{k|k-1}$ is given by

$$\tilde{x}_{k+1|k} = (a - K(k))\tilde{x}_{k|k-1} + v_k - K(k)w_k.$$

For large k, $K(k)$ is close to K^* (or exactly equal to K^* if $P_0 = P^*$) so that this equation becomes

$$\tilde{x}_{k+1|k} = 0.382\,\tilde{x}_{k|k-1} + v_k - 1.618\,w_k$$

which expresses $\tilde{x}_{k|k-1}$ in the form of a *stable* autoregression. The point is that $\mathrm{var}(x_k) = \mathrm{var}(\hat{x}_{k|k-1}) + \mathrm{var}(\tilde{x}_{k|k-1})$, and $\mathrm{var}(\tilde{x}_{k|k-1})$ remains bounded even through the other two terms do not. Intuitively, the observer has enough information to 'track' x_k successfully although it is generated by an unstable system.

Computation of P(k)

For the general system (3.3.1)–(3.3.2) with time-varying coefficients, the Kalman filter is implemented by precomputing the gain sequence $K(k)$ and this of course involves calculating the sequence of covariances $P(k)$. In principle this can be done by direct recursion of the Riccati equation (3.3.6) but that is not in fact a very good way of doing it, since (3.3.6) is numerically ill-conditioned. The three terms on the right of (3.3.6) are symmetric and non-negative definite, but the last one is *subtracted*, so there is nothing preventing non-negative definiteness of $P(k+1)$ from being lost, and if this ever happens the Riccati equation can become completely unstable. Consider for instance the example in Section 3.2. Taking the scalar case with $N = H = 1$, the Riccati equation (3.2.9) becomes

$$P(k) = P(k-1) - \frac{P^2(k-1)}{1 + P(k-1)} = \frac{P(k-1)}{1 + P(k-1)}.$$

Thus $\quad q(k) := P^{-1}(k)$ satisfies

$$q(k) = 1 + q(k-1).$$

Now suppose $P_0 = -0.1$; what happens? The moral is that $P(k)$ must be computed in such a way that successive terms of the recursion are intrinsically non-negative definite. The simplest way to do this is to use (3.3.5) and (3.3.13):

$$K(k) = [AP(k)H^{\mathrm{T}} + CG^{\mathrm{T}}][HP(k)H^{\mathrm{T}} + GG^{\mathrm{T}}]^{-1}$$

$$\begin{aligned} P(k+1) = {} & (A - K(k)H)P(k)(A - K(k)H)^{\mathrm{T}} \\ & + (C - K(k)G)(C - K(k)G)^{\mathrm{T}}. \end{aligned}$$

This is much better, as now $P(k+1)$ is expressed as a *sum* of non-negative definite terms.

An alternative approach is to propagate a square root of $P(k)$, i.e. calculate matrices $W(k)$ such that $P(k) = W(k)W^{\mathrm{T}}(k)$, an idea that has been the subject of considerable research. The situation is complicated by the fact that such a factorization of $P(k)$ is not unique and therefore a variety of different algorithms is possible. Some references to this subject are given in the Notes at the end of this chapter.

3.3.2 Time-invariant systems

Suppose that the coefficient matrices A, B, C, H, G in the system model (3.3.1), (3.3.2) are time-invariant (do not depend on k). The results of the above example suggest that we should study the *algebraic Riccati equation*

$$\begin{aligned} P = {} & APA^{\mathrm{T}} + CC^{\mathrm{T}} - [APH^{\mathrm{T}} + CG^{\mathrm{T}}] \\ & \cdot [HPH^{\mathrm{T}} + GG^{\mathrm{T}}]^{-1}[APH^{\mathrm{T}} + CG^{\mathrm{T}}]^{\mathrm{T}}. \end{aligned} \qquad (3.3.19)$$

If the initial covariance P_0 satisfies this equation then evidently, from (3.3.6), $P(k) = P_0$ for all k, and the Kalman filter (3.3.4) is time-invariant since now $K(k)$, as well as the other coefficients, is a constant matrix. Notice that this does *not* imply that the state process x_k (with $u_k = 0$) is wide-sense stationary. The condition for this was given in Proposition 2.4.1 and is

$$P_0 = AP_0A^{\mathrm{T}} + CC^{\mathrm{T}}.$$

Equation (3.3.19) represents a trade-off between two opposing effects: on the one hand the observer is learning more about x_k as more data accumulates, but on the other hand the position of x_k may become less certain as it moves away from its initial position. The initial covariance which satisfies the algebraic Riccati equation is the value at which these factors exactly balance, leaving a precisely constant degree of uncertainty as to the position of x_0 as measured by the estimation error covariance.

Under what conditions does the algebraic Riccati equation have a solution? Can there be more than one solution? It seems clear that the answers to these questions must be related to stabilizability and detectability properties. Suppose for example that one of the states x_k^i is completely unobserved by the output. Then the best estimate for x_k^i is just its mean Ex_k^i, and the mean square estimation error is $E(x_k^i - Ex_k^i)^2 = \text{var}(x_k^i)$. This converges only if x_k^i is stable. The questions of existence of solutions to the algebraic Riccati equation, and convergence of the sequence $P(0), P(1), \ldots$ of matrices generated by the Riccati equation (3.3.6), are studied in detail in Appendix B. It is a fundamental feature of linear system theory that the same Riccati equations appear in connection with a certain optimal control problem, the *linear regulator problem*, which is discussed in Section 6.1. The properties of these equations are most readily obtained from control-theoretic considerations, and we therefore limit ourselves here to stating the results and giving some interpretation of them in the filtering context.

We require matrices $\breve{A}$, $\breve{C}$ defined by

$$\breve{A} = A - CG^{\mathrm{T}}(GG^{\mathrm{T}})^{-1}H$$
$$\breve{C} = C[I - G^{\mathrm{T}}(GG^{\mathrm{T}})^{-1}G].$$

Theorem 3.3.3

(a) If the pair (H, A) is detectable then there exists at least one non-negative solution to the algebraic Riccati equation (3.3.19).

(b) If further the pair $(\breve{A}, \breve{C})$ is stabilizable then this solution P is unique and $P(k) \to P$ as $k \to \infty$ where $P(k)$ is the sequence generated by (3.3.6) with arbitrary initial covariance P_0. The matrix $A - KH$ is stable, where K is the Kalman gain corresponding to P, i.e.

$$K = [APH^{\mathrm{T}} + CG^{\mathrm{T}}][HPH^{\mathrm{T}} + GG^{\mathrm{T}}]^{-1}.$$

PROOF This is Theorem B.1 of Appendix B. In Appendix B the Riccati equation appears in different notation, appropriate to its role in the control problems of Chapter 6. One obtains (3.3.19) by identifying coefficients as in Table 6.1. It will be found that $\breve{A}$ corresponds to $\hat{A}^{\mathrm{T}}$ and $\breve{C}$ to $\hat{D}^{\mathrm{T}}$. □

Note that the Kalman filter equation (3.3.4) can be written

$$\hat{x}_{k+1|k} = (A - K(k)H)\hat{x}_{k|k-1} + Bu_k + K(k)y_k. \tag{3.3.20}$$

The above results say that, under the stated conditions, this will be a time-invariant system (i.e. $K(k) = K$ for all k) if $P(0) = P$, the solution to the algebraic Riccati equation. If $P(0) \neq P$ then the gain sequence $K(k)$ tends to the stationary value K as $k \to \infty$. Thus the filter is almost time-invariant for large K and furthermore the filter is stable in that the system matrix $A - KH$ in (3.3.20) is stable. Convergence to the stationary state is often rapid, and this justifies the widely employed practice of using the time-invariant filter even when $P(0) \neq P$.

The following remarks are intended to illuminate the role of the matrices $(\check{A}, \check{C})$ which appear in the conditions of Theorem 3.3.3. Returning to the state-space model (3.3.1), (3.3.2), let us denote

$$e_k = Cw_k$$
$$f_k = Gw_k = y_k - Hx_k.$$

These are the 'noise' terms appearing in the state and observation equations respectively. They are not uncorrelated – in fact $\operatorname{cov}(e_k, f_k) = CG^{\mathrm{T}}$ – but e_k and f_l are uncorrelated for $k \neq l$ since w_k is white noise. The best estimate $\hat{e}_k$ of e_k given f_k is give by the general formula (3.1.6) as

$$\hat{e}_k = CG^{\mathrm{T}}(GG^{\mathrm{T}})^{-1} f_k$$

and the covariance of the error $\tilde{e}_k = e_k - \hat{e}_k$ is

$$\begin{aligned}\operatorname{cov}(\tilde{e}_k) &= C[I - G^{\mathrm{T}}(GG^{\mathrm{T}})^{-1}G][I - G^{\mathrm{T}}(GG^{\mathrm{T}})^{-1}G]^{\mathrm{T}}C^{\mathrm{T}}\\ &= \check{C}\check{C}^{\mathrm{T}}.\end{aligned}$$

Thus we can express $\tilde{e}_k$ in the form

$$\tilde{e}_k = \check{C}v_k$$

where $\operatorname{cov}(v_k) = I$ and v_k is (like $\tilde{e}_k$) uncorrelated with f_l for all l (including $l = k$). The state equation (3.3.1) now becomes

$$\begin{aligned}x_{k+1} &= Ax_k + Bu_k + \hat{e}_k + \tilde{e}_k\\ &= Ax_k + Bu_k + CG^{\mathrm{T}}(GG^{\mathrm{T}})^{-1}(y_k - Hx_k) + \check{C}v_k\end{aligned}$$

i.e.

$$\begin{aligned}x_{k+1} &= \check{A}x_k + Bu_k + CG^{\mathrm{T}}(GG^{\mathrm{T}})^{-1}y_k + \check{C}v_k\\ y_k &= Hx_k + Gw_k.\end{aligned}$$

This expresses the system in a form in which the noises appearing in the state and observation equations are uncorrelated (at the expense of adding an extra 'feedback' term from y_k to x_{k+1}). The stabilizability and detectability conditions stated above refer to the system in this form, involving matrices $(\breve{A}, \breve{C}, H)$ rather than the original (A, C, H). (Note that detectability of $(H, \breve{A})$ is equivalent to detectability of (H, A).) If $CG^{\mathrm{T}} = 0$ then e_k and f_k are already uncorrelated and $\breve{A} = A$, $\breve{C} = C$.

Computation of the solution of the algebraic Riccati equation has been the subject of active research and the best algorithms are of comparatively recent vintage. It is true that $P(k)$ generated by (3.3.6) converges to P but as an algorithm this is not numerically robust. We do not discuss this subject here; see the Notes at the end of the chapter for further information.

3.4. Innovations representation of state-space models

In Section 2.4 it was shown that the state-space model and the ARMAX model

$$A(z^{-1})y_k = B(z^{-1})u_k + C(z^{-1})w_k$$

are interchangeable in the sense that the ARMAX model can be realized in state-space form, while a state-space model can be recast as an ARMAX system by calculating its transfer function. The two forms are complementary: the ARMAX model is preferred in *system identification* because of its 'parsimonious parametrization' (as Box and Jenkins (1976) put it), while *optimal control theory* has been developed primarily for state-space systems. In this section we discuss further the concept of an innovations model, which was already briefly introduced in Section 2.5. Recall that the state-space model

$$\begin{aligned} x_{k+1} &= Ax_k + Bu_k + Cw_k \\ y_k &= Hx_k + Gw_k \end{aligned} \tag{3.4.1}$$

is in *innovations form* if the matrix G is non-singular, and that the standard state-space realization of the ARMAX model has this property. The main result of this section, Theorem 3.4.1 below, states that to every state-space model (subject to mild restrictions) there corresponds an innovations model with the same external behaviour. This result is an important by-product of Kalman filtering theory. It implies that, in terms of input–output modelling, the class of innovations-form state-space models is just as rich as that of state-space models as a whole, a fact which is by no means apparent since the innovations model involves fewer parameters and a reduced number of noise inputs.

We consider the system (3.4.1) with time-invariant coefficients A, B, etc., and made the following assumptions:

(a) The matrix A is stable. (3.4.2)
(b) $GG^{\mathrm{T}} > O$.

Condition (a) does not imply that $(\check{A}, \check{C})$ is stabilizable, or conversely; see the example at the end of this section.

The input–output properties of (3.4.1) under stationary conditions were derived in Section 2.4. The state x_k has covariance Q satisfying

$$Q = AQA^{\mathrm{T}} + CC^{\mathrm{T}} \tag{3.4.3}$$

and the equations represent a linear system as shown in Fig. 3.3. Here $\mathscr{L}$ is a linear system with transfer function

$$T_{\mathrm{L}}(z^{-1}) = z^{-1}H(I - z^{-1}A)^{-1}B$$

and $\bar{y}_k$ is a stationary process with covariance function

$$\operatorname{cov}(\bar{y}_k, \bar{y}_l) = \begin{cases} HQH^{\mathrm{T}} + GG^{\mathrm{T}} & k = l \\ HA^{l-k}QH^{\mathrm{T}} + HA^{l-k-1}CG^{\mathrm{T}} & l > k. \end{cases} \tag{3.4.4}$$

In accordance with the introductory remarks in Chapter 2, we describe this as an *external model.*

Now consider the Kalman filter (3.3.3)–(3.3.6) for this system with the constant gain K corresponding to a solution P of the algebraic Riccati equation (such a solution exists since (3.4.2) (a) implies that (H, A) is detectable) P and K satisfy

$$\begin{aligned} P = APA^{\mathrm{T}} + CC^{\mathrm{T}} - (APH^{\mathrm{T}} + CG^{\mathrm{T}}) \\ \cdot(HPH^{\mathrm{T}} + GG^{\mathrm{T}})^{-1}(APH^{\mathrm{T}} + CG^{\mathrm{T}})^{\mathrm{T}} \end{aligned} \tag{3.4.5}$$

$$K = (APH^{\mathrm{T}} + CG^{\mathrm{T}})(HPH^{\mathrm{T}} + GG^{\mathrm{T}})^{-1} \tag{3.4.6}$$

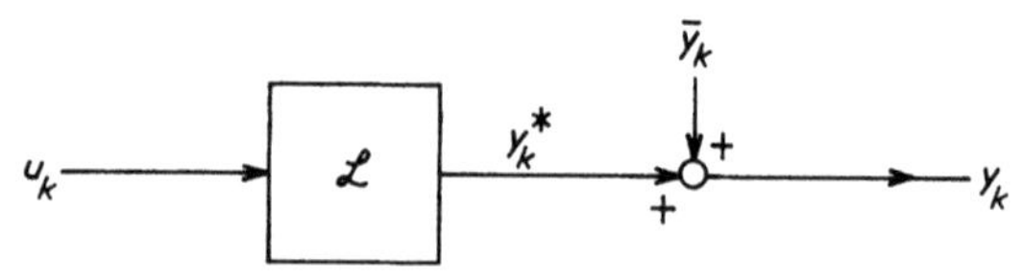

Fig. 3.3

Recall that the 'innovations process'

$$\tilde{y}_{k|k-1} = y_k - H\hat{x}_{k|k-1} \tag{3.4.7}$$

consists of uncorrelated random vectors with covariance

$$\operatorname{cov}(\tilde{y}_{k|k-1}) = HPH^{\mathrm{T}} + GG^{\mathrm{T}}. \tag{3.4.8}$$

Let D denote any $r \times r$ square root of this covariance, i.e. any $r \times r$ matrix satisfying

$$DD^{\mathrm{T}} = HPH^{\mathrm{T}} + GG^{\mathrm{T}}. \tag{3.4.9}$$

Now consider the following state-space model with state ξ_k, output η_k and normalized white-noise input e_k ($\operatorname{cov}(e_k) = I_r$):

$$\xi_{k+1} = A\xi_k + Bu_k + KDe_k \tag{3.4.10}$$

$$\eta_k = H\xi_k + De_k. \tag{3.4.11}$$

Note that this is obtained from the Kalman filter equation (3.3.4) and the definition (3.4.7) of the innovations process, with $\xi_k = \hat{x}_{k|k-1}$, $\eta_k = y_k$ and $De_k = \tilde{y}_{k|k-1}$, but in (3.4.10) we are thinking of e_k as an exogenous white noise driving a state-space model in the standard form. This system is called the *innovations representation* of the state-space model (3.4.1).

Theorem 3.4.1

Suppose conditions (3.4.2) hold; then the state-space models (3.4.1) and (3.4.10), (3.4.11) are alternative realizations of the *same* external model.

PROOF It is clear that the input-to-output transfer functions of the two systems coincide since these involve only the matrices A, B, H. It remains to show that the output noise covariance of the innovations representation coincides with (3.4.4). To calculate this let $\bar{\xi}_k$, $\bar{\eta}_k$ be the state and output processes in (3.4.10), (3.4.11) when $u_k \equiv 0$. From Proposition 2.4.2, the stationary covariance $\bar{P}$ of $\bar{\xi}_k$ satisfies

$$\begin{aligned}\bar{P} &= A\bar{P}A^{\mathrm{T}} + KDD^{\mathrm{T}}K^{\mathrm{T}} \\ &= A\bar{P}A + (APH^{\mathrm{T}} + GG^{\mathrm{T}})(HPH^{\mathrm{T}} \\ &\quad + GC^{\mathrm{T}})^{-1}(APH^{\mathrm{T}} + GG^{\mathrm{T}})^{\mathrm{T}}.\end{aligned}$$

Adding this equation and the algebraic Riccati equation (3.4.5) we see that $\bar{P}+P$ satisfies

$$(\bar{P}+P)=A(\bar{P}+P)A^{\mathrm{T}}+CC^{\mathrm{T}}.$$

Thus in view of (3.4.3),

$$\bar{P}+P=Q.$$

Now the two terms on the right-hand side of (3.4.11) are uncorrelated, and therefore

$$\begin{aligned}\operatorname{cov}(\bar{\eta}_k)&=H\bar{P}H^{\mathrm{T}}+DD^{\mathrm{T}}\\&=H\bar{P}H^{\mathrm{T}}+(HPH^{\mathrm{T}}+GG^{\mathrm{T}})\\&=HQH^{\mathrm{T}}+GG^{\mathrm{T}}.\end{aligned}$$

This shows that $\operatorname{cov}(\bar{\eta}_k)=\operatorname{cov}(\bar{y}_k)$. It remains to show that $\operatorname{cov}(\bar{\eta}_k,\bar{\eta}_l)=\operatorname{cov}(\bar{y}_k,\bar{y}_l)$ for $k\neq l$. By direct recursion of (3.4.10) we see that for $l>k$,

$$\bar{\xi}_l=A^{l-k}\bar{\xi}_k+A^{l-k-1}KDe_k+f(e_{k+1},\ldots,e_{l-1})$$

where $f(\cdot)$ is a linear function of the indicated random variables, all of which are uncorrelated with $\bar{\xi}_k, e_k$. Thus

$$\begin{aligned}E[\bar{\eta}_l\bar{\eta}_k^{\mathrm{T}}]&=E[(HA^{l-k}\bar{\xi}_k+HA^{l-k-1}KDe_k+De_l)(\bar{\xi}_k^{\mathrm{T}}H^{\mathrm{T}}+e_k^{\mathrm{T}}D^{\mathrm{T}})]\\&=HA^{l-k}\bar{P}H^{\mathrm{T}}+HA^{l-k-1}KDD^{\mathrm{T}}\\&=HA^{l-k}\bar{P}H^{\mathrm{T}}+HA^{l-k-1}(APA^{\mathrm{T}}+CG^{\mathrm{T}})\\&=HA^{l-k}QH^{\mathrm{T}}+HA^{l-k-1}CG^{\mathrm{T}}.\end{aligned}$$

But this agrees with the expression (3.4.4) for $\operatorname{cov}(\bar{y}_k,\bar{y}_l)$ when $l>k$. Thus models (3.4.1) and (3.4.10) are both represented by Fig. 3.3 with additive noise whose covariance function is given by (3.4.4); i.e. they are the same system in terms of their external behaviour. □

It was shown in Section 2.4 that by calculating the transfer functions of the models (3.4.1) and (3.4.10), (3.4.11) we can express them in general stochastic dynamical model form as

$$y_k=P(z^{-1})u_k+Q(z^{-1})w_k \tag{3.4.12}$$

$$\eta_k=P(z^{-1})u_k+\tilde{Q}(z^{-1})e_k \tag{3.4.13}$$

respectively, where $P, Q, \tilde{Q}$ are matrices of rational functions. Q and $\tilde{Q}$ are not the same since the dimension of w_k is possibly greater than that of e_k (it cannot be less, in view of condition (c) of (3.4.2)). None the less, Theorem 3.4.1 implies that the spectral densities of y_k and η_k with

$u_k = 0$ are the same, so that

$$Q(z^{-1})Q^{\mathrm{T}}(z) = \tilde{Q}(z^{-1})\tilde{Q}^{\mathrm{T}}(z).$$

This shows that the innovations representation is the *most efficient parametrization* of the state-space model in the sense of giving a desired output spectral density with a minimal number of uncorrelated noise inputs.

Finally, let us consider representing the (common) input/output behaviour of our state-space models in ARMAX form. As shown in Section 2.4 this can always be done by factoring out the lowest common multiples of the denominators of P, Q and P, $\tilde{Q}$ respectively in the general models (3.4.12), (3.4.13). Model (3.4.13) is the better choice since the noise dimension is reduced, but the factorization procedure is in general a laborious one. We now show that *in the single input–single output case* one can *read off* the coefficients of the corresponding ARMAX model after a simple change of coordinates. We need to introduce the additional assumption that (H, A) is observable. In a sense this assumption entails no loss of generality since, as discussed in Section 1.2, if the system is not observable then it is possible to construct a reduced-order observable system with the same transfer function.

Suppose then that (3.4.1) is a single input–single output system satisfying conditions (3.4.2) and that the pair (H, A) is observable. Referring to Section 2.4, note that the state-space representation of the ARMAX model given in proposition 2.4.2 is identical in structure to (3.4.10), (3.4.11) but has the additional feature that the A and H matrices take a particular form (the so-called transposed companion form). However, the general model (3.4.10), (3.4.11) can always be put in this form by a change of basis in the state space. Indeed, suppose T is any non-singular matrix and define

$$\tilde{\xi}_k = T^{-1}\xi_k.$$

Then $\tilde{\xi}_k$ satisfies

$$\begin{aligned}\tilde{\xi}_{k+1} &= \tilde{A}\tilde{\xi}_k + \tilde{B}u_k + \tilde{K}De_k\\ \eta_k &= \tilde{H}\tilde{\xi}_k + De_k\end{aligned}$$

where

$$\begin{aligned}\tilde{A} &= TAT^{-1}\\ \tilde{B} &= T^{-1}B\\ \tilde{K} &= T^{-1}K\\ \tilde{H} &= HT.\end{aligned}$$

We claim that it is possible to choose T so that $\tilde{A}$, $\tilde{H}$ are in transposed companion form. To achieve this, let w be any n-vector such that the matrix

$$T := [w \mid Aw \mid A^2w \mid \dots \mid A^{n-1}w] \qquad (3.4.14)$$

is non-singular. Then

$$AT = [Aw \quad A^2w \quad \dots \quad A^nw]$$

and hence

$$T^{-1}AT = \left[\begin{array}{ccccc|c} 0 & & & & 0 & \\ 1 & 0 & & & \cdot & \\ 0 & 1 & & & \cdot & \\ \cdot & & \cdot & & \cdot & T^{-1}A^nw \\ \cdot & & & \cdot & & \\ \cdot & & & \cdot & 0 & \\ 0 & \cdot & \cdot & \cdot & 1 & \end{array}\right].$$

This gets $\tilde{A}$ in the appropriate form. We also require $HT = [0, 0, \dots, 0, 1]$. If this is to be satisfied, then by the definition of T, and recalling that H is now a row n-vector, the vector w must satisfy

$$\begin{aligned} Hw &= 0 \\ HAw &= 0 \\ &\vdots \\ HA^{n-2}w &= 0 \\ HA^{n-1}w &= 1, \end{aligned} \qquad (3.4.15)$$

i.e.

$$(\Gamma w)^{\mathrm{T}} = [0, 0, \dots, 0, 1] \qquad (3.4.16)$$

where Γ is the observability matrix:

$$\Gamma = \begin{bmatrix} H \\ HA \\ \vdots \\ HA^{n-1} \end{bmatrix}.$$

By assumption this is non-singular, and therefore (3.4.16) states that (3.4.15) is satisfied if w is the last column of Γ^{-1}. It remains to show that T defined by (3.4.14) is non-singular with this choice of w. But in view of (3.4.15)

$$J := \Gamma T = \begin{bmatrix} H \\ HA \\ \vdots \\ HA^{n-1} \end{bmatrix} [w \,\vdots\, Aw \,\vdots\, \cdots \,\vdots\, A^{n-1}w]$$

$$= \begin{bmatrix} 0 & \cdots & 1 \\ \vdots & 1 & * \\ \vdots & & \vdots \\ 1 & * \cdots & * \end{bmatrix}$$

(where the stars denote possibly non-zeros elements). Thus J has rank n, so that $T = \Gamma^{-1}J$ is non-singular. This completes the identification of the state space and ARMAX models: once the innovations representation of the state-space model has been transformed to transposed companion form, the coefficients of the corresponding ARMAX model can be read off by referring to the state-space realization of the ARMAX model given in Proposition 2.4.3. In terms of the original model (3.4.1), these coefficients are given by

$$\begin{aligned} a_i &= -[T^{-1}A^n w]^{n-i+1} \\ b_i &= [T^{-1}B]^{n-i+1} \\ c_0 &= \sqrt{(HPH^{\mathrm{T}} + GG^{\mathrm{T}})} \\ c_i &= c_0 a_i + [T^{-1}K]^{n-i+1} \end{aligned}$$

for $i = 1, 2, \ldots, n$, where $[x]^i$ denotes the ith component of the n-vector x.

Finally, it is instructive to consider what happens when (3.4.1) is already in 'innovations form'. This occurs when y_k and w_k have the same dimension and G is non-singular, so that (3.4.1) becomes

$$\begin{aligned} w_k &= G^{-1}(y_k - Hx_k) \\ x_{k+1} &= Ax_k + Bu_k + CG^{-1}(y_k - Hx_k) \\ &= (A - CG^{-1}H)x_k + Bu_k + CG^{-1}y_k. \end{aligned} \tag{3.4.17}$$

If the initial state x_0 is known then the states $x_1, x_2, \ldots$ can be recovered *exactly* from the observations by recursion of (3.4.17). If

x_0 is unknown, start (3.4.17) with an arbitrary initial state ξ and let $\bar{x}_k$ be the resulting sequence of states, i.e.

$$\begin{aligned} \bar{x}_1 &= (A - CG^{-1}H)\xi + Bu_0 + CG^{-1}y_0 \\ \bar{x}_2 &= (A - CG^{-1}H)\bar{x}_1 + Bu_1 + CG^{-1}y_1 \\ &\text{etc.} \end{aligned} \tag{3.4.18}$$

Then $\varepsilon_k := x_k - \bar{x}_k$ satisfies

$$\begin{aligned} \varepsilon_{k+1} &= (A - CG^{-1}H)\varepsilon_k \\ \varepsilon_0 &= x_0 - \xi, \end{aligned}$$

so that $\varepsilon_k \to 0$ as $k \to \infty$ as long as $A - CG^{-1}H$ is stable.

The following result is obvious by inspection but is worth pointing out explicitly.

Proposition 3.4.2

Suppose that, in the model (3.4.1), y_k and w_k have the same dimension and G is non-singular. Then $P = 0$ is a solution of the algebraic Riccati equation (3.4.5).

These results help to evaluate the relationship between the various conditions used above. Indeed, if G is invertible, then $\check{C} = 0$ and $\check{A} = A - CG^{-1}H$, so that the pair $(\check{A}, \check{C})$ is stabilizable if and only if $A - CG^{-1}H$ is stable. Under this condition the algebraic Riccati equation has a unique non-negative definite solution, and we know that $P = 0$ is a solution if G is non-singular. The corresponding values of K and D are, from (3.4.6) and (3.4.9), $K = CG^{-1}$ and $D = G$. Thus the innovations model (3.4.10), (3.4.11) coincides with the original model (3.4.1), as it should. For any ξ, the sequence $\bar{x}_k$ given by (3.4.18) with $\bar{x}_0 = \xi$ forms a sub-optimal estimate of x_k which nevertheless has asymptotically zero error covariance.

This example also helps to elucidate the relationship between the conditions of Section 3.3 and those of (3.4.2). If (H, A) is observable then *pole placement* is possible, i.e. the eigenvalues of the matrix $A + SH$ can be assigned to arbitrary locations by suitable choice of the matrix S. Thus observability of (H, A) does not imply stabilizability of $(\check{A}, \check{C})$, which is equivalent to stability of $A - CG^{-1}H$, in the absence of any restrictions on C and G.

Notes

The idea of representing stochastic processes in terms of innovations or orthogonal components goes back at least to Wold (1938) and reaches its furthest development in the papers of Wiener and Masani (1958). Prediction problems were tackled simultaneously by Wiener (1949) and Kolmogorov (1941). (Wiener's book contains the results of previously classified wartime research.) Both of these authors were concerned with stationary processes. The time-domain approach based on state-space models was initiated by Kalman and Bucy (1960; 1961).

The literature on Kalman filtering is now immense. Textbook accounts that we have found particularly informative are Anderson and Moore (1979), Gelb (1974) and Maybeck (1979). All of these are valuable colateral reading in that they cover applications issues not discussed in this book. In particular, square root algorithms for propagating the conditional covariance matrix are discussed in detail in Anderson and Moore and in Maybeck. Anderson and Moore also discuss solution of the algebraic Riccati equation; for some of the latest work in this area, see Pappas, Laub and Sandell (1980).

References

Anderson, B. D. O. and Moore, J. B. (1979) *Optimal Filtering*, Prentice-Hall, Englewood Cliffs, NJ.

Box, G. E. P. and Jenkins, G. M. (1976) *Time Series Analysis, Forecasting and Control*, 2nd edn, Holden-Day, San Francisco.

Gelb, A. (ed.) (1974) *Applied Optimal Estimation*, MIT Press, Cambridge, Mass.

Kailath, T. (1980) *Linear Systems*, Prentice-Hall, Englewood Cliffs, NJ.

Kalman, R. E. (1960) A new approach to linear filtering and prediction problems. *ASME Transactions*, Part D (*J. of Basic Engineering*), **82**, 35–45.

Kalman, R. E. and Bucy, R. S. (1961) New results in linear filtering and prediction theory. *ASME Transactions*, Part D (*J. of Basic Engineering*), **83**, 95–108.

Kolmogorov, A. N. (1941) Interpolation und Extrapolation von Stationären Zufälligen Folgen. *Bull. Acad. Sci. URSS, Sér. Math.* **5**, 3–14.

Maybeck, P. S. (1979) *Stochastic Models, Estimation and Control*, Vol. 1, Academic Press, NY.

Pappas, T., Laub, A. J. and Sandell, N. R. (1980) On the numerical solution of the algebraic Riccati equation. *IEEE Trans. Automatic Control* **AC-25**, 631–641.

Wiener, N. (1949) *Extrapolation, Interpolation and Smoothing of Stationary Time Series*, MIT Press.

Wiener, N. and Masani, P. The prediction theory of multivariable stochastic processes I and II. *Acta Math.* (1957) **98**, 111–150; (1958) **99**, 93–137.

Wold, H. (1938) *A Study in the Analysis of Stationary Time Series*, Almqvist and Wiksell, Stockholm.

CHAPTER 4

System identification

An implicit assumption in the theory of optimal filtering and control is the availability of a mathematical model which adequately describes the behaviour of the system concerned. We pointed out in Chapter 2 that such models can be obtained from the physical laws governing the system or alternatively by some form of data analysis. The latter approach, known as 'system identification', is discussed in this chapter and is appropriate in cases where the physical mechanisms of the system are either highly complex or imprecisely understood, but where the input/output behaviour of the system is sufficiently regular to be represented adequately by fairly simple models.

The methodology of system identification involves a number of steps:

(a) Selection of a class of models from which a model to represent the system is to be chosen.
(b) Experiment design: choice of the inputs to be supplied and the readings to be taken in the identification experiment.
(c) Selection of a model on the basis of the experimental data.
(d) Model validation: this involves checking the adequacy of the chosen model in relation to some specific task such as prediction or use as the basis of control system design.

In this chapter we are concerned with the techniques of system identification when the models considered are linear discrete-time stochastic models of the sort described in Chapter 2. These models represent linear time-invariant systems with stationary additive random disturbances. Data analysis is then necessarily based on statistical techniques. The field of statistical identification is however a large one, and it is possible to treat only certain topics in the space available here. Attention will be given almost exclusively to the problem of how to analyse data from an identification experiment and thereby choose a suitable model from some given, finitely parametrized, class of models. This aspect of identification is usually called *parameter estimation*, even though this terminology misleadingly suggests that there is some 'true' parameter value which provides a perfect description of the system and which it is our task to estimate. In practice we can never achieve a perfect description, and the object of identification is merely to furnish a model whose response adequately approximates that of the system in significant respects.

Even on the subject of parameter estimation we have been selective. Emphasis is given to methods which admit a 'prediction error' formulation and consequently there is no mention of correlation techniques (such as that of 'instrumental variables') which do not fit into this framework. Our models all involve stationary disturbances, so we do not discuss non-stationary behaviour – trends and seasonal variations – which is so important in econometric time series. Nor do we investigate issues of numerical stability. Some references to the literature on these and other omitted topics are provided in the Notes at the end of this chapter.

Our object in this chapter is to describe certan important parameter estimation methods, and to investigate the quality of the estimates in some cases where the analysis is relatively straightforward. The task of analysing the asymptotic properties of the estimates in a general context is undertaken in Chapter 5.

4.1 Point estimation theory

Here we describe some classical concepts from point estimation theory of relevance to identification. The problem considered in point estimation theory is that of estimating the value of some function of an unknown parameter given an observation of a random variable x whose statistical properties depend on the parameter.

To be more specific, suppose that $f(\cdot, \theta)$ is a collection of probability densities in n variables, parameterized by vectors $\theta \in D$ where D is some set of q-vectors, and suppose $d: \mathbb{R}^q \to \mathbb{R}^r$ is some function of the parameter we are interested in, the parameter itself say. A random variable x, which has density $f(\cdot, \theta^*)$ for some unknown $\theta^* \in D$, is observed. An *estimator* for $d(\theta^*)$ is a function $g: \mathbb{R}^n \to \mathbb{R}^r$. It supplies an *estimate* $g(x)$ of $d(\theta^*)$. We view an estimate $g(x)$ of $d(\theta^*)$ either as a random variable, defined as a function of a random variable, or as the function g evaluated at the observation of the random variable x, depending on context. It is understood that an estimate of $d(\theta^*)$ in some sense approximates $d(\theta^*)$. Estimates of θ^* are of primary interest and we shall often refer to these simply as 'estimates'.

What is a good estimator of $d(\theta^*)$? In order to ask this question precisely, we introduce the following definitions:

$g(\cdot)$ is an *unbiased* estimator for $d(\theta^*)$ if

$$E_\theta g(x) = d(\theta), \qquad \text{for all } \theta$$

where $E_\theta g(x)$ is the expected value of $g(x)$ given than x has the probability density $f(\cdot, \theta)$, that is

$$E_\theta g(x) = \int g(\xi) f(\xi, \theta)\, d\xi.$$

An unbiased estimate averaged over independent experiments gives the correct parameter value, whatever this is.

It is also desirable that the covariance matrix of $g(x)$ be 'small'. This property in itself, however, gives little indication of the quality of an estimator: the estimate $g(x) = \psi$, where ψ is some fixed vector, has covariance matrix the zero matrix and yet it is useless as an estimator since it will be biased except in the fortuitous circumstances that $\psi = d(\theta^*)$. For this reason, bounds on the covariance of unbiased estimates are of particular interest.

Theorem 4.1.1 (Cramér–Rao lower bound)

Suppose the function $f(\cdot,\cdot)$ defining the collection of probability density functions $f(\cdot, \theta), \theta \in D$, is sufficiently regular. Define the matrix $M_\theta = \{m_{ij}\}$ by

$$m_{ij} = E_\theta\left(\frac{\partial}{\partial \theta_i} \log f(x, \theta) \frac{\partial}{\partial \theta_j} \log f(x, \theta)\right) \tag{4.1.1}$$

and suppose that M_θ is non-singular. Then for an arbitrary unbiased estimator $g(\cdot)$ of θ, we have

$$\text{cov}_\theta\, g(x) \geq M_\theta^{-1} \qquad \text{for all } \theta \in D.$$

PROOF Let $g(\cdot)$ be an unbiased estimator of θ. Since $g(\cdot)$ is unbiased,

$$E_\theta\{g(x)\} = \theta \qquad \text{for all } \theta. \tag{4.1.2}$$

This means

$$\int g(\xi) f(\xi, \theta) \, \mathrm{d}\xi = \theta \qquad \text{for all } \theta$$

whence

$$\frac{\partial}{\partial \theta_i} \int g_j(\xi) f(\xi, \theta) \mathrm{d}\xi = \delta_{ij} \qquad \text{for all } \theta.$$

Under suitable conditions on f and g, we can carry the differentiation with respect to θ_i operator under the integral sign. There results the equation†

$$\int g(\xi) \frac{\partial}{\partial \theta} f(\xi, \theta) \, \mathrm{d}\xi = I \qquad \text{for all } \theta$$

which can be written

$$\int g(\xi) \frac{\partial}{\partial \theta} (\log f(\xi, \theta)) f(\xi, \theta) \, \mathrm{d}\xi = I \qquad \text{for all } \theta.$$

or, in terms of the expectation operator corresponding to $f(\cdot, \theta)$,

$$E_\theta \left\{ g(x) \frac{\partial}{\partial \theta} (\log f(x, \theta)) \right\} = I \qquad \text{for all } \theta. \tag{4.1.3}$$

We now use the fact that $f(\cdot, \theta)$ is a probability density to derive another relationship. Since

$$\int f(\xi, \theta) \, \mathrm{d}\xi = 1$$

we can write, under suitable conditions on f,

$$\int \frac{\partial}{\partial \theta} f(\xi, \theta) \, \mathrm{d}\xi = \frac{\partial}{\partial \theta} \int f(\xi, \theta) \, \mathrm{d}\xi = 0^{\mathrm{T}} \qquad \text{for all } \theta.$$

Here 0 denotes a column vector of zeros. But then

$$\int \frac{\partial}{\partial \theta} (\log f(\xi, \theta)) f(\xi, \theta) \, \mathrm{d}\xi = 0^{\mathrm{T}} \qquad \text{for all } \theta.$$

† $(\partial/\partial\theta) f(\xi, \theta)$ denotes the *row vector* with components $(\partial/\partial\theta_i) f(\xi, \theta)$. We adhere to this convention throughout.

and this equation can be expressed

$$E_\theta\left\{\frac{\partial}{\partial\theta}\log f(x,\theta)\right\} = 0^{\mathrm{T}} \qquad \text{for all } \theta. \tag{4.1.4}$$

Now let us examine the covariance matrix Q_θ of the composite random variable $[g^{\mathrm{T}}(x), \partial/\partial\theta \log f(x,\theta)]^{\mathrm{T}}$ when x is taken to have probability density $f(\cdot,\theta)$. By (4.1.2) and (4.1.4) this random variable has mean col $[\theta, 0]$, so

$$Q_\theta = E_\theta\left\{\left[\begin{bmatrix} g(x)-\theta \\ \left[\dfrac{\partial}{\partial\theta}\log f(x,\theta)\right]^{\mathrm{T}}\end{bmatrix}\right]\left[(g(x)-\theta)^{\mathrm{T}} \vdots \frac{\partial}{\partial\theta}\log f(x,\theta)\right]\right\}. \tag{4.1.5}$$

It follows from (4.1.3), (4.1.4) and the definition of M_θ that

$$Q_\theta = \begin{bmatrix} \mathrm{cov}_\theta\, g & I \\ I & M_\theta \end{bmatrix}.$$

Now suppose that M_θ is non-singular. Since Q_θ is a covariance matrix, Q_θ is non-negative and therefore

$$[I \;\vdots\; -M_\theta^{-1}]\begin{bmatrix} \mathrm{cov}\, g & I \\ I & M_\theta \end{bmatrix}\begin{bmatrix} I \\ -M_\theta^{-1} \end{bmatrix} = \mathrm{cov}_\theta\, g - M_\theta^{-1} \tag{4.1.6}$$

is non-negative. It follows that

$$\mathrm{cov}_\theta\, g(x) \geq M_\theta^{-1}. \qquad \square$$

The regularity hypotheses on the function $f(\cdot,\cdot)$ referred to in Theorem 4.1.1, and those which we tacitly assume concerning the 'arbitrary' unbiased estimator $g(\cdot)$, are such as to justify differentiating $f(\cdot,\cdot)$ with respect to the θ variable and, where necessary in the proof, carrying the θ-derivative operator under the integral sign.

M_θ is called *Fisher's information matrix*. We remark that there is an alternative and often more convenient formula than that given in Theorem 4.1.1 for the entries m_{ij}, namely

$$m_{ij} = -E_\theta\left(\frac{\partial^2}{\partial\theta_i\partial\theta_j}\log f(x,\theta)\right).$$

To check the validity of this formula, note that

$$f\frac{\partial^2}{\partial\theta_i\partial\theta_j}\log f = -f\frac{\partial}{\partial\theta_i}\log f\frac{\partial}{\partial\theta_j}\log f + \frac{\partial^2 f}{\partial\theta_i\partial\theta_j}.$$

Integrating over values of x, we obtain

$$E_\theta\left(\frac{\partial^2}{\partial\theta_i\partial\theta_j}\log f\right) = -E_\theta\left(\frac{\partial}{\partial\theta_i}\log f\frac{\partial}{\partial\theta_j}\log f\right)$$

since

$$\int\frac{\partial^2 f}{\partial\theta_i\partial\theta_j}\mathrm{d}x = \frac{\partial^2}{\partial\theta_i\partial\theta_j}\int f\mathrm{d}x = 0.$$

The quality of an unbiased estimator can be assessed by comparing its covariance to the lower bound provided by Theorem 4.1.1. An unbiased estimate $g(\cdot)$ is said to be *efficient* if

$$E_\theta[(g(y)-\theta)(g(y)-\theta)^{\mathrm{T}}] = M_\theta^{-1} \qquad \text{for all } \theta\in D.$$

Notice that if $g^*(\cdot)$ is an efficient estimator and $g(\cdot)$ is an unbiased estimator then for any q-vector $c, E[c^{\mathrm{T}}g(y)] = c^{\mathrm{T}}\theta$ and we have by Theorem 4.1.1,

$$\begin{aligned}\operatorname{var}\{c^{\mathrm{T}}g(y)\} &= c^{\mathrm{T}}E_\theta[(g(y)-\theta)(g(y)-\theta)^{\mathrm{T}}]c \geq c^{\mathrm{T}}M_\theta^{-1}c\\ &= \operatorname{var}\{c^{\mathrm{T}}g^*(y)\}.\end{aligned}$$

It follows from this inequality that if an unbiased estimator $g^*(\cdot)$ is efficient, it provides an estimate $c^{\mathrm{T}}g^*(y)$ of an arbitrary linear combination $c^{\mathrm{T}}\theta$ of the components of θ with variance which is a minimum as compared with that provided by other unbiased estimators.

Consider now the situation when the parameter θ^* is to be estimated on the basis of observations of a sequence of random variables $x_1, x_2, \ldots$ For $n = 1, 2, \ldots$, let x^n denote the composite random variable $x^n = \operatorname{col}[x_1, \ldots, x_n]$ and let $\hat{g}_n(\cdot)$ be an estimate for θ^* given x^n. We say the sequence of estimators $\{g_n(\cdot)\}$ is *consistent* if $\hat{g}_n(\cdot) \to \theta^*$ almost surely.

We now introduce a particularly important kind of estimator. Take $f(\cdot,\theta), \theta\in D$ as above. A function $\hat{\theta}(\cdot)$ is a *maximum likelihood estimator* if for every x

$$f(x, \hat{\theta}(x)) = \max_{\theta\in D} f(x, \theta).$$

The maximum may be attained at more than one point, so maximum likelihood estimators are not necessarily unique. $\hat{\theta}(x)$ has the interpretation that it is the value of θ which maximizes the probability that the random variable x will be in an infinitesimal region about the observed value; very roughly, it chooses the value of θ which makes the observed x 'as likely as possible'. This interpretation is in itself no justification for introducing the maximum likelihood estimator, but there are in fact excellent reasons for doing so, one of which is the following result.

Proposition 4.1.2

Suppose as before that the family $f(\cdot,\theta), \theta\in D$ satisfies appropriate regularity conditions and that M_θ is non-singular for all θ. Then any efficient estimator is a maximum likelihood estimator.

PROOF Let $g(\cdot)$ be an efficient estimator, so that

$$\operatorname{cov}(g(x)) = M_\theta^{-1}.$$

In view of (4.1.5) and (4.1.6) above this means that for any q-vector a,

$$E_\theta\left[a^{\mathrm{T}}[I \,\vdots\, -M_\theta^{-1}]\begin{bmatrix} g(x)-\theta \\ \dfrac{\partial}{\partial\theta}\log f(x,\theta))^{\mathrm{T}}\end{bmatrix}\right.$$

$$\left.\cdot\left[(g(x)-\theta)^{\mathrm{T}} \,\vdots\, \frac{\partial}{\partial\theta}\log f(x,\theta)\right]\begin{bmatrix} I \\ -M_\theta^{-1}\end{bmatrix} a\right] = 0,$$

i.e.

$$E_\theta|b|^2 = 0 \tag{4.1.7}$$

where

$$b = a^{\mathrm{T}}\left[g(x)-\theta-M_\theta^{-1}\left(\frac{\partial}{\partial\theta}\log f(x,\theta)\right)^{\mathrm{T}}\right].$$

Now (4.1.7) implies that $b=0$ a.s. for any θ, and hence, since a is arbitrary, that

$$g(x)-\theta = M_\theta^{-1}\left(\frac{\partial}{\partial\theta}\log f(x,\theta)\right)^{\mathrm{T}} \qquad \text{a.s.,}$$

i.e. that

$$\frac{\partial}{\partial\theta}\log f(x,\theta) = (g(x)-\theta)^{\mathrm{T}} M_\theta \qquad \text{a.s.}$$

Now suppose $\hat{\theta}(x)$ is a maximum likelihood estimator. Then

$\log f(x,\theta)$ is maximized at $\theta = \hat{\theta}(x)$, so that

$$\frac{\partial}{\partial\theta}\log f(x,\hat{\theta}(x)) = (g(x)-\hat{\theta}(x))^{\mathrm{T}}M_\theta = 0 \qquad \text{a.s.}$$

But this implies that $g(x) = \hat{\theta}(x)$, i.e. the efficient estimator $g(\cdot)$ coincides almost surely with the maximum likelihood estimator $\theta(\cdot)$.

□

Proposition 4.1.2 is less far-reaching than it seems at first, since efficient estimators only exist in very special circumstances. The main justification for the maximum likelihood method lies in its *large sample* properties, which we discuss next.

Suppose the situation is the same as before except that we now observe the values of n independent 'samples' of x, i.e. we observe $\{x_1,\ldots,x_n\}$ which are independent, each x_i having density function $f(\cdot,\theta)$ (the same θ for all i). The joint density function is

$$f_n(x_1,\ldots,x_n,\theta) = \prod_{i=1}^{n} f(x_i,\theta).$$

An estimator $\hat{\theta}_n(\cdot)$ of θ is a function of all the available data $\{x_1,\ldots,x_n\}$. Since

$$E_\theta\left[\frac{\partial^2}{\partial\theta_i\partial\theta_j}\log f_n(x_1,\ldots,x_n,\theta)\right] = nE_\theta\left[\frac{\partial^2}{\partial\theta_i\partial\theta_j}\log f(x_1,\theta)\right],$$

the Fisher information matrix for the n-observation case is just

$$M_\theta^n = nM_\theta$$

where M_θ is defined by (4.1.1). Thus for any unbiased estimator $\hat{\theta}_n(x_1,\ldots,x_n)$,

$$\operatorname{cov}(\hat{\theta}_n(x_1,\ldots,x_n)) \geq \frac{1}{n}M_\theta^{-1}. \tag{4.1.8}$$

With increasing n, more accurate estimation of θ is in principle possible, as indicated by the decreasing lower bound. As before, $\hat{\theta}_n$ is efficient if equality holds in (4.1.8). A more useful concept, however, is that of *asymptotic efficiency*. Here we consider a sequence of estimators $\hat{\theta}_1,\hat{\theta}_2,\ldots$ based on increasing numbers of observations. The sequence $\{\hat{\theta}_n\}$ is said to be *asymptotically unbiased* if $E_\theta\hat{\theta}_n \to \theta$ as $n\to\infty$ for any θ. $\{\theta_n\}$ is *asymptotically efficient* if for any θ

$$n\operatorname{cov}_\theta(\hat{\theta}_n) \to M_\theta^{-1}, \qquad n\to\infty.$$

Consider for example the normal distribution with parameters $\theta^T = (\mu, v)$ (the mean and variance respectively) so that

$$f(x, \theta) = \frac{1}{(2\pi v)^{1/2}} \exp\left(-\frac{1}{2v}(x-\mu)^2 \right).$$

The inverse of the Fisher information matrix is

$$(M_\theta^n)^{-1} = \frac{1}{n}\begin{pmatrix} v & 0 \\ 0 & 2v^2 \end{pmatrix}$$

and it is easily shown that the maximum likelihood estimators of μ and v are the sample mean and variance $\bar{x}$ and s^2 respectively, given by

$$\bar{x} = \frac{1}{n}\sum_{i=1}^{n} x_i$$

$$s^2 = \frac{1}{n}\sum_{i=1}^{n} (x_i - \bar{x})^2$$

These statistics have mean and variance as follows (see Kendall and Stuart (1979)):

$$E_\theta(\bar{x}) = \mu \qquad \text{var}_\theta(\bar{x}) = \frac{1}{n}v,$$

$$E_\theta(s^2) = \frac{n-1}{n}v \qquad \text{var}_\theta(s^2) = \frac{n-1}{n}\frac{2v^2}{n}.$$

Thus $\bar{x}$ is unbiased and efficient while s^2 is asymptotically unbiased and efficient. (Note that $\text{var}_\theta(s^2)$ is less than the Cramér–Rao lower bound, but this is not a contradiction since the lower bound only applies to unbiased estimators). It is a remarkable fact that similar properties apply to sequences of estimators based on independent samples with *any* family $f(\cdot, \theta), \theta \in D$ of density functions, subject only to regularity conditions similar to those assumed earlier. A full statement and proof of the following proposition will be found in Kendall and Stuart (1979).

Proposition 4.1.3

Let $x_1, x_2 \ldots$ be independent random variables with density function $f(\cdot, \theta)$ and let $\hat{\theta}_n$ be the maximum likelihood estimator for θ based on $\{x_1, \ldots, x_n\}$. Then subject to regularity hypotheses, the sequence $\{\hat{\theta}_n\}$

is consistent, asymptotically unbiased and efficient. Further, $\hat{\theta}_n$ is asymptotically normally distributed; more precisely the distribution of $\sqrt{n}(\theta_n - \theta)$ converges to the $N(0, M_\theta^{-1})$ distribution as $n \to \infty$.

The last statement means that if F_n denotes the distribution function of the random variable $\sqrt{n}(\hat{\theta}_n - \theta)$ and F the normal $N(0, M_\theta^{-1})$ distribution function then $F_n(a) \to F(a)$ as $n \to \infty$ for each a at which $F(\cdot)$ is continuous[†]. Thus for large n the distribution of $\hat{\theta}_n$ is very close to $N(\theta, (1/n)M_\theta^{-1})$. This is useful information as it provides a precise measure of the accuracy of parameter estimates, at least when the sample size is large.

Our main concern in this chapter is to estimate the parameters of dynamical systems such as the ARMA model introduced in Chapter 2. It is of course an essential property of the ARMA model that the successive outputs $y_1, y_2 \ldots$ are not independent, and analysis of the large sample behaviour of maximum likelihood estimates presents a much more delicate problem than the 'classical' case considered in Proposition 4.1.3. Nonetheless it has been shown in a number of papers listed in the Notes at the end of this chapter that properties of asymptotic efficiency and normality similar to those of the classical case continue to hold. We discuss certain of these results in Section 4.4.

For simplicity we have adopted a framework in this section in which a family of probability density functions $f(\cdot, \theta), \theta \in D$, is specified. We remark that it still makes sense to speak of 'estimators', 'unbiased estimators', 'consistent sequences of estimators', etc., even if $\{f(\cdot, \theta), \theta \in D\}$ is replaced by a family of distribution functions.

4.2 Models

In system identification we observe input and output sequences, $\{u_0, \ldots, u_{N-1}\}$ and $\{y_0, \ldots, y_N\}$, of our system and attempt to 'fit' a model which best represents the data. Invariably the models considered are *parametric*, i.e. selection of a parameter vector $\theta \in \mathbb{R}^q$ fully specifies a model $M(\theta)$. Thus the 'model set' is $\{M(\theta): \theta \in D\}$ where D is a set of allowable parameter values. Often θ will simply list the entries of the matrices involved in the model, but it is possible that these

[†]In the present case $F(\cdot)$ is continuous everywhere, being the distribution function of a non-degenerate normal distribution. We need the extra generality later.

matrices could be given as functions of some (say lower-dimensional) parameter θ. The parameter θ will usually not be 'free'; it will be restricted to maintain, for example, certain stability properties of the models.

We shall be primarily concerned in our treatment of system identification with the (linear, time-invariant) stochastic dynamical models introduced in Section 2.4. For purposes of algorithm description and analysis, however, we consider also static models (see below) and the predictor models of Section 2.6.

4.2.1 Static models

The observed p-vector random variable y is taken to satisfy the equation

$$y = X\theta + e.$$

Here X is a given deterministic $p \times q$ matrix and e is a p-vector random variable with zero mean. Choice of θ specifies the mean value of y since $Ey = X\theta$. This set-up is known as the 'general linear model' in the statistics literature. It covers in particular moving-average type stochastic dynamical models of the form

$$y_k = B(z^{-1})u_k + w_k$$

where $\{w_k\}$ is a white noise sequence, i.e. ARMAX models with $A(z^{-1}) = C(z^{-1}) = I$, when the entries of the matrix coefficients of the polynomial $B(\sigma)$ are treated as the unknown parameters (see Example 4.3.1 below and subsequent remarks).

4.2.2 Stochastic dynamical models

These are the stochastic dynamical models of Section 2.4 (general stochastic dynamical models, ARMAX models and stochastic state-space models), parametrized by the parameter vector θ.

The general stochastic dynamical model equations relating inputs $\{u_k\}$ and outputs $\{y_k\}$ are

$$y_k = P_\theta(z^{-1})u_{k-1} + Q_\theta(z^{-1})e_k. \qquad (4.2.1)$$

Here $P_\theta(\sigma)$, $Q_\theta(\sigma)$ are $r \times m$, $r \times r$, matrices of rational functions in σ expressible as

$$P_\theta(\sigma) = [p_\theta(\sigma)]^{-1}\bar{P}_\theta(\sigma), \qquad Q_\theta(\sigma) = [q_\theta(\sigma)]^{-1}\bar{Q}_\theta(\sigma).$$

In these expressions $p_\theta(\sigma), q_\theta(\sigma)$ are polynomials with coefficients real-valued functions of the parameter θ which satisfy $p_\theta(0) \neq 0$, $q_\theta(0) \neq 0$. $\bar{P}_\theta(\sigma)$, $\bar{Q}_\theta(\sigma)$ are polynomials in σ with coefficients matrix-valued functions of θ. $\{e_k\}$ is an r-vector white-noise sequence. (We refer back to Section 2.1 for interpretation of the output generated by these equations, discussion of initial conditions, etc.)

The ARMAX model equations are

$$A_\theta(z^{-1})y_k = B_\theta(z^{-1})u_{k-1} + C_\theta(z^{-1})e_k. \tag{4.2.2}$$

In these equations, $A_\theta(\sigma)$, $B_\theta(\sigma)$, $C_\theta(\sigma)$ are polynomials in σ with coefficients $r \times r$, $r \times m$, $r \times r$ matrix functions of the parameter θ, and $A_\theta(\sigma)$ satisfies det $A_\theta(0) \neq 0$. $\{e_k\}$ is an r-vector white-noise sequence.

Finally, the stochastic state-space model equations considered are

$$\left.\begin{aligned} x_{k+1} &= A(\theta)x_k + B(\theta)u_k + K(\theta)e_k \\ y_k &= H(\theta)x_k + e_k. \end{aligned}\right\} \tag{4.2.3}$$

(an innovations representation has been adopted). Here $A(\theta)$, $B(\theta)$, $K(\theta)$, $H(\theta)$, are $n \times n$, $n \times m$, $n \times r$, $r \times n$ matrix-valued functions of θ. Again $\{e_k\}$ is an r-vector white-noise sequence.

From the point of view of analysis, general stochastic dynamical models, ARMAX models and stochastic state-space models are interchangeable, except for details involved in the specification of initial conditions (see Section 2.4). Notice, however, that a change from one model set description is accompanied by modification of the definition of the parameter set D in terms of the coefficients in the new description. A model set expressed, say, in terms of stable space equations in which one matrix entry ranges over an interval will give rise to an ARMAX model in which the description of the permissible coefficients in P and Q is rather complicated. Simplicity of the parameter constraint set will affect ease of implementation, and performance, of identification methods. So there may be grounds for choosing one model set rather than another.

4.2.3 Predictor models

We consider the predictor models of Section 2.6, but we now suppose that the predictor function at time k, f_k, depends on a parameter θ. Thus we take the r-vector output y_k to be related to past outputs y_{k-1}, $y_{k-1}, \ldots$, and past m-vector inputs $u_{k-1}, u_{k-2}, \ldots$, by the equations

$$y_k = f_k(\theta; y^{k-1}, u^{k-1}) + e_k \qquad k = 0, 1, \ldots$$

Here y^{k-1} and u^{k-1} denote (as before) col$[y_{k-1}, y_{k-2}, \ldots, y_0]$ and

$\text{col}[u_{k-1}, u_{k-2}, \ldots, u_0]$ respectively. $f_k: \mathbb{R}^q \times \mathbb{R}^{kr} \times \mathbb{R}^{km} \to \mathbb{R}^r$, $k = 0, 1, \ldots$, are given deterministic functions of the parameter and past outputs and inputs, and $\{e_k\}$ is a sequence of independent, zero-mean random variables.

We recall that $f_k(\theta; y^{k-1}, u^{k-1})$ is the conditional expectation of y_k given y^{k-1}, u^{k-1}, and hence the best 'one-step-ahead predictor' in the mean square sense under the assumption, of course, that $M(\theta)$ is the true model. A predictor model then is basically a rule (evaluation of the function f_k) for predicting the value of y_k given y^{k-1}, u^{k-1}. As we have seen, predictor models essentially subsume the stochastic dynamical models of the previous subsection provided the driving noise vectors are independent. When we reformulate a stochastic dynamical model as a predictor model we replace it, in effect, by an algorithm for calculating predictions. A typical identification procedure involves selection of a parameter value θ to minimize in some sense the prediction errors, namely the discrepancy between the observed output and the prediction of the output supplied by the algorithm corresponding to parameter value θ. Identification procedures formulated in terms of predictor models, on which we concentrate in our study of identification, can be viewed then as identification procedures for stochastic dynamical models reduced to a family of algorithms, parametrized by θ, each of which supplies a predictor.

We have already determined (Section 2.6) the predictors associated with the stochastic dynamical models considered here. For the general stochastic dynamical model (4.2.1), the predictor functions f_k take the form

$$f_k(\theta; y^{k-1}, u^{k-1}) = [I - Q_\theta^{-1}(z^{-1})]y_k + Q_\theta^{-1}(z^{-1})P_\theta(z^{-1})u_{k-1}, k \geq 0$$

(where we assume zero initial data, $u_k = 0$, $y_k = 0$, $k < 0$, and take $Q_\theta(\sigma)$ such that $Q_\theta(0) = I$). For the ARMAX models (4.2.2)

$$f_k(\theta; y^{k-1}, u^{k-1}) = \hat{y}_k \qquad k = 0, 1, \ldots$$

where $\hat{y}_k$ is calculated form the recursive equations

$$C_\theta(z^{-1})\hat{y}_i = [C_\theta(z^{-1}) - A_\theta(z^{-1})]y_i + B_\theta(z^{-1})u_{i-1} \qquad i = 0, 1, \ldots$$

(We assume zero initial data, $y_k = 0, \hat{y}_k = 0$, $e_k = 0$, $k < 0$, and take $A_\theta(\sigma)$, $C_\theta(\sigma)$ to be such that $A_\theta(0) = C_\theta(0) = I$.) For the stochastic state-space models (4.2.3),

$$f_k(\theta; y^{k-1}, u^{k-1}) = \hat{y}_k \qquad k = 0, 1, \ldots$$

where now

$$\hat{y}_k = H(\theta)\hat{x}_k$$

and $\hat{x}_k$ is calculated from the recursive equations

$$\hat{x}_{i+1} = A(\theta)\hat{x}_i + B(\theta)u_i + K(\theta)(y_i - H(\theta)\hat{x}_i), \qquad i \geq 0$$

(again we assume zero initial data, $\hat{x}_0 = 0$)

4.3 Parameter estimation for static systems

In this section we describe and analyse techniques for identifying a static system. The analysis suffers from the limitation that it is based on the hypothesis that the model set considered contains a model which perfectly describes the system. We cannot expect in practice that this hypothesis is valid. The analysis is none the less significant since it is reasonable to suppose that the analysis will give some indication of the quality of an estimator when the model set, if it does not actually contain a true description, comes close to doing so.

Let X be a given $p \times q$ matrix and let e be a zero-mean p-vector random variable. Suppose that the p-vector random variable y satisfies the equation

$$y = X\theta^* + e \tag{4.3.1}$$

for some (unknown) q-vector θ^*. Further statistical information about e may, or may not, be available. In this section we study the problem of estimating the parameter θ^* (and also, possibly, statistical properties of e), given an observation of y. The problem then is to choose a model from the model set described by the equations

$$y = X\theta + e$$

as the parameter θ ranges over $\mathbb{R}^q$, when it is known that some parameter value ($\theta = \theta^*$) provides a true description of the system.

Of course, parameter estimation for dynamical systems is of primary interest in this chapter and, before proceeding, we give an example illustrating the extent to which consideration of static models is relevant to dynamical systems.

Example 4.3.1

Consider scalar ARMAX models of the form

$$\begin{aligned} y_k + a_1 y_{k-1} + \cdots + a_n y_{k-n} \\ = b_1 u_{k-1} + \cdots b_m u_{k-m} + e_k \qquad k = 1, \ldots, N. \end{aligned} \tag{4.3.2}$$

Here we treat $y_0,\ldots,y_{-n+1}$ and $u_0,\ldots,u_{-m+1}$ as initial conditions. The e_i are zero-mean random variables. The model parameter is the vector of coefficients $a_1,\ldots,a_n,b_1,\ldots,b_m$.

The equations (4.3.2) can be expressed as a single vector equation

$$y = X\theta + e$$

in which $p = N$, $q = n + m$, $y = \mathrm{col}(y_1,\ldots,y_N)$, $e = \mathrm{col}(e_1,\ldots,e_N)$, $\theta = \mathrm{col}(a_1,\ldots,a_n,b_1,\ldots b_m)$ and

$$X = \left[\begin{array}{ccc|ccc} -y_0 & \cdots & -y_{-n+1} & u_0 & \cdots & u_{-m+1} \\ -y_1 & \cdots & -y_{-n} & u_1 & \cdots & u_{-m} \\ \vdots & & \vdots & \vdots & & \vdots \\ -y_{N-1} & \cdots & -y_{N-n} & u_{N-1} & \cdots & u_{N-m} \end{array}\right]. \qquad (4.3.3)$$

We have limited ourselves here to treatment of the scalar case. A reduction of a vector ARMAX model to a static model of the form (4.3.1) can be performed along similar lines.

It is clear from this example that estimators, procedures for selection of model order, etc., devised for static models translate into corresponding estimates, etc., for dynamical models of the sort described in Section 2.4.

Notice that the matrix X, given by (4.3.3), depends on the random variable $y_1,\ldots,y_N$ and is therefore, in general, random. X is deterministic, however, in those situations when $a_1,\ldots,a_n$ can be taken zero, i.e. when the dynamical system has a moving-average description. Much of the analysis of this section is based on the assumption that X is a known deterministic matrix (or at least that X is the realization of a matrix random variable which is independent of e). It should be borne in mind, then, the analysis is directly relevant only to rather special dynamical systems.

4.3.1 Least squares estimation of static systems

A natural approach to the problem of estimating θ^* in the model (4.3.1) is to choose an estimate which minimizes some measure of the discrepancy, or error, between the observation of y and the value of y which the model predicts in the absence of disturbances. A particularly appealing estimate is one which minimizes the sum of squares of the components of the error, because we can expect it to be analytically and computationally tractable. Such an estimate minimizes

$$\theta \to \frac{1}{2}\sum_{i=1}^{p}\left|y_i - \sum_{j=1}^{q} x_{ij}\theta_j\right|^2$$

(the x_{ij} are the components of the matrix X). In vector notation this function becomes

$$\theta \to \tfrac{1}{2}(y - X\theta)^{\mathrm{T}}(y - X\theta),$$

A slight refinement of such an estimate is one which minimizes the function f:

$$f(\theta) = \tfrac{1}{2}(y - X\theta)^{\mathrm{T}}Q(y - X\theta), \tag{4.3.4}$$

in which Q is some symmetric, non-negative matrix. Choice of Q will depend on our judgement about the relative importance of different components of the error, or other considerations.

A least squares estimate $\hat{\theta}$ of θ^* (corresponding to Q) is one which achieves the minimum of f defined by (4.3.4).

Notice that, once the matrix Q is fixed, the problem of determining a least squares estimate is a purely deterministic one and does not involve statistical information about y.

We observe that the gradient of the function f at θ is

$$f'(\theta) = [X^{\mathrm{T}}QX\theta - X^{\mathrm{T}}Qy]^{\mathrm{T}}.$$

It is now shown that the condition $f'(\hat{\theta}) = 0$ fully characterizes the least squares estimate $\hat{\theta}$.

Proposition 4.3.2

Let Q be an arbitrary symmetric, non-negative $p \times p$ matrix. A least squares estimate of θ^* (corresponding to Q) exists. $\hat{\theta}$ is a least squares estimate if and only if

$$X^{\mathrm{T}}QX\hat{\theta} = X^{\mathrm{T}}Qy. \tag{4.3.5}$$

PROOF For any $\theta, \hat{\theta} \in \mathbb{R}^q$ we can write

$$\begin{aligned} f(\theta) - f(\hat{\theta}) &= -y^{\mathrm{T}}QX(\theta - \hat{\theta}) + \tfrac{1}{2}\theta^{\mathrm{T}}X^{\mathrm{T}}QX\theta - \tfrac{1}{2}\hat{\theta}^{\mathrm{T}}X^{\mathrm{T}}QX\hat{\theta} \\ &= (\theta - \hat{\theta})^{\mathrm{T}}[-X^{\mathrm{T}}Qy + X^{\mathrm{T}}QX\hat{\theta}] \\ &\quad + \tfrac{1}{2}(\theta - \hat{\theta})^{\mathrm{T}}X^{\mathrm{T}}QX(\theta - \hat{\theta}), \end{aligned} \tag{4.3.6}$$

after some rearrangement. (The right-hand side of (4.3.6) will be recognized as the expansion of f as a finite Taylor series about $\hat{\theta}$, namely

$$f(\theta)-f(\hat{\theta})=f'(\hat{\theta})(\theta-\hat{\theta})+\tfrac{1}{2}(\theta-\hat{\theta})^{\mathrm{T}}f''(\hat{\theta})(\theta-\hat{\theta})).$$

Suppose that $\hat{\theta}$ satisfies (4.3.5). Since $X^{\mathrm{T}}QX\geq 0$, it follows from (4.3.6) that $f(\theta)\geq f(\hat{\theta})$, for all $\theta\in\mathbb{R}^q$. In other words, (4.3.5) is a sufficient condition for $\hat{\theta}$ to be a least squares estimator.

Suppose that $\hat{\theta}$ does not satisfy (4.3.5); then $[-X^{\mathrm{T}}Qy+X^{\mathrm{T}}QX\hat{\theta}]=\xi$ for some non-zero vector ξ. Choose $\theta=\hat{\theta}-\alpha\xi$, for $\alpha>0$. From (4.3.6),

$$f(\theta)-f(\hat{\theta})=-\alpha\|\xi\|^2+\tfrac{1}{2}\alpha^2\xi^{\mathrm{T}}X^{\mathrm{T}}QX\xi.$$

It is clear that for α sufficiently small, $f(\theta)-f(\hat{\theta})<0$. So $\hat{\theta}$ does not minimize f. We have shown that (4.3.5) is also a necessary condition.

It remains to show that there exists some $\hat{\theta}$ satisfying (4.3.5). For this purpose we introduce the symmetric, non-negative square root of the matrix Q (see Appendix D). Suppose in contradiction that (4.3.5) does not have a solution. This means that the vector $X^{\mathrm{T}}Qy$ does not lie in the subspace $\{X^{\mathrm{T}}QX\theta\colon\theta\in\mathbb{R}^q\}$. Then there exist a q-vector ξ such that

$$\xi^{\mathrm{T}}X^{\mathrm{T}}Qy(=\xi^{\mathrm{T}}X^{\mathrm{T}}(Q^{1/2})^2y)\neq 0. \tag{4.3.7}$$

But

$$\xi^{\mathrm{T}}X^{\mathrm{T}}QX\theta=0,\qquad \text{all } \theta\in\mathbb{R}^q. \tag{4.3.8}$$

We conclude from (4.3.7) that $Q^{1/2}X\xi\neq 0$. It follows that

$$\xi^{\mathrm{T}}X^{\mathrm{T}}QX\xi=\|Q^{1/2}X\xi\|^2\neq 0.$$

This contradicts (4.3.8). Equation (4.3.5) therefore has a solution.

□

The equations (4.3.5), are called the *normal equations* for the least squares estimate. They have a unique solution

$$\hat{\theta}=(X^{\mathrm{T}}QX)^{-1}X^{\mathrm{T}}Qy$$

if and only if $X^{\mathrm{T}}QX$ is non-singular. A sufficient condition for non-singularity of $X^{\mathrm{T}}QX$ is that Q is positive definite and X has full column rank. In this case, for arbitrary, non-zero $\xi\in\mathbb{R}^q$, $X\xi\neq 0$. But then $\xi^{\mathrm{T}}X^{\mathrm{T}}QX\xi=(X\xi)^{\mathrm{T}}Q(X\xi)>0$. It follows that $X^{\mathrm{T}}QX$ is a positive definite, and therefore a non-singular matrix.

Suppose that Q is positive definite. A least squares estimate $\hat{\theta}$ then has a geometric interpretation in terms of orthogonal projections. The function $(u, v) \to u^{\mathrm{T}}Qv$ defines an inner product on $\mathbb{R}^q$ which we write $\langle \cdot, \cdot \rangle_Q$. The normal equations (4.3.5) may be written in terms of the inner product

$$\langle x_i, y - X\hat{\theta} \rangle_Q = 0, \qquad i = 1, \ldots, q. \tag{4.3.9}$$

In these equations $x_1, \ldots, x_q$ are the columns of X. The x_i span the range of X and $X\hat{\theta}$ lies in the range of X. Equations (4.3.9) mean therefore that $X\hat{\theta}$, the value of y predicted by $\hat{\theta}$ in the absence of disturbances, is the orthogonal projection of the observation y onto the range of X, with respect to $\langle \cdot, \cdot \rangle_Q$.

4.3.2 Statistical properties of least squares estimates

We now examine statistical properties of least squares estimates $\hat{\theta}$ of the parameter θ^* in the system equations under the assumption that e is a zero-mean, second-order random variable.

The following results establish that least squares estimators are very good estimators (at least when the weighting matrix Q is suitably chosen).

Proposition 4.3.3

Suppose that the matrix $X^{\mathrm{T}}QX$ is non-singular. Then

$$E_\theta\{\hat{\theta}(y)\} = \theta, \qquad \text{for all } \theta, \tag{4.3.10}$$

where E_θ denotes expectation under the hypothesis that θ is the 'true' parameter value.

PROOF Under the assumptions, $\hat{\theta}$ is unique and is given by

$$\hat{\theta} = (X^{\mathrm{T}}QX)^{-1}X^{\mathrm{T}}Q(X\theta + e).$$

But e has zero mean so, $E\{\hat{\theta}\} = (X^{\mathrm{T}}QX)^{-1}X^{\mathrm{T}}QX\theta = \theta$. □

The proposition asserts that under the conditions which make it uniquely defined, the least squares estimate is unbiased.

The covariance of the least squares estimate is easily calculated:

Proposition 4.3.4

Suppose that the matrix Q is non-singular. Then

$$\operatorname{cov}\{\hat{\theta}(y)\} = (X^{\mathrm{T}}QX)^{-1}X^{\mathrm{T}}QRQX(X^{\mathrm{T}}QX)^{-1} \tag{4.3.11}$$

in which R is the covariance matrix of e.

PROOF $\hat{\theta}(y)-\theta^*=(X^TQX)^{-1}X^TQ(X\theta^*+e)-\theta^*$
$$=(X^TQX)^{-1}X^TQe.$$
Since $\hat{\theta}$ is an unbiased estimator, it follows that
$$\text{cov}\{\hat{\theta}(y)\}=E[X^TQX]^{-1}X^TQee^TQX(X^TQX)^{-1}$$
$$=(X^TQX)^{-1}X^TQRQX(X^TQX)^{-1}.$$
□

Given an estimate $\hat{\theta}$ of the parameter θ^* it is natural to estimate the arbitrary linear combination $c^T\theta^*$ of the components of θ^*, defined by the vector c, by the same linear combination of the components of $\hat{\theta}$, namely $c^T\hat{\theta}$. The procedure yields the estimate $\hat{\theta}_i$ of θ_i, in particular. The next two results assert that estimates induced in this way by the least squares estimates for θ^* have minimum variance, provided the weighting matrix is suitably chosen; the variance is minimal compared with that of arbitrary linear unbiased estimates or, in the case that the disturbance vector e is normally distributed, compared with arbitrary unbiased estimates.

Theorem 4.3.5 (Gauss–Markov)

Suppose that
$$\text{cov}\{e\}=\sigma^2\Sigma$$

for some positive number σ^2 and some non-singular $p\times p$ matrix Σ. Suppose also that X has linearly independent columns. Let $\hat{\theta}$ be the least squares estimate for θ^* in the system (4.3.1), corresponding to a choice of weighting matrix
$$Q=\Sigma^{-1}.$$

Then for any q-vector $c, c^T\hat{\theta}$ has minimum variance in the class of linear unbiased estimators for $c^T\theta^*$.

PROOF Notice first of all that, under the hypotheses on X and Σ, $X^T\Sigma^{-1}X$ is a non-singular matrix, so we can speak unambiguously of the least squares estimate.

The estimator $c^T\hat{\theta}(\cdot)$ is obviously linear. It is unbiased since $E_\theta c^T\hat{\theta}(y)=c^TE_\theta\hat{\theta}(y)=c^T\theta$ for all θ, by (4.3.10).

We must show then, given $\psi(\cdot)$ any other linear, unbiased estimator for $c^T\theta$,
$$\text{var}\{\psi(y)\}-\text{var}\{c^T\hat{\theta}(y)\}\geq 0.$$

Since the estimator ψ is linear, there exists a p-vector ξ such that

$$\psi(y) = \xi^{\mathrm{T}} y, \qquad \text{for all } y \tag{4.3.12}$$

and since it is unbiased

$$E_\theta \psi(y) = c^{\mathrm{T}}\theta, \qquad \text{for all } \theta. \tag{4.3.13}$$

From (4.3.12) and (4.3.13)

$$E_\theta\{\xi^{\mathrm{T}}(X\theta + e)\} = c^{\mathrm{T}}\theta, \qquad \text{for all } \theta.$$

Since e has zero mean,

$$\xi^{\mathrm{T}} X\theta = c^{\mathrm{T}}\theta, \qquad \text{for all } \theta.$$

This is only possible if

$$\xi^{\mathrm{T}} X = c^{\mathrm{T}}. \tag{4.3.14}$$

Now,

$$\operatorname{var}\{\psi\} = E[\xi^{\mathrm{T}} y - c^{\mathrm{T}}\theta^*]^2$$

since $\psi(\cdot)$ is an unbiased estimator for $c^{\mathrm{T}}\theta^*$,

$$= E[\xi^{\mathrm{T}} X\theta^* + \xi^{\mathrm{T}} e - \xi^{\mathrm{T}} X\theta^*]^2$$

by (4.3.12)

$$= E\xi^{\mathrm{T}} e e^{\mathrm{T}} \xi = \sigma^2 \xi^{\mathrm{T}} \Sigma \xi.$$

It follows from Proposition 4.3.4 that

$$\operatorname{var}\{\psi(y)\} - \operatorname{var}\{c^{\mathrm{T}}\hat{\theta}(y)\} = \sigma^2[\xi^{\mathrm{T}}\Sigma\xi - \hat{\xi}^{\mathrm{T}}\Sigma\hat{\xi}] \tag{4.3.15}$$

in which

$$\hat{\xi}^{\mathrm{T}} = c^{\mathrm{T}}(X^{\mathrm{T}}\Sigma^{-1}X)^{-1}X^{\mathrm{T}}\Sigma^{-1}. \tag{4.3.16}$$

Substitution of (4.3.14) and (4.3.16) into (4.3.15) gives

$$\operatorname{var}\{\psi(y)\} - \operatorname{var}\{c^{\mathrm{T}}\hat{\theta}(y)\} = \sigma^2\xi^{\mathrm{T}}[\Sigma - D]\xi$$

in which

$$D = X(X^{\mathrm{T}}\Sigma^{-1}X)^{-1}X^{\mathrm{T}}.$$

We can check, however, by direct expansion that

$$[\Sigma - D] = [\Sigma - D]^{\mathrm{T}}\Sigma^{-1}[\Sigma - D].$$

We have shown that

$$\operatorname{var}\{\psi(y)\} - \operatorname{var}\{c^{\mathrm{T}}\hat{\theta}(y)\} = \sigma^2\xi^{\mathrm{T}}[\Sigma - D]^{\mathrm{T}}\Sigma^{-1}[\Sigma - D]\xi.$$

This last expression is non-negative, since Σ^{-1} is a non-negative, symmetric matrix. We have shown that $c^{\mathrm{T}}\hat{\theta}(y)$ has minimum variance.

□

If the disturbance vector is normally distributed, then the least squares estimator has 'minimum variance' over the class† of unbiased estimators, linear or not, for a suitable choice of weighting matrix. This is proved by showing that the least squares estimator is efficient.

Theorem 4.3.6

Suppose that $e \sim N(0, \sigma^2\Sigma)$ for some number σ^2 and some non-singular $p \times p$ matrix Σ. It is assumed that X has linearly independent columns. Let $\hat{\theta}^*$ be the least squares estimate for θ^* in the system (4.3.1) corresponding to a choice of weighting matrix $Q = \Sigma^{-1}$, and let $\psi(\cdot)$ be an arbitrary unbiased estimator (not necessarily linear).

Then for any q-vector c,

$$\mathrm{var}\{c^{\mathrm{T}}\hat{\theta}(y)\} \leq \mathrm{var}\{c^{\mathrm{T}}\psi(y)\}.$$

PROOF Under the hypothesis that θ is the true parameter value, we have that $y = X\theta + e, e \sim N(0, \sigma^2\Sigma)$, and consequently the probability density $p(y|\theta)$ of y is:

$$p(y|\theta) = \frac{1}{(2\pi\sigma^2)^{p/2}(\det \Sigma)^{1/2}} \exp\left\{-\frac{1}{2\sigma^2}(y - X\theta)^{\mathrm{T}}\Sigma^{-1}(y - X\theta)\right\}.$$

So

$$\log p(y|\theta) = -\log\{(2\pi\sigma^2)^{p/2}(\det \Sigma)^{1/2}\} - (2\sigma^2)^{-1}(y - X\theta)^{\mathrm{T}}\Sigma^{-1}(y - X\theta)$$

Fisher's information matrix $M_\theta = -\partial^2/\partial\theta^2 \log p(y|\theta)$ can now be calculated:

$$M_\theta = \sigma^{-2}(X^{\mathrm{T}}\Sigma^{-1}X)$$

(see Section 4.1). By theorem 4.1.1,

$$\mathrm{cov}\{\psi\} \geq \sigma^2(X^{\mathrm{T}}\Sigma^{-1}X)^{-1}. \tag{4.3.17}$$

However, since the weighting matrix is Σ^{-1}, we see from formula

† We are somewhat vague here about the class of comparison estimators; it comprises those estimators for which the Cramér–Rao lower bound is valid.

(4.3.11) that

$$\operatorname{cov}\{\hat{\theta}(y)\} = \sigma^2(X^{\mathrm{T}}\Sigma^{-1}X)^{-1}. \tag{4.3.18}$$

We deduce from (4.3.17) and (4.7.18) that

$$\operatorname{var}(c^{\mathrm{T}}\hat{\theta}(y)) = c^{\mathrm{T}} \operatorname{cov}\{\hat{\theta}(y)\}c \leq c^{\mathrm{T}} \operatorname{cov}\{\psi(y)\}c = \operatorname{var}\{c^{\mathrm{T}}\psi(y)\},$$

the desired inequality. □

4.3.3 Estimation of the variance of the noise components

Suppose that the observed random variable y satisfies the static system equation

$$y = X\theta^* + e$$

in which e is a p-vector with components uncorrelated, zero-mean random variables with common variance σ^2. As usual y is a p-vector random variable, and θ^* is a q-vector. It is assumed that X has full column rank and $p > q$.

Suppose σ^2 is unknown. A plausible estimate for σ^2 is

$$\frac{1}{p}(y - X\hat{\theta})^{\mathrm{T}}(y - X\hat{\theta})$$

in which $\hat{\theta}$ is the least squares estimate of θ^* given y:

$$\hat{\theta} = (X^{\mathrm{T}}X)^{-1}X^{\mathrm{T}}y. \tag{4.3.19}$$

Indeed the expression (4.3.19) is the sample variance of the e_i under the assumption that $\hat{\theta}$ coincides with the true parameter value. The fact that $\hat{\theta}$ is used in the expression introduces a bias into the estimate. This can however be corrected by a simple scaling:

Proposition 4.3.7

$\hat{\sigma}^2$ defined by

$$\hat{\sigma}^2 = (p - q)^{-1}(y - X\hat{\theta})^{\mathrm{T}}(y - X\hat{\theta}), \tag{4.3.20}$$

in which $\hat{\theta}$ is given by (4.3.19), is an unbiased estimate of σ^2.

PROOF By (4.3.19)

$$\begin{aligned}(y - X\hat{\theta})^{\mathrm{T}}(y - X\hat{\theta}) &= (y - X(X^{\mathrm{T}}X)^{-1}X^{\mathrm{T}}y)^{\mathrm{T}}(y - X(X^{\mathrm{T}}X)^{-1}X^{\mathrm{T}}y)\\ &= y^{\mathrm{T}}(I_p - X(X^{\mathrm{T}}X)^{-1}X^{\mathrm{T}})(I_p - X(X^{\mathrm{T}}X)^{-1}X^{\mathrm{T}})y\end{aligned}$$

(I_p denotes the $p \times p$ identity matrix)

$$= y^{\mathrm{T}}(I_p - X(X^{\mathrm{T}}X)^{-1}X^{\mathrm{T}})y$$
$$= \operatorname{trace}\{(I_p - X(X^{\mathrm{T}}X)^{-1}X^{\mathrm{T}})yy^{\mathrm{T}}\}$$

by properties of the trace operator

$$= \operatorname{trace}\{(I_p - X(X^{\mathrm{T}}X)^{-1}X^{\mathrm{T}})(X\theta^* + e)(X\theta^* + e)^{\mathrm{T}}\}.$$

Since the trace operator is linear it commutes with the expectation operator and so

$$E[y - X\hat{\theta}]^{\mathrm{T}}[y - X\hat{\theta}]$$
$$= \operatorname{trace}\{(I_p - X(X^{\mathrm{T}}X)^{-1}X^{\mathrm{T}})(X\theta^*\theta^{*\mathrm{T}}X^{\mathrm{T}} + \sigma^2 I_p)\}.$$

However, $(I_p - X(X^{\mathrm{T}}X)^{-1}X^{\mathrm{T}})(X\theta^*\theta^{*\mathrm{T}}X^{\mathrm{T}}) = 0$, so

$$\begin{aligned} E[y - X\hat{\theta}]^{\mathrm{T}}[y - X\hat{\theta}] &= \sigma^2 \operatorname{trace}\{I_p - X(X^{\mathrm{T}}X)^{-1}X^{\mathrm{T}}\} \\ &= \sigma^2[p - \operatorname{trace}\{X(X^{\mathrm{T}}X)^{-1}X^{\mathrm{T}}\}] \\ &= \sigma^2[p - \operatorname{trace}\{(X^{\mathrm{T}}X)^{-1}X^{\mathrm{T}}X\}] \\ &= \sigma^2[p - \operatorname{trace} I_q] \\ &= \sigma^2[p - q]. \end{aligned}$$

It is clear from this equation that $\hat{\sigma}^2$ given by (4.3.20) is an unbiased estimate of σ^2. □

4.3.4 Maximum likelihood estimation for static systems

Suppose again that the observed variable y satisfies the static system equation

$$y = X\theta^* + e.$$

We derive equations satisfied by maximum likelihood estimates of θ^* and of the unknown satistics of the disturbance e, when $e \sim N(0, \Sigma)$ and Σ is non-singular.

If θ were the true parameter value we would have $y \sim N(X\theta, \Sigma)$. The likelihood function $p(y|\theta, \Sigma)$ is therefore

$$p(y|\theta, \Sigma) = [(2\pi)^p \det \Sigma]^{-1/2} \exp\{-\tfrac{1}{2}(y - X\theta)^{\mathrm{T}}\Sigma^{-1}(y - X\theta)^{\mathrm{T}}\}.$$

The log likelihood function is

$$\log p(y|\theta, \Sigma) = -\frac{p}{2}\log 2\pi - \tfrac{1}{2}\log\det\Sigma - \tfrac{1}{2}(y - X\theta)^{\mathrm{T}}\Sigma^{-1}(y - X\theta). \tag{4.3.21}$$

The maximum likelihood estimates are obtained by maximizing the log likelihood function over the unknown parameters following substitution of the observation in place of the y variable.

Case 1 (Σ known)

In this case the maximum likelihood estimate $\hat{\theta}$ coincides with the least squares estimate, corresponding to weighting Σ^{-1}:

$$\hat{\theta} = (X^{\mathrm{T}}\Sigma^{-1}X)^{-1}X^{\mathrm{T}}\Sigma^{-1}y.$$

To see this, we need merely note that $\hat{\theta}$, which maximizes the expression (4.3.21) for given Σ and observation y, minimizes the least squares criterion $\theta \to \frac{1}{2}(y - X\theta)^{\mathrm{T}}\Sigma^{-1}(y - X\theta)$.

Case 2 (Σ partially known)

When only partial information about Σ is available we can expect that estimation of the unknown parameters and statistics (θ^* and Σ) will no longer reduce to a least squares problem and numerical methods will be required. In illustrating the kind of analysis which is possible, we consider here only the case when $X = \mathrm{col}\{X_1, \ldots, X_N\}$ and $\Sigma = \mathrm{diag}\{\Lambda^*, \ldots, \Lambda^*\}$, in which Λ^* is an unknown non-singular matrix. When $X_1 = X_2 = \cdots = X_N$, this case corresponds to estimating θ^* and Λ^* in the model

$$\bar{y} = X_1\theta^* + e$$

when it is known $e \sim N(0, \Lambda^*)$, given N independent observations $\bar{y}_1, \bar{y}_2, \ldots, \bar{y}_N$ of the vector $\bar{y}$. Here it is possible at least to derive coupled equations satisfied by estimates $\hat{\theta}$, $\hat{\Lambda}$, if they exist and $\hat{\Lambda}$ is invertible, which have a rather natural form. These express $\hat{\theta}$ as the maximum likelihood estimate of the unknown parameter when the covariance matrix is taken to be $\hat{\Lambda}$ and they express $\hat{\Lambda}$ as the sample covariance of e, based on the assumption that the unknown parameter is $\hat{\theta}$:

$$\hat{\theta} = \left[\sum_{k=1}^{N} X_k^T \hat{\Lambda}^{-1} X_k\right]^{-1} \sum_{k=1}^{N} X_k \hat{\Lambda}^{-1} y_k$$
$$\hat{\Lambda} = \frac{1}{N}\sum_{k=1}^{N} (y_k - X_k\hat{\theta})(y_k - X_k\hat{\theta})^{\mathrm{T}}. \tag{4.3.22}$$

(We have partitioned $y = \mathrm{col}\{y_1, \ldots, y_N\}$ compatibly with

$\text{col}\{X_1,\ldots,X_N\}$.) Indeed, by equation (4.3.21), $(\hat{\theta},\hat{\Lambda})$ minimizes

$$J(\theta,\Lambda)=\tfrac{1}{2}N\log\det\Lambda+\tfrac{1}{2}\sum_{k=1}^{N}(y_k-X_k\theta)^{\mathrm{T}}\Lambda^{-1}(y_k-X_k\theta).$$

The fact that $\hat{\theta}$ minimizes the least squares criterion $\theta\to J(\theta,\hat{\Lambda})$ leads to the first equation in (4.3.22). The estimates also satisfy

$$\frac{\partial}{\partial\Sigma}J(\hat{\theta},\hat{\Sigma})=0. \tag{4.3.23}$$

We can evaluate this partial Jacobian with the help of the following identities from matrix calculus (see Appendix D.4):
$(\mathrm{d}/\mathrm{d}Q)\log\det Q=Q^{-1}$, on the space of $n\times n$ non-singular matrices Q and, for any vector a,
$(\mathrm{d}/\mathrm{d}Q)a^{\mathrm{T}}Q^{-1}a=-Q^{-1}aa^{\mathrm{T}}Q^{-1}$, on the space of $n\times n$ non-singular matrices Q. From (4.3.23) we deduce that

$$\tfrac{1}{2}N\Lambda^{-1}-\tfrac{1}{2}\Lambda^{-1}\left[\sum_{k=1}^{N}(y_k-X_k\hat{\theta})(y_k-X_k\hat{\theta})^{\mathrm{T}}\right]\Lambda^{-1}=0$$

which implies the second equation in (4.3.22).

Case 3 ($\Sigma=\sigma^2I$, σ^2 unknown)

This is an instance of Case 2 in which the estimates can be determined analytically. We deduce from (4.3.22) that

$$\hat{\theta}=(X^{\mathrm{T}}X)^{-1}X^{\mathrm{T}}y\quad\text{and}\quad\hat{\sigma}^2=\frac{1}{p}(y-X\hat{\theta})^{\mathrm{T}}(y-X\hat{\theta}).$$

In view of Proposition 4.3.7, the maximum likelihood estimate of σ^2 in case 3 is biased, though the percentage bias will be small for p much larger than q.

4.3.5 Model order selection

Suppose that a dynamical system is described by scalar ARMAX model equations of the form

$$y_k=b_1u_{k-1}+\cdots+b_qu_{k-q}+e_k\qquad k=1,\ldots,p \tag{4.3.24}$$

together with initial data $u_0,\ldots,u_{-q+1}$. Here the e_k are zero-mean, independent gaussian random variables with common variance σ^2.

As we have observed (Example 4.3.1), the problem of estimating the parameters $b_1,\ldots,b_q$ can be reformulated as that of estimating the vector parameter θ^* in the static model

$$y = x\theta^* + e$$

given observations of y, when we take $y = \mathrm{col}(y_1,\ldots,y_p)$, $\theta^* = \mathrm{col}(b_1,\ldots,b_q)$, $e = \mathrm{col}(e_1,\ldots,e_p)$ and

$$X = \begin{bmatrix} u_0 & \ldots u_{1-q} \\ u_1 & \ldots u_{2-q} \\ \vdots & \vdots \\ u_{p-1} & \ldots u_{p-q} \end{bmatrix}.$$

We refer to the integer q as the *order* of the model. So far we have concerned ourselves with estimating the unknown parameters when the model order is pre-set. What model order should we adopt when this is not fixed beforehand? We now study this question.

One obvious procedure is to fix the model order at some large number. This might seem reasonable since, if q is increased in equation (4.3.24), the equation still describes the response of the dynamical system (when the added parameters are set to zero.)

There are disadvantages in this procedure, however. The models that result will be unnecessarily complicated. Also, we can expect that an increase in model order will lead to an increase in the variances of the significant components of the least squares parameter estimates and consequently to a reduction in the reliance we can place upon them. These considerations make desirable more sophisticated procedures in which the model order is estimated from the observations.

Henceforth we study the model order selection problem only in relation to the static model (4.3.1). It is assumed that $E\{e\} = 0$, $\mathrm{cov}\{e\} = \sigma^2 I$, X is a known $p \times q$ matrix with linearly independent columns, $p > q$ and θ^* is the vector of parameters to be estimated from observations of y.

Let an integer d, $0 \leq d \leq q$, be given. Our problem is that of deciding, on the basis of observations, when the hypothesis

$$\theta_q^* = \theta_{q-1}^* = \cdots = \theta_{q-d+1}^* = 0 \qquad (4.3.25)$$

should be rejected.

We shall describe some statistical tests of hypothesis (4.3.25). These involve the χ^2 and F distributions, defined as follows.

Let k_1, k_2 be positive numbers. A random variable is said to have a $\chi^2(k_1)$ distribution if it can be expressed as a sum of squares of k_1 independent random variables each with distribution $N(0, 1)$.

A random variable v is said to have a $F(k_1, k_2)$ distribution if it can be expressed:

$$v = \frac{s_1}{k_1} \bigg/ \frac{s_2}{k_2}$$

in which s_1, s_2 are independent random variables with distributions $\chi^2(k_1)$ and $\chi^2(k_2)$ respectively.

It is convenient also to define $\chi^2(0)$ to be the distribution of the degenerate random variable taking value 0, almost surely.

Analytical definitions of the distributions can be given, but we shall find these implicit definitions easier to work with. Percentiles of these distributions are tabulated in books of statistical tables. Some representative functions are illustrated in Fig. 4.1.

For large values of k_2, if v has an $F(k_1, k_2)$ distribution, then to a good approximation $k_1 v$ has a $\chi^2(k_1)$ distribution.

Let $\hat{\theta}$ be the least squares estimate of θ^*, and let $\hat{\hat{\theta}}$ be the least squares estimate under hypothesis (4.3.25). $\hat{\hat{\theta}}$ is calculated as

$$\hat{\hat{\theta}} = \mathrm{col}\,(\theta_0, 0, \ldots, 0), \quad \theta_0 = (X_0^{\mathrm{T}} X_0)^{-1} X_0^{\mathrm{T}} y$$

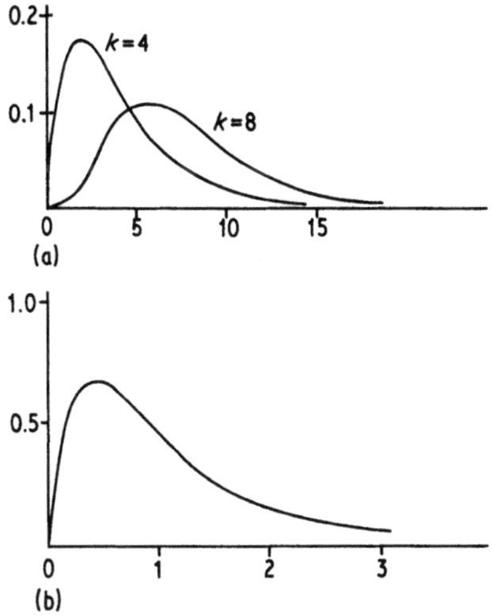

Fig. 4.1 *(a) Probability density function of $\chi^2(k)$ for $k = 4, 8$; (b) Probability density function of $F(4, 8)$.*

in which X_0 is the $p \times (q-d)$ matrix obtained from X by removing the last d columns.

Define

$$S(\theta) = \varepsilon^{\mathrm{T}}(\theta)\varepsilon(\theta) \tag{4.3.26}$$

in which

$$\varepsilon(\theta) = y - X\theta, \quad \theta \in \mathbb{R}^q.$$

We refer to a function of the observed data as a *statistic*. The statistic $S(\hat{\hat{\theta}}) - S(\hat{\theta})$ measures the increase in the minimum of the least squares criterion when we decrease the number of parameters from q to $q-d$. It is natural to reject hypothesis (4.3.25) if $S(\hat{\hat{\theta}}) - S(\hat{\theta})$ is large, since then a significantly better fit to the data can be achieved by the higher-order model. The following proposition brings together results necessary for formulating a test along these lines.

Proposition 4.3.8

Suppose that for some integer d, $0 \leq d \leq q$, hypothesis (4.3.25) is true. (In the case $d = 0$, no restrictions are placed upon θ^*). Let $\hat{\theta}$ be the least squares estimate of θ^*, and let $\hat{\hat{\theta}}$ be the least squares estimate under hypothesis (4.3.25). Let $S(\theta)$ be defined by (4.3.26). Then $\hat{\hat{\theta}}$, $S(\hat{\theta})$, $S(\hat{\hat{\theta}}) - S(\hat{\theta})$ are independent, and

$$(\sigma^2)^{-1}S(\hat{\theta}) \sim \chi^2(p-q), (\sigma^2)^{-1}(S(\hat{\hat{\theta}}) - S(\hat{\theta})) \sim \chi^2(d).$$

Notice that, when d takes the value 0, the proposition says that, if no hypothesis (4.3.25) is imposed, $\hat{\theta}$ and $S(\hat{\theta})$ are independent and $(\sigma^2)^{-1}s^2(\hat{\theta}) \sim \chi^2(p-q)$. We shall make use of this fact when we come to the calculation of confidence regions.

PROOF We shall deal first of all with the case $0 < d < q$. Take once again X_0 to be the matrix X in which the last d columns have been replaced by zero columns.

We observe that, by the nature of least squares estimates,

$$(y - X\hat{\theta}) \quad \text{is orthogonal to the range of } X \tag{4.3.27}$$

and

$$(y - X\hat{\hat{\theta}}) \quad \text{is orthogonal to the range of } X_0. \tag{4.3.28}$$

Properties (4.3.27) and (4.3.28) imply that

$$X(\hat{\theta} - \hat{\hat{\theta}}) \quad \text{is orthogonal to the range of } X_0, \tag{4.3.29}$$

since $X(\hat{\theta}-\hat{\hat{\theta}})$ can be expressed as the sum of $(y-X\hat{\hat{\theta}})$ and $-(y-X\hat{\theta})$ and both terms in the sum are orthogonal to the range of X_0.

We have, from (4.3.27),

$$\|y-X\hat{\hat{\theta}}\|^2=\|(y-X\hat{\theta})+X(\hat{\theta}-\hat{\hat{\theta}})\|^2=\|y-X\hat{\theta}\|^2+\|X(\hat{\theta}-\hat{\hat{\theta}})\|^2.$$

This equation can be expressed as

$$S(\hat{\hat{\theta}})-S(\hat{\theta})=\|X(\hat{\theta}-\hat{\hat{\theta}}\|^2. \tag{4.3.30}$$

Let $b_1,\ldots,b_p$ be an orthonormal basis for $\mathbb{R}^p$ with the properties:

(a) The vectors $b_1,\ldots,b_{q-d}$ span the range of X_0,
(b) The vector $b_{q-d+1},\ldots,b_q$ are orthogonal to the range of X_0 and such that $b_1,\ldots,b_q$ span the range of X; and,
(c) The vectors $b_{q+1},\ldots,b_p$ are orthogonal to the range of X.

Such vectors can be chosen since the columns of X are linearly independent.

Let $B:=(b_1\vdots\ldots\vdots b_p)$ and set

$$v=B^{\mathrm{T}}e.$$

We deduce from the fact that the columns of B form an orthonormal basis for $\mathbb{R}^p$, that

$$B^{\mathrm{T}}=B^{-1}. \tag{4.3.31}$$

Notice also that, since $e\sim N(0,\sigma^2 I)$,

$$\operatorname{cov}\{v\}=E[B^{\mathrm{T}}ee^{\mathrm{T}}B]=\sigma^2B^{\mathrm{T}}B=\sigma^2 I$$

and so $v\sim N(0,\sigma^2 I)$.

Consider the following decomposition of e:

$$e=y-X\theta^*=(y-X\hat{\theta})+X(\hat{\theta}-\hat{\hat{\theta}})+X(\hat{\hat{\theta}}-\theta^*).$$

Multiplying through by B^{T} and using (4.3.31), we obtain

$$v=B^{\mathrm{T}}e=B^{-1}(y-X\hat{\theta})+B^{-1}X(\hat{\theta}-\hat{\hat{\theta}})+B^{-1}\,X(\hat{\hat{\theta}}-\theta^*). \tag{4.3.32}$$

Now the mapping $x\to B^{-1}x$ transforms the coordinates w.r.t. the standard basis into a coordinates w.r.t. the basis $b_1,\ldots,b_p$. By (4.3.27), $(y-X\hat{\theta})$ is orthogonal to $b_1,\ldots,b_q$ and so

$$B^{-1}(y-X\hat{\theta})\in\{\xi\in\mathbb{R}^p:\xi_1=\cdots\xi_q=0\}.$$

By (4.3.29), $X(\hat{\theta}-\hat{\hat{\theta}})$ lies in the range of X but is orthogonal to the range of X_0 and so

$$B^{-1}x(\hat{\theta}-\hat{\hat{\theta}})\in\{\xi\in\mathbb{R}^p:\xi_1=\cdots=\xi_{q-d}=\xi_{q+1}=\cdots=\xi_p=0\}.$$

Since $X(\hat{\hat{\theta}}-\theta^*)$ lies in the range of X_0

$$B^{-1}X(\hat{\hat{\theta}}-\theta^*)\in\{\xi\in\mathbb{R}^p:\xi_{q-d+1}=\cdots=\xi_p=0\}.$$

It follows from these properties, together with (4.3.31) and (4.3.32), that

$$B^{\mathrm{T}}(y-X\hat{\theta})=(0,\ldots,0,v_{q+1},\ldots,v_p)^{\mathrm{T}} \tag{4.3.33}$$

$$B^{\mathrm{T}}X(\hat{\theta}-\hat{\hat{\theta}})=(0,\ldots,0,v_{q-d+1},\ldots,v_q,0,\ldots,0)^{\mathrm{T}} \tag{4.3.34}$$

$$B^{\mathrm{T}}X(\hat{\hat{\theta}}-\theta^*)=(v_1,\ldots,v_{q-d},0,\ldots)^{\mathrm{T}}. \tag{4.3.35}$$

By (4.3.31) and (4.3.34)

$$S(\hat{\hat{\theta}})-S(\hat{\theta})=\|X(\hat{\theta}-\hat{\hat{\theta}})\|^2=\|B^{\mathrm{T}}X(\hat{\theta}-\hat{\hat{\theta}})\|^2=\sum_{i=q-d+1}^{q}v_i^2. \tag{4.3.36}$$

By (4.3.31) and (4.3.33),

$$S(\hat{\theta})=\|y-X\hat{\theta}\|^2=\|B^{\mathrm{T}}(y-X\hat{\theta})\|^2=\sum_{i=q+1}^{p}v_i^2. \tag{4.3.37}$$

By (4.3.31) and (4.3.35)

$$\begin{aligned}\hat{\hat{\theta}}&=\theta^*+(X^{\mathrm{T}}X)^{-1}X^{\mathrm{T}}X(\hat{\hat{\theta}}-\theta^*)=\theta^*+(X^{\mathrm{T}}X)^{-1}X^{\mathrm{T}}BB^{\mathrm{T}}X(\hat{\hat{\theta}}-\theta^*)\\&=(X^{\mathrm{T}}X)^{-1}X^{\mathrm{T}}B(v_1,\ldots,v_q,0\ldots0)^{\mathrm{T}}.\end{aligned} \tag{4.3.38}$$

Since the v_i are independent and have common distribution $N(0,\sigma^2)$ we deduce from (4.3.36) and (4.3.37) that

$$\sigma^{-2}(S(\hat{\hat{\theta}})-S(\hat{\theta}))\sim\chi^2(d)\quad\text{and}\quad\sigma^{-2}S(\hat{\theta})\sim\chi^2(p-q).$$

Finally, we note that (4.3.36), (4.3.37) and (4.3.38) imply that $S(\hat{\hat{\theta}})-S(\hat{\theta})$, $S(\hat{\theta})$ and $\hat{\hat{\theta}}$ are independent.

This completes the proof when $0<d<q$. It remains to consider $d=0$ and $d=q$. Obvious modifications to our earlier arguments, in which we now select $v_1,\ldots,v_q$ to span the range of X, give the assertions of the properties in these cases also. □

The case when σ^2 is known

The proposition tells us that the statistic $\sigma^{-2}[S(\hat{\hat{\theta}})-S(\hat{\theta})]$ has the distribution $\chi^2(d)$ under hypothesis (4.3.25). Let k_α be the upper α-percentile of the $\chi^2(d)$ distribution, i.e. if $x\sim\chi^2(d)$, the event

$x \geq k_\alpha$ has probability α. Then the probability that the event

$$S(\hat{\hat{\theta}}) - S(\hat{\theta}) > \sigma^2 k_\alpha \tag{4.3.39}$$

will occur when (4.3.25) is true, is α.

There is therefore good evidence for rejecting hypothesis (4.3.25), if the inequality (4.3.39) holds for some pre-set value of α (0.05, say).

The case when σ^2 is not known

In this case we use the property that

$$\frac{S(\hat{\hat{\theta}}) - S(\hat{\theta})}{d} \Bigg/ \frac{S(\hat{\theta})}{(p-q)} \sim F(d, p-q), \tag{4.3.40}$$

if hypothesis (4.3.25) is true.

Let k_α now denote the α-percentile for the $F(d, p-q)$ distribution. The event

$$\frac{S(\hat{\hat{\theta}}) - S(\hat{\theta})}{d} \Bigg/ \frac{S(\hat{\theta})}{(p-q)} > k_\alpha \tag{4.3.41}$$

has probability α if (4.3.25) is true, and there is good evidence for rejecting hypothesis (4.3.25) if (4.3.41) holds for some pre-selected value for α.

In typical applications to modelling of dynamical systems, p (which is related to the number of data points) will be large and the model orders considered will be small. Tests based on the property (4.3.40) suggest a procedure for selecting model order in such situations.

Let S_n be the minimum of the least squares criterion over vectors of parameters of dimension $n, n = 1, 2, \ldots$

Since p is assumed large and q/p small, the distribution $F(1, p-q)$ closely approximates $\chi^2(1)$. We deduce from Proposition 4.3.8 that, if n is a possible model order,

$$(S_n - S_{n+1}) \Bigg/ \frac{S_{n+1}}{p}$$

has approximately the $\chi^2(1)$ distribution.

The 0.05 percentile for $\chi^2(1)$ is approximately 4. There are grounds then for rejecting n as a possible model order, at approximately a 5% risk level, if the inequality

$$S_n - S_{n+1} > \kappa \frac{S_{n+1}}{p}$$

is satisfied in which the coefficient κ takes value 4.

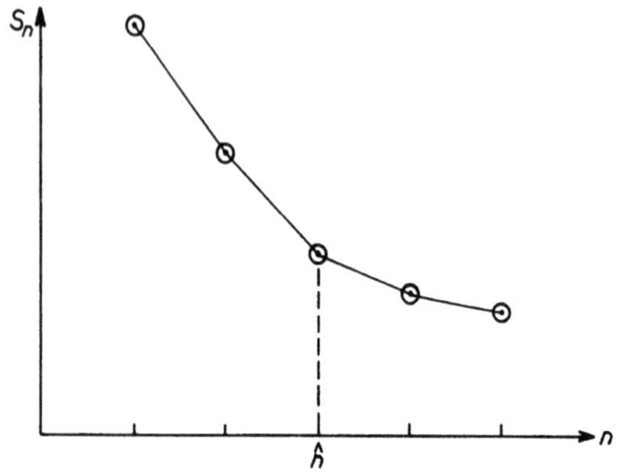

Fig. 4.2

Consider now the graph of S_n against n (see Fig. 4.2). These observations suggest that an estimate $\hat{n}$ of the model order be chosen to satisfy

$$S_{\hat{n}-1} - S_{\hat{n}} > \kappa \frac{S_{\hat{n}}}{p}, \quad S_{\hat{n}} - S_{\hat{n}+1} \leq \kappa \frac{S_{\hat{n}+1}}{p}, \tag{4.3.42}$$

for some pre-set value of κ (4, for example).

Before we leave the topic of model order selection, we point out an interesting interpretation of the inequalities (4.3.42). We can view S_n, $n = 1, \ldots, p$ as a uniform discretization of a continuously differentiable function $g: [0, 1] \to \mathbb{R}$. By this we mean

$$g\left(\frac{n}{p}\right) = S_n, \qquad n = 1, 2, \ldots, p.$$

Now $(S_{\hat{n}-1} - S_{\hat{n}})/(1/p)$ is a finite difference approximation to

$$-\frac{\mathrm{d}}{\mathrm{d}x} g$$

at $\hat{x} = \hat{n}/p$. The condition (4.3.42) can be expressed approximately in terms of g:

$$-\frac{\dfrac{\mathrm{d}}{\mathrm{d}x} g(x)}{g(x)} - \kappa \Big|_{x = \hat{n}/p} = 0$$

or

$$\frac{\mathrm{d}}{\mathrm{d}x} [\log g(x) + \kappa x] \Big|_{x = \hat{n}/p} = 0.$$

We recognize this last equation as a necessary condition for the function

$$x \to \log\{g(x)\} + \kappa x \tag{4.3.43}$$

to achieve a minimum at $x = \hat{n}/p$. The property that the function (4.3.43) achieves its minimum at $x = \hat{n}/p$ can be expressed in terms of S_n: $\hat{n}$ minimizes

$$n \to \log\{S_n\} + \frac{\kappa}{p} n.$$

Since p is fixed, we can alternatively take $\hat{n}$ to minimize $A(n)$ where

$$A(n) = \log\left\{\frac{1}{p} S_n\right\} + \tilde{\kappa} n$$

in which $\tilde{\kappa} = \kappa/p$.

These formal calculations justify a loose interpretation of $\hat{n}$ as the model order minimizing the criterion function $A(n)$ for a pre-set value of $\tilde{\kappa}$. The function $A(n)$ is customarily referred to as a criterion of Akaike type for selection of model order. See Section 4.8 for further discussion.

4.3.6 Accuracy of estimates

Let the observed vector y satisfy the static system equation

$$y = X\theta^* + e$$

in which we assume $e \sim N(0, \sigma^2 I)$. Suppose that the $p \times q$ matrix X has full column rank, and $p > q$.

Let $\hat{\theta}$ be the least squares estimate of θ^* given y:

$$\hat{\theta} = (X^T X)^{-1} X^T y.$$

The trustworthiness of the estimated components $\hat{\theta}_i$ can be gauged from an α-confidence region for $\hat{\theta}_i$; this is an interval $I(\hat{\theta})$, which depends on the estimate $\hat{\theta}$, and has the property that the event $\{\theta_i^* \in I_i(\hat{\theta})\}$ occurs with probability α.

We provide α-confidence regions in the cases that σ^2 is, and is not, known.

Case 1 (σ^2 known)

Since the estimate $\hat{\theta}$ is linear, unbiased and has covariance matrix $\sigma^2(X^T X)^{-1}$, and since e is a vector of jointly normally distributed random variables,

$$\hat{\theta} \sim N(\theta, \sigma^2(X^T X)^{-1}).$$

It follows that, for $i = 1, \ldots, q$,

$$\hat{\theta}_i \sim N(\theta_i, \sigma^2 c_{ii})$$

in which $\{c_{ij}\} = (X^T X)^{-1}$. Since $X^T X$ is a positive definite matrix, the c_{ii} will all be positive, and

$$(\hat{\theta}_i - \theta_i^*)/\sqrt{(\sigma^2 c_{ii})} \sim N(0, 1).$$

We can use this property to construct a confidence region. Let k_β be the upper $\beta/2$ percentile for the distribution $N(0, 1)$. Then

$$(\hat{\theta}_i - k_\beta \sqrt{(\sigma^2 c_{ii})}, \hat{\theta}_i + k_\beta \sqrt{(\sigma^2 c_{ii})})$$

is a $(1 - \beta)$ confidence region for θ_i^*, $i = 1, \ldots, q$. This is the case since the normal density function is symmetric.

Case 2 (σ^2 unknown)

When σ^2 is unknown, a realization of $\hat{\theta}_i/\sqrt{(\sigma^2 c_{ii})}$ is no longer available. It is natural in this situation to construct regions from the statistic $\hat{\theta}_i/\sqrt{(\hat{\sigma}^2 c_{ii})}$, in which $\hat{\sigma}^2$ is the unbiased estimate of σ^2,

$$\hat{\sigma}^2 = (p - q)^{-1} \|y - X\hat{\theta}\|^2$$

provided in Section 4.2. (See Proposition 4.3.7.)

At this stage we must introduce another distribution: given a positive integer k, a random variable v is said to have the $t(k)$ distribution if it can be expressed

$$v = \frac{d}{\sqrt{(e/k)}}$$

in which d and e are independent random variables with distributions $N(0, 1)$ and $\chi^2(k)$ respectively. Percentiles for these distributions are tabulated in books of statistical tables.

Now, for $i = 1, \ldots, q$,

$$\frac{\hat{\theta}_i - \theta_i^*}{\sqrt{c_{ii}}} \Big/ \sqrt{\hat{\sigma}^2} = \frac{\hat{\theta}_i - \theta_i^*}{\sqrt{(\sigma^2 c_{ii})}} \Big/ \sqrt{\frac{\|y - X\hat{\theta}\|^2}{(p-q)\sigma^2}}.$$

We know that

$$\frac{\hat{\theta}_i - \theta_i^*}{\sqrt{(\sigma^2 c_{ii})}} \sim N(0, 1).$$

Proposition 4.3.8, applied when $d = 0$, tells us that $\hat{\theta}_i$ is independent of $\| y - X\hat{\theta} \|^2$ and

$$\frac{1}{\sigma^2} \| y - X\hat{\theta} \|^2 \sim \chi^2(p - q).$$

It follows from these properties and definitions of the t-distribution that

$$\frac{\hat{\theta}_i - \theta_i^*}{\sqrt{c_{ii}}} \Big/ \sqrt{\hat{\sigma}^2} \sim t(p - q). \tag{4.3.44}$$

Let k_β be the upper $\beta/2$ percentile for the distribution $t(p - q)$. Since the associated probability density function is symmetric, it follows from (4.3.44) that

$$(\hat{\theta}_i - k_\beta \sqrt{(\hat{\sigma}^2 c_{ii})}, \hat{\theta}_i + k_\beta \sqrt{(\hat{\sigma}^2 c_{ii})})$$

is a $(1 - \beta)$ confidence region for $\hat{\theta}_i$, $i = 1, \ldots, q$.

4.4 Parameter estimation for dynamical systems

A great variety of parameter estimation techniques for dynamical systems have been proposed. Most of these share the following ingredients:

Observations are available of the r-vector output y_k, $k = 0, 1, \ldots, N$ and the m-vector input u_k, $k = 0, 1, \ldots, N - 1$ of a dynamical system.

A set M of models is specified. The models in M are parametrized by a q-vector θ, which ranges over a set D. The model in M corresponding to choice of parameter θ is denoted by $M(\theta)$.

A real-valued function V_N of the parameter θ and of the data y^N, u^{N-1} is also specified. V_N, which is a measure of the discrepancy between the observed outputs and those predicted by the model on the basis of earlier inputs and outputs, is called the identification criterion.

The parameter estimation problem is that of selecting a model from M which best matches the data according to the identification criterion. It amounts to finding a parameter value which minimizes $\theta \to V_N(\theta, y^N, u^{N-1})$. Different parameter estimation techniques will result from changing the specification of M, V_N and choice of numerical scheme for determining the minimizing value of θ.

4.4.1 Prediction error formulation

A particularly important class of parameter estimation techniques is formulated in terms of the predictor models of Section 4.2:

$$y_k = f_k(\theta; y^{k-1}, u^{k-1}) + e_k, \qquad k = 0, 1, \ldots. \tag{4.4.1}$$

To emphasize the point that, if the data were realizations of the processes generated by application of the input to the model (4.4.1) then $f_k(\theta; y^{k-1}, u^{k-1})$ would be the conditional expectation of y_k, given y^{k-1}, u^{k-1}, we adopt the hat notation '$\hat{\ }$', customarily used to denote estimates given past outputs and inputs and write

$$\hat{y}_k(\theta) = f_k(\theta; y^{k-1}, u^{k-1}).$$

We also write $\varepsilon_k(\theta)$ for the error in the prediction of y_k provided by $\hat{y}_k(\theta)$, namely

$$\varepsilon_k(\theta) = y_k - \hat{y}_k(\theta). \tag{4.4.2}$$

The sequence $\{\varepsilon_k(\theta)\}$ is commonly called the sequence of 'residuals' or 'prediction errors' associated with the model (4.4.1).

It is natural to assess the quality of the model according to the accuracy of its predictors $y_k(\theta)$, and to choose therefore the identification criterion to be a function of the prediction errors. A versatile identification criterion is defined in terms of:

(a) A sequence of functions $\{l_k(\cdot,\cdot)\}$ from the space $\mathbb{R}^q \times \mathbb{R}^r$ to the space of $d \times d$ matrices, and,
(b) A real valued function $h(\cdot)$ with domain the space of $d \times d$ matrices.

The identification criterion is

$$V_N(\theta; y^N, u^{N-1}) = h(Q_N(\theta; y^N, u^{N-1})) \tag{4.4.3}$$

in which

$$Q_N(\theta; y^N, u^{N-1}) = \frac{1}{N} \sum_{k=1}^{N} l_k(\theta; \varepsilon_k(\theta)). \tag{4.4.4}$$

Common choices of identification criterion of this form are:

$$V_N(\theta; y^N, u^{N-1}) = \frac{1}{N} \sum_{k=1}^{N} \varepsilon_k^{\mathrm{T}}(\theta) W_k \varepsilon_k(\theta) \tag{4.4.5}$$

in which $\{W_k\}$ is a given set of positive definite weighting matrices, and

$$V_N(\theta; y^N, u^{N-1}) = \det\left[\frac{1}{N} \sum_{k=1}^{N} \varepsilon_k(\theta) \varepsilon_k^{\mathrm{T}}(\theta)\right]. \tag{4.4.6}$$

Choice (4.4.5) corresponds to selecting a model to minimize a weighted sum of squares of the prediction errors, and choice (4.4.6) to selecting a model by the maximum likelihood method, as we shall see.

Recall that the predictor model description covers the stochastic dynamical models of Section 2.4 when the noise vectors are assumed independent. Let us illustrate compution of the prediction errors (and hence the identification criterion via (4.4.3) and (4.4.4)), in the case that (4.4.1) is a reformulation of the ARMAX model equations

$$A_\theta(z^{-1}) y_k = B_\theta(z^{-1}) u_{k-1} + C_\theta(z^{-1}) e_k, \qquad k = 0, 1, \ldots \tag{4.4.7}$$

with zero initial conditions ($y_k = 0, u_k = 0, e_k = 0$, for $k < 0$). We suppose that the polynomials $A_\theta(\sigma)$, $C_\theta(\sigma)$ in σ are such that $A_\theta(0) = C_\theta(0) = I$. In these circumstances, as was shown in Section 2.6, the predictors $\hat{y}_k(\theta)$ are given by

$$\hat{y}_k(\theta) = [I - C_\theta^{-1} A_\theta] y_k - C_\theta^{-1} B_\theta u_k, \qquad k = 0, 1, \ldots$$

The prediction errors $\varepsilon_k(\theta) = y_k - \hat{y}_k(\theta)$ corresponding to (4.4.7) are therefore

$$\varepsilon_k(\theta) = C_\theta^{-1} A_\theta y_k + C_\theta^{-1} B_\theta u_k, \qquad k = 0, 1, \ldots$$

The prediction errors can be computed then by recursive solution of the difference equations

$$C_\theta(z^{-1}) \varepsilon_k(\theta) = A_\theta(z^{-1}) y_k + B_\theta(z^{-1}) u_k, \qquad k = 0, 1, \ldots$$

with zero initial conditions ($\varepsilon_k(\theta) = 0$, $y_k = 0$, $u_k = 0$ for $k < 0$).

4.4.2 Least squares parameter estimation

Least squares parameter estimation methods for dynamical systems are methods in which a model is chosen to minimize a weighted sum of squares of the prediction errors, $\{\varepsilon_k\}$, defined by (4.4.2). We seek then a value of θ which minimizes the identification criterion V_N:

$$V_N(\theta; y^N, u^{N-1}) = \frac{1}{N} \sum_{k=1}^{N} \varepsilon_k^T(\theta) W_k \varepsilon_k(\theta).$$

Here the weighting matrices W_k are symmetric, positive definite matrices chosen to reflect the relative importance attached to the different components of the prediction.

The least squares method for given weighting matrices admits a prediction error formulation in which we choose the functions h and $l_1, \ldots, l_N$ defining the identification criterion

$$h\left(\frac{1}{N} \sum_{k=1}^{N} l_k(\theta; \varepsilon_k(\theta))\right)$$

to be

$$l_k(\theta, \varepsilon) = \varepsilon_k \varepsilon_k^T W_k \qquad \text{for } k = 1, 2, \ldots \text{ and } h(\cdot) = \text{trace}\,\{\cdot\}.$$

Consider now scalar ARMAX models of the form

$$A_\theta(z^{-1}) y_k = B_\theta(z^{-1}) u_k + e_k, \qquad k = 0, 1, \ldots$$

with zero initial conditions ($y_k = 0$, $u_k = 0$, $k < 0$). Here

$$A_\theta(z^{-1}) = 1 + a_1 z^{-1} + \cdots + a_n z^{-n}, \quad B_\theta(z^{-1}) = b_1 z^{-1} + \cdots + b_n z^{-n}$$

and e_k, $k = 1, 2, \ldots$, is a sequence of zero-mean independent random, variables. The vector θ of unknown parameters, made up of the coefficients $a_1, \ldots, a_n$, $b_1, \ldots, b_n$, is to be estimated from observations of y^N, u^{N-1}. As we have already noted, the prediction errors are

$$\varepsilon_k(\theta) = A_\theta(z^{-1}) y_k - B_\theta(z^{-1}) u_k, \qquad k = 0, 1, \ldots \tag{4.4.8}$$

The problem of minimizing the identification criterion $V_N(\cdot, y^N, u^{N-1})$:

$$V_N(\theta; y^N, u^{N-1}) = \sum_{k=1}^{N} \varepsilon_k^T(\theta) \varepsilon_k(\theta)$$

in which $\varepsilon_k(\theta)$ is given by (4.4.8) can be expressed explicitly as that of minimizing $\|y - X\theta\|^2$ over θ, where

$$y = \begin{bmatrix} y_1 \\ \vdots \\ y_N \end{bmatrix}$$

and

$$X = \left[\begin{array}{ll|ll} -y_0 & -y_{-1}\cdots -y_{-n+1} & u_0 & u_{-1}\cdots u_{-m+1} \\ -y_1 & -y_0\cdots -y_{-n+2} & u_1 & u_0\cdots u_{-m+2} \\ \vdots & & \vdots & \vdots \quad\quad \vdots \\ -y_{N-1} & -y_{N-2}\cdots -y_{N-n} & u_{N-1} & u_{N-2}\cdots u_{N-m} \end{array}\right].$$

In this case the parameter estimation procedure is equivalent to application of least squares estimation, as described in Section 4.3, to the static model

$$y = X\theta + e$$

obtained from reformulation of the model equations as in Example 4.3.1. If X has full column rank, the least squares estimate is given then by

$$\hat{\theta} = (X^{\mathrm{T}}X)^{-1}X^{\mathrm{T}}y.$$

4.4.3 Maximum likelihood estimation for dynamical systems

Consider again predictor models:

$$y_k = f_k(\theta; y^{k-1}, u^{k-1}) + e_k \qquad k = 1, \ldots, N. \tag{4.4.9}$$

We now permit $e_1, \ldots, e_N$ to be random variables whose joint probability density functions are specified functions of the unknown parameter θ. Assume that the inputs $u_1, \ldots, u_N$ are independent of $e_1, \ldots, e_N$. Let $p(x_N, \ldots, x_1 | \theta, u^{N-1})$ be the joint probability density function of y^N given θ and u^{N-1} (assumed to exist). Maximum likelihood estimates of the unknown parameter are those values of θ which maximize $V_N(\cdot, y^N, u^{N-1})$:

$$V_N(\theta, y^N, u^{N-1}) = p(y_N, \ldots, y_1 | \theta, u^{N-1})$$

(in which $y_1, \ldots, y_N$ are observations of the output).

Let us examine maximum likelihood estimates in more detail under the following additional assumptions: for $k = 1, 2, \ldots$

(a) e_k has zero-mean and is normally distributed with non-singular

covariance matrix (we write the covariance matrix Λ_k), (4.4.10)
and,

(b) e_k is independent of e_l, $l \neq k$, for $l = 1,\ldots,N$. (4.4.11)

Denote by $p(x_k,\ldots,x_j|y^{j-1},u^{N-1},\theta)$ (or, briefly, $p(x_k,\ldots,x_j|y^{j-1})$) the probability density function of $y_k,\ldots,y_j$ given y^{j-1}, u^{N-1}, $1 \leq j \leq k \leq N$. It follows from our assumptions that $p(x_k|y^{k-1})$ exists for $k = 1,\ldots,N$, and is given by

$$p(x_k|y^{k-1}) = [(2\pi)^r \det \Lambda_k]^{-1/2} \exp\{-\tfrac{1}{2}\|x_k - \hat{y}_k(\theta)\|^2_{\Lambda_k^{-1}}\}$$

in which, as usual, $\hat{y}_k(\theta) = f_k(\theta; y^{k-1}, u^{k-1})$. We have used the notation "$\|x\|_P$" for "$x^T P x$". Bayes' rule tells us that $p(x_N, x_{N-1}|y^{N-2})$ also exists and is given by

$$\begin{aligned} p(x_N, x_{N-1}|y^{N-2}) &= p(x_N|y_{N-1} = x_{N-1}, y^{N-2})p(x_{N-1}|y^{N-2}) \\ &= p(x_N|y^{N-1})p(x_{N-1}|y^{N-2}) \end{aligned} \tag{4.4.12}$$

(when x_{N-1} replaces y_{N-1} in y^{N-1}).

We deduce from (4.4.12) and repeated application of Bayes' rule that $p(x_N,\ldots,x_j|y^{j-1})$ exists for $j = N-1,\ldots,1$ and the likelihood function $p(y_N,\ldots,y_1|u^{N-1},\theta)$ is given by

$$\begin{aligned} p(y_N,\ldots,y_1|u^{N-1},\theta) &= \prod_{k=1}^{N} p(y_k|y^{k-1},u^{k-1},\theta) \\ &= (2\pi)^{-Nr/2}(\det \Lambda_N \ldots \det \Lambda_1)^{-1/2} \\ &\quad \cdot \exp\left(-\frac{1}{2}\sum_{k=1}^{N} \|y_k - \hat{y}_k(\theta)\|^2_{\Lambda_k^{-1}}\right). \end{aligned}$$

The log likelihood function is therefore

$$\begin{aligned} \log p(y_N,\ldots,y_1|u^{N-1},\theta) = &-\frac{Nr}{2}\log 2\pi - \frac{1}{2}\sum_{k=1}^{N} \log \det \Lambda_k \\ &-\frac{1}{2}\sum_{k=1}^{N} \|y_k - \hat{y}_k(\theta)\|^2_{\Lambda_k^{-1}}. \end{aligned}$$

Since maximizing the likelihood function is equivalent to minimizing minus the log likelihood function, we conclude that maximum likelihood estimates are those values of the parameter θ which minimize the identification criterion

$$L(\theta) = \sum_{k=1}^{N} \|\varepsilon_k(\theta)\|^2_{\Lambda_k^{-1}(\theta)} + \sum_{k=1}^{N} \log \det \Lambda_k(\theta)$$

where once again the $\varepsilon_k(\theta)$ are the prediction errors. We have added an argument θ to Λ_k to emphasize that it too can depend on the unknown parameter.

We shall now give further information about maximum likelihood estimates in three cases.

Case 1 (Λ_k known)

Assume that Λ_k is known and $\Lambda_k > 0$, $k = 1, \ldots, N$. In this case, maximum likelihood estimates are values of θ which minimize

$$V_1(\theta) = \frac{1}{N} \sum_{k=1}^{N} \|\varepsilon_k(\theta)\|^2_{\Lambda_k^{-1}}. \tag{4.4.13}$$

Case 2 ($\Lambda_k = \sigma^2\Sigma$, Σ known)

Assume that the distributions have common covariance matrix $\sigma^2\Sigma$, in which Σ is a fixed positive definite matrix and σ^2 is an unknown positive parameter to be estimated. We can arrange that the vector of unknown parameters takes the form $\mathrm{col}(\psi, \sigma^2)$ in which ψ is a vector comprising the other unknown parameters. Assume that the parameter constraint set has the form $\tilde{D} \times (0, \infty)$ for some set $\tilde{D}$. In this case maximum likelihood estimates are values of ψ and σ^2 which minimize

$$V_2(\psi, \sigma^2) = (\sigma^2)^{-1}\left[\frac{1}{N} \sum_{k=1}^{N} \|\varepsilon_k(\psi)\|^2_{\Sigma^{-1}}\right] + \log\det\{\sigma^2\Sigma\}. \tag{4.4.14}$$

Case 3 ($\Lambda_k = \Sigma, \Sigma$ unknown)

Let the disturbances have common covariance matrix Σ. Suppose that the unknown parameter vector comprises the components of Σ and other unknown parameters assembled to form the vector ψ. We assume that the parameter constraint set is of the form $\{(\psi, \Sigma)$: $\theta \in \tilde{D}$, $\Sigma > 0\}$ for some set $\tilde{D}$. In this case the maximum likelihood estimates are values of (θ, Σ) which minimize

$$V_3(\psi, \Sigma) = \frac{1}{N} \sum_{k=1}^{N} \|\hat{\varepsilon}_k(\psi)\|^2_{\Sigma^{-1}} + \log\det\Sigma. \tag{4.4.15}$$

The following result tells us that, in each case, the parameter estimation problem reduces to that of minimizing an identification criterion of the form

$$\psi \to h\left(\frac{1}{N}\sum^{N} l_k(\psi; \varepsilon_k(\psi))\right) \tag{4.4.16}$$

for appropriate choices of the functions $h(\cdot)$ and $l(\cdot,\cdot,\cdot)$. In other words, the problem admits a prediction error formulation.

Proposition 4.4.1

Consider the problem of obtaining maximum likelihood estimates of the unknown parameters in the model (4.4.9), under the assumptions (4.4.10) and (4.4.11). Let the special cases 1, 2, and 3 be as described above. We have:

Case 1 (Λ_k known) $\hat{\theta}$ is a maximum likelihood estimate if and only if $\hat{\theta}$ minimizes the identification criterion (4.4.16), when we choose $h(\cdot) = \text{trace}\{\cdot\}$ and $l_k(\theta, \varepsilon) = \varepsilon\varepsilon^{\mathrm{T}}\Lambda_k^{-1}$, $k = 1, \ldots, N$.

Case 2 ($\Lambda_k = \sigma^2\Sigma$, Σ known) ($\hat{\psi}$, $\hat{\sigma}^2$) is a maximum likelihood estimate if and only if

(a) $\hat{\psi}$ minimizes the identification criterion (4.4.16) when we choose $h(\cdot) = \text{trace}\{\cdot\}$, $l_k(\psi, \varepsilon) = \varepsilon\varepsilon^{\mathrm{T}}\Sigma^{-1}$, $k = 1, \ldots, N$; and,

(b)

$$\hat{\sigma}^2 = \frac{1}{Nr}\sum_{k=1}^{N} \|\varepsilon_k(\hat{\psi})\|^2_{\Sigma^{-1}} \quad \text{and} \quad \hat{\sigma}^2 \neq 0.$$

(r is the dimension of the output vector.)

Case 3 ($\Lambda_k = \Sigma$, Σ unknown) ($\hat{\psi}$, $\hat{\Sigma}$) is a maximum likelihood estimate if and only if

(a) $\hat{\psi}$ minimizes the identification criterion (4.4.16) when we choose $h(\cdot) = \det\{\cdot\}$, $l_k(\psi, \varepsilon) = \varepsilon\varepsilon^{\mathrm{T}}$, $k = 1, \ldots, N$, and,

(b)

$$\hat{\Sigma} = \frac{1}{N}\sum_{k=1}^{N} \varepsilon_k(\hat{\psi})\varepsilon_k^{\mathrm{T}}(\hat{\psi}) \quad \text{and} \quad \det\{\hat{\Sigma}\} > 0.$$

PROOF *Case* 1 In view of the fact that $\varepsilon^{\mathrm{T}}\Lambda\varepsilon = \text{trace}\{\varepsilon\varepsilon^{\mathrm{T}}\Lambda\}$, the proposition (in this case) merely restates the property that $\hat{\theta}$ minimizes the function V_1 defined by (4.4.13).

Case 2 Since $\log\det(\sigma^2\Sigma) = r\log\sigma^2 + \log\det\Sigma$, we see that $(\hat{\psi}, \hat{\sigma}^2)$ minimizes V_2 given by (4.4.14) if and only if $\hat{\psi}$, $\hat{\sigma}^2$ minimizes the function $(\psi, \sigma^2) \to (\sigma^2)^{-1}\left[\frac{1}{N}\sum_k \|\varepsilon_k(\psi)\|^2_{\Sigma^{-1}}\right] + r\log\sigma^2$ over

$\tilde{D} \times (0, \infty)$. This last property is equivalent to

(a) $\hat{\psi}$ minimizes $\psi \to \left[\frac{1}{N} \sum_k \| \varepsilon_k(\psi) \|^2 \right]$; and,

(b) $\hat{\sigma}^2$ minimizes $\sigma^2 \to (\sigma^2)^{-1} \left[\frac{1}{N} \sum_k \| \varepsilon_k(\hat{\psi}) \|^2_{\Sigma^{-1}} \right] + r \log \sigma^2$.

However the function $\sigma^2 \to (\sigma^2)^{-1} c + r \log(\sigma^2) : (0, \infty) \to \mathbb{R}$ (for $c \geq 0$, $r \geq 0$) has no stationary point if $c = 0$, and a unique stationary point $\sigma^2 = c/r$, at which the function achieves its minimum value, if $c > 0$. The proof is completed by applying this result when $c = \frac{1}{N} \sum_k \| \varepsilon_k(\hat{\psi}) \|^2_{\Sigma^{-1}}$.

Case 3 $(\hat{\psi}, \hat{\Sigma})$ minimizes V_3 given by (4.4.15) if and only if $\hat{\Sigma} = \hat{Q}^{-1}$ and $(\hat{\psi}, \hat{Q})$ minimizes the function

$$g(\psi, Q) := \text{trace}\{Q V(\psi)\} - \log \det Q$$

over $(\psi, Q) \in \tilde{D} \times \{Q : Q = Q^{\mathrm{T}}, Q > 0\}$. Here

$$V(\psi) := \frac{1}{N} \sum_k \varepsilon_k(\psi) \varepsilon_k^{\mathrm{T}}(\psi).$$

These conditions are equivalent to

$$\begin{aligned} &\hat{\psi} \text{ minimizes } J(\psi) := \min\{g(\psi, Q) : Q = Q^{\mathrm{T}}, Q > 0\} \\ &\text{over the subset of } \tilde{D} \text{ on which } J(\cdot) \text{ is defined,} \\ &Q \text{ minimizes } Q \to g(\hat{\psi}, Q) \text{ over } \{Q : Q = Q^{\mathrm{T}}, Q > 0\} \text{ and} \\ &\hat{\Sigma} = \hat{Q}^{-1} \end{aligned} \tag{4.4.17}$$

Let us investigate $J(\cdot)$ introduced in (4.4.17). The domain of $J(\cdot)$ comprises those ψ such that the minimum of $Q \to g(\psi, Q)$ is achieved over $\{Q : Q = Q^{\mathrm{T}}, Q > 0\}$. Fix such a ψ.

We shall require the following identities from matrix calculus (see Appendix D.4):

$$\frac{\partial}{\partial S} \log \det S = S^{-1}$$

on the space of $n \times n$ non-singular matrices S, and, given any $n \times n$ matrix D,

$$\frac{\partial}{\partial S} \text{trace}\{SD\} = D$$

on the space of $n \times n$ matrices S. Using these identities we deduce that the stationary points of the function $Q \to g(\hat{\psi}, Q)$ on $\{Q : Q = Q^{\mathrm{T}},$

$Q > 0\}$ are those $\bar{Q}$ which satisfy

$$V(\hat{\psi}) - \bar{Q}^{-1} = 0.$$

It follows that $\bar{Q} = (V(\hat{\psi}))^{-1}$ and $V(\hat{\psi}) > 0$ are necessary conditions for $\bar{Q}$ to minimize $Q \to g(\hat{\psi}, Q)$ over $\{Q: Q^{\mathrm{T}}, Q > 0\}$. Since these necessary conditions define a unique matrix $\bar{Q}$ (provided $V(\hat{\psi}) > 0$) we will be able to conclude that the matrix $(V(\hat{\psi}))^{-1}$ actually minimizes $Q \to g(\psi, Q)$ on $\{Q: Q = Q^{\mathrm{T}}, Q > 0\}$ if we can show that (when $V(\psi) > 0$), $Q \to g(\psi, Q)$ achieves a minimum on $\{Q: Q = Q^{\mathrm{T}}, Q > 0\}$. We prove existence of a minimizing element. Bearing in mind that an arbitrary symmetric positive definite matrix can be factored as a product of positive definite symmetric matrices (Appendix D.1) we see that it suffices to show that, given a symmetric positive definite matrix W, the minimization problem:

$$\begin{array}{l} \text{minimize } F(H) := \operatorname{trace}\{WHH\} - \log\det\{HH\} \\ \text{over symmetric positive definite-matrices } H \end{array} \tag{4.4.18}$$

has a solution.

Let $\|\cdot\|_{\mathrm{tr}}$ and $\|\cdot\|$ denote the trace and spectral matrix norms respectively, (see Appendix D.2).

By the equivalence of norms, there exists $\alpha > 0$ such that $\|P\|_{\mathrm{tr}} \geq \alpha\|P\|$, for all matrices P, of fixed dimension. We shall use the facts that $\|P\|^2$ is the maximum eigenvalue of $P^{\mathrm{T}}P$ and that, for P_1 a non-singular matrix, $\|P_1 P\| \geq \|P_1^{-1}\|^{-1}\|P\|$.

Take H an arbitrary symmetric $r \times r$ matrix. We have

$$\begin{aligned} \operatorname{trace}\{WHH\} &= \operatorname{trace}\{W^{1/2}HHW^{1/2}\} \\ &= \operatorname{trace}\{(W^{1/2}H)(W^{1/2}H)^{\mathrm{T}}\} = \|W^{1/2}H\|_{\mathrm{tr}}^2 \\ &\geq \alpha^2\|W^{1/2}H\|^2 \geq \alpha^2\|W^{-1/2}\|^{-2}\|H\|^2. \end{aligned}$$

Since $\det\{HH\}$ is the product of eigenvalues $\{\lambda_i\}$ of HH,

$$\begin{aligned} \log\det\{HH\} &= \log\{\prod_i \lambda_i\} \leq \log|\max_i \lambda_i|^r \\ &= r\log\|H\|^2. \end{aligned}$$

It follows from these inequalities that

$$F(H) \geq \alpha\|W^{-1/2}\|^2\|H\|^2 - r\log\|H\|^2.$$

We see that

$$\lim_{\|H\| \to 0} F(H) = +\infty \quad \text{and} \quad \lim_{\|H\| \to \infty} F(H) = +\infty.$$

In consequence, there exists $\delta > 0$ such that the infimum of the minimization problem (4.4.18) is unaltered by addition of the constraint:

$$\delta^{-1} \leq \|H\| \leq \delta.$$

But the $r \times r$ matrices H which are symmetric, positive definite, and satisfy these inequalities form a closed bounded set. The function $F(\cdot)$ is continuous and, by Weierstrass' theorem, achieves its minimum on this set. Problem (4.4.18) has a solution then.

Summing up, we have shown that, given any $\psi \in \tilde{D}$, the function $Q \to g(\psi, Q)$ achieves its minimum on $\{Q: Q = Q^{\mathrm{T}}, Q > 0\}$ if and only if $V(\psi)$ is non-singular, in which case the minimum is achieved at the unique point $(V(\psi))^{-1}$. It follows that the domain of the function $J(\cdot)$ of (4.4.17) is $\{\psi \in \tilde{D}: V(\psi) \text{ is non-singular}\}$ and

$$J(\psi) = \operatorname{trace}\{I\} - \log\det(V(\psi))^{-1}.$$

Since, however, $-\log\det(V(\psi))^{-1} = \log\det V(\psi)$ and the logarithmic function is monotone, minimization of J is equivalent to minimization of $\bar{J}(\psi) := \det V(\psi)$ over $\{\psi \in \tilde{D}: V(\psi) > 0\}$. The infimum of $\bar{J}$ is clearly unaltered if we take the domain of $\bar{J}$ to be to all of $\tilde{D}$. It follows that conditions (4.4.17) can be expressed in the desired form:

$$\hat{\psi} \text{ minimizes } \det\left(\frac{1}{N}\sum_k \varepsilon_k(\psi)\varepsilon_k^{\mathrm{T}}(\psi)\right),$$

$$\frac{1}{N}\sum_k \varepsilon_k(\hat{\psi})\varepsilon_k^{\mathrm{T}}(\psi) \text{ is non-singular}$$

and

$$\hat{\Sigma} = \frac{1}{N}\sum_k \varepsilon_k(\psi)\varepsilon_k^{\mathrm{T}}(\psi).$$

4.4.4 Asymptotic distributions of parameter estimates

Consider again selection procedures which admit a prediction error formulation. Here we select a model defined by a parameter $\hat{\theta}_N$ which minimizes the identification criterion

$$\theta \to h\left(\frac{1}{N}\sum_{k=1}^{N} l_k(\theta, e_k(\theta))\right)$$

Let us examine how we might assess the quality of the estimate $\hat{\theta}_N$. Recall that, since the outputs are random variables, the estimate which is some function of the inputs and outputs is also a random variable. Ideally then we would like to know the probability distribution of $\hat{\theta}_N$.

While the task of calculating the probability distribution of $\hat{\theta}_N$ is a formidable one except in highly restrictive circumstances (for example when the dynamic models can be reformulated as static models and conditions are satisfied under which the theory of Section 4.3 is applicable), we might hope at least to obtain estimates of the asymptotic distribution of $\hat{\theta}_N$ as $N \to \infty$. It turns out that this is possible; parameter estimates supplied by prediction error schemes have similar properties (notably consistency and asymptotic normality) to those of maximum likelihood estimates based on independent samples of a random variable, summarized in Proposition 4.1.3.

Suppose that the limit

$$\bar{V}(\theta) = \lim_{N \to \infty} h\left(\frac{1}{N} \sum_{k=1}^{N} El_k(\theta, \varepsilon_k(\theta)) \right) \tag{4.4.19}$$

exists for each θ and that, for N sufficiently large, the estimates $\hat{\theta}_N$ are confined to some closed ball B in parameter space such that†:

$$\frac{\partial^2}{\partial \theta^2} \bar{V}(\theta) > \delta I \qquad \text{for all } \theta \in B. \tag{4.4.20}$$

Then, provided certain mild conditions are satisfied (we shall be precise about such conditions in Chapter 5),

$$\hat{\theta}_N \to \theta^* \qquad \text{a.s.}$$

where θ^* minimizes $\bar{V}(\theta)$ over B. (This follows from Theorem 5.2.1 since the limit (4.4.19) is assumed to exist, and the convexity hypothesis (4.4.20) ensures the $\bar{V}(\theta)$ has at most one minimizer over B).

We can interpret θ^* as a parameter value associated with a model which best approximates the system as measured by some kind of average value of the identification criterion in the limit as $N \to \infty$. An asymptotic analysis of the probability distribution of $\hat{\theta}_N$ is possible if θ^* merely provides an approximation of the system (see Ljung and Caines, 1979). But we examine here the limiting distribution only in situations where θ^* provides a true description of the system in the following sense:

$$\{\varepsilon_k(\theta^*)\} \text{ is a sequence of zero mean, independent random variables with common covariance matrix } \Sigma_0. \tag{4.4.21}$$

† $\frac{\partial^2}{\partial \theta^2} \bar{V}$ denotes the matrix $\left\{ \frac{\partial^2}{\partial \theta_i \partial \theta_j} \bar{V} \right\}$.

We focus attention on the least squares identification criterion

$$J_1(\theta; N) = \frac{1}{N} \sum_{k=1}^{N} \varepsilon_k^{\mathrm{T}}(\theta) W \varepsilon_k(\theta) \tag{4.4.22}$$

(here W is a given weighting matrix) and the identification criterion which results from formulation of maximum likelihood estimation as a prediction error scheme:

$$J_2(\theta; N) = \det\left[\frac{1}{N} \sum_{k=1}^{N} \varepsilon_k(\theta)\varepsilon_k^{\mathrm{T}}(\theta)\right]. \tag{4.4.23}$$

Proposition 4.4.2

Consider either the identification criterion (4.4.22) or (4.4.23) and let θ^* be as above. Let $\psi_k(\theta)(=\{(\psi_k(\theta))_{ij}\})$ be the gradient of the predictions:

$$(\psi_k(\theta))_{ij} = \frac{\partial}{\partial \theta_j}(\hat{y}_k(\theta))_i,$$

$k = 1, 2, \ldots$, and suppose that $\{\psi_k(\theta^*)\}$ is a stationary process. Define

$$P_1(W) = [E\psi_k^{\mathrm{T}}(\theta^*)W\psi_k(\theta^*)]^{-1}[E\psi_k^{\mathrm{T}}(\theta^*)W\Sigma_0 W\psi_k(\theta^*)] \cdot [E\psi_k^{\mathrm{T}}(\theta^*)W\psi_k(\theta^*)]^{-1} \tag{4.4.24}$$

and

$$P_2 = [E\psi_k^{\mathrm{T}}(\theta^*)\Sigma_0^{-1}\psi_k(\theta^*)]^{-1}. \tag{4.4.25}$$

Then under certain conditions, described in (Ljung and Caines (1979)), the distribution of $N^{1/2}(\hat{\theta}_N - \theta)$ converges to the $N(0, G)$ distribution† as $N \to \infty$, where $G = P_1(w)$ if the identification criterion is $J_1(\theta; N)$ and $G = P_2$ if the criterion is $J_2(\theta; N)$.

PROOF See Ljung and Caines (1979).

We have seen in Section 4.3 how knowledge of the probability distribution of the parameter estimate permits us to construct confidence regions for the true parameter value. In the same spirit we can use properties such as those described in Proposition 4.4.2 to estimate confidence regions here too in a dynamic setting. Of course, since these estimates of confidence regions are based on the asympto-

†This mode of convergence is defined following Proposition 4.1.3.

tic behaviour of $\hat{\theta}_N$ we can expect them to be useful only when N is large.

The expressions (4.4.24) and (4.4.25) for the limiting covariance matrices $P_1(W)$ and P_2 cannot be evaluated exactly, but we can approximate them by related expressions whose values can be computed. For example we can replace Σ_0 by the estimate $\tilde{\Sigma}_0$:

$$\tilde{\Sigma}_0 = \frac{1}{N} \sum_{k=1}^{N} \varepsilon_k(\hat{\theta}_N)\varepsilon_k^{\mathrm{T}}(\hat{\theta}_N)$$

and replace the action of the expectation operator by sample averages about $\hat{\theta}_N$. Thus we use in place of $P_1(W)$ and P_2 the matrices $\tilde{P}_1(W)$ and $\tilde{P}_2$:

$$P_1(W) = Q(W)^{-1} Q(W\tilde{\Sigma}_0 W) Q(W)^{-1}$$

and

$$P_2 = (Q(\tilde{\Sigma}_0^{-1}))^{-1}.$$

Here $Q(\cdot)$ is defined by

$$Q(S) = \frac{1}{N} \sum_{k=1}^{N} \psi_k^{\mathrm{T}}(\hat{\theta}_N) S \psi_k(\hat{\theta}_N).$$

It is convenient that the gradients of the prediction $\{\psi_k(\hat{\theta}_N)\}$ are often available anyway as a byproduct from application of the algorithm used for numerical minimization of the identification criterion.

Results on the asymptotic distribution of estimates are significant not only as regards error analysis, but have a bearing on experiment design and questions of identification criterion selection too.

Consider the least squares identification criterion (4.4.24). We might ask, what is the best choice of W in the sense that the variances of linear combinations of components of the estimates are minimized in the limit as $N \to \infty$? Here we are helped by the following lemma.

Lemma 4.4.3

Let Z be an $n \times m$ matrix of second order random variables, and let Σ and W be symmetric $n \times n$ matrices. Suppose that Σ, $E(Z^{\mathrm{T}}WZ)$ and $E(Z^{\mathrm{T}}\Sigma^{-1}Z)$ are positive definite. Then

$$(EZ^{\mathrm{T}}WZ)^{-1}(EZ^{\mathrm{T}}W\Sigma WZ)(EZ^{\mathrm{T}}WZ)^{-1} \geq (EZ^{\mathrm{T}}\Sigma^{-1}Z)^{-1},$$

and equality holds if $W = \Sigma^{-1}$.

PROOF That equality holds when $W=\Sigma^{-1}$ is obvious. Since Σ (and therefore Σ^{-1}) is positive definite, it follows that

$$E[(a^TZ^T+b^TZ^TW\Sigma)\Sigma^{-1}(Za+\Sigma WZb)]\geq 0$$

for all n vectors a and b. Now, for arbitrary b, the left hand side is minimized by $a=-(EZ^T\Sigma^{-1}Z)^{-1}(EZ^TWZ)b$. From this choice of a there results

$$b^T[(EZ^TW\Sigma WZ)-(EZ^TWZ)(EZ^T\Sigma^{-1}Z)^{-1}(EZ^TWZ)]b\geq 0.$$

The lemma is proved by setting $b=(EZ^TWZ)^{-1}c$, for arbitrary c.

□

It is evident from the lemma and equation (4.4.24) that, for the least squares identified criterion (4.4.22), greatest accuracy is achieved, in the sense that the covariance matrix of $N^{1/2}(\theta_N-\theta^*)$ is minimized (with respect to the usual partial ordering of positive semidefinite matrices) in the limit as $N\to\infty$, if the weighting matrix is chosen to be the inverse of the covariance matrix of the innovations, Σ_0. What is more, estimates provided by the identification criterion (4.4.22), which arises in maximum likelihood estimation and for which knowledge of Σ_0 is not required, have accuracy, in the limit as $N\to\infty$, that of least squares estimates corresponding to a best choice of weighting matrix. These properties, somewhat akin to the asymptotic efficiency of maximum likelihood estimates for independent observations, make maximum likelihood estimates very attractive in dynamical system identification.

We conclude this section by indicating why we can expect the limiting covariance matrix of the estimate to be as given in Proposition 4.4.2 in one special case. The case considered is that when the output is scalar valued, $\operatorname{var}\{e_k^2(\theta^*)\}=\sigma_0^2$, and the following least squares identification criterion is adopted:

$$V_N(\theta)=N^{-1}\left\{\sum_{k=1}^{N}\varepsilon_k^2(\theta)\right\}. \tag{4.4.26}$$

For simplicity we take θ to be scalar valued.

Provided $\hat{\theta}_N$ is interior to the parameter constraint set, we have

$$\frac{\partial}{\partial\theta}V_N(\hat{\theta}_N)=0.$$

By the mean value theorem applied to $\frac{\partial}{\partial\theta}V_N(\theta)$ then,

$$\frac{\partial}{\partial\theta}V_N(\theta^*)=0+\frac{\partial^2}{\partial\theta^2}V_N(\gamma_N)(\hat{\theta}_N-\theta^*).$$

Here γ_N is a point on the line segment joining θ^* and $\hat{\theta}_N$. It follows that

$$N^{1/2}(\hat{\theta}_N-\theta^*))^2=\left(\frac{\partial^2}{\partial\theta^2}V_N(\gamma_N)\right)^{-2}N\left(\frac{\partial}{\partial\theta}V_N(\theta^*)\right)^2.$$

In view of this equation, and since $\hat{\theta}_N\to\theta^*$ a.s., it is not implausible that $N^{1/2}(\hat{\theta}_N-\theta^*)$ should have variance, in the limit as $N\to\infty$,

$$\left[\frac{\partial^2}{\partial\theta^2}\bar{V}(\theta^*)\right]^{-2}\lim_{N\to\infty}NE\left(\frac{\partial}{\partial\theta}V_N(\theta^*)\right)^2.$$

However, for the identification criterion here considered (4.4.26),

$$\begin{aligned}\frac{\partial^2}{\partial\theta^2}\bar{V}(\theta^*)&=\frac{\partial^2}{\partial\theta^2}\lim_{N\to\infty}\frac{1}{N}\sum_{k=1}^{N}E(\varepsilon_k^2(\theta^*))\\&=\lim_{N\to\infty}\frac{2}{N}\sum_{k=1}^{N}\left[E\left(\varepsilon_k(\theta^*)\frac{\partial^2}{\partial\theta^2}\varepsilon_k(\theta^*)\right)\right.\\&\qquad\left.+E\left(\left(\frac{\partial}{\partial\theta}\varepsilon_k(\theta^*)\right)^2\right)\right]\end{aligned}$$

(It is assumed that the operations just carried out are valid). The first term under the summation is zero by assumption (4.4.21) and since

$$\frac{\partial^2}{\partial\theta^2}\varepsilon_k(\theta^*)$$

is a function of $\{\varepsilon_j(\theta^*), j<k\}$. By stationarity then,

$$\frac{\partial^2}{\partial\theta^2}\bar{V}(\theta^*)=2E\psi_k^2(\theta^*)$$

where ψ_k is as defined in Proposition 4.4.2. Note also that, for N a positive integer,

$$\begin{aligned}NE\left(\frac{\partial}{\partial\theta}V_N(\theta^*)\right)^2&=4E\left[\frac{1}{N}\sum\sum_{j,k=1}^{N}\varepsilon_j(\theta^*)\frac{\partial}{\partial\theta}\varepsilon_j(\theta^*)\varepsilon_k(\theta^*)\frac{\partial}{\partial\theta}\varepsilon_k(\theta^*)\right]\\&=4E(\varepsilon_k^2(\theta^*))E(\psi_k^2(\theta^*))=4\sigma_0^2E(\psi_k^2(\theta^*)).\end{aligned}$$

Once again we have appealed to assumption (4.4.21) and noted that θ

derivatives of $\varepsilon_k^2(\theta^*)$ are functions of $\{\varepsilon_j(\theta^*), j < k\}$. The variance of $N^{1/2}(\hat{\theta}_N - \theta^*)$, in the limit as $N \to \infty$, is therefore

$$\sigma_0^2/E[\psi_k^2(\theta^*)]$$

in accordance with Proposition 4.4.2.

4.5 Off-line identification algorithms

Prediction error parameter estimation techniques involve minimization of an identification criterion $\theta \to J(\theta)$. In special cases (see Example 4.3.1) a closed-form solution can be found to this minimization problem. For others, typically those involving models with correlated disturbances, we must resort to numerical search procedures to find a minimizing parameter value.

4.5.1 A modified Newton–Raphson algorithm

Suppose that the identification criterion J is twice continuously differentiable, and that the parameter values are unconstrained. The Newton–Raphson algorithm generates a sequence of parameter values $\{\theta^{(k)}\}$, given a starting value $\theta^{(0)}$, by means of the recursion

$$\theta^{(k+1)} = \theta^{(k)} - \left[\frac{\partial^2 J}{\partial \theta^2}(\theta^{(k)})\right]^{-1} \frac{\partial J^{\mathrm{T}}}{\partial \theta}(\theta^{(k)}). \qquad k = 1, 2, \ldots \quad (4.5.1)$$

Here the row vector $\partial J/\partial \theta$ denotes, as usual, the gradient of J. $\partial^2 J/\partial \theta^2$ is the matrix of second partial derivatives $\{\partial^2 J/\partial \theta_i \partial \theta_j\}$ (the Hessian of J). If the Hessian is positive definite at $\bar{\theta}$, a minimizing value of the parameter, then it is known that

$$\limsup_{k \to \infty} \| \theta^{(k)} - \bar{\theta} \| / \| \theta^{(k-1)} - \bar{\theta} \|^2 < \infty$$

for any starting value $\theta^{(0)}$ sufficiently close to $\bar{\theta}$; that is, the method has local 'second-order convergence' properties. A natural variant on this algorithm which can be expected to perform satisfactorily even when $\theta^{(0)}$ is not close to a minimizing value is the following: let $J''(\theta)$ be an approximation to the Hessian at θ which is symmetric and positive definite. The recursion (4.5.1) is replaced by a one-dimensional search in an approximate 'Newton–Raphson' direction, namely

$$\theta^{(k+1)} = \theta^{(k)} - \alpha_k [J''(\theta^{(k)})]^{-1} \frac{\partial J^{\mathrm{T}}}{\partial \theta}(\theta^{(k)}),$$

where α_k is chosen to minimize

$$\alpha \to J\left[\theta^{(k)} - \alpha(J''(\theta^{(k)}))^{-1}\frac{\partial J}{\partial \theta}(\theta^{(k)})\right]$$

over $\alpha \geq 0$. Second-order expansion of J about $\theta^{(k)}$ reveals that, whatever $\theta^{(0)}$, the value of identification criterion is reduced at each step, provided the gradient

$$\frac{\partial J}{\partial \theta}(\theta^{(k)}),$$

is non-zero. Since $J''(\theta^{(k)})$ approximates the Hessian, it is reasonable to suppose that the method will have the desirable second-order convergence properties associated with Newton–Raphson algorithms, and that (4.4.2) will rapidly generate a parameter value at which $\partial J/\partial \theta$ is small.

This scheme is particularly attractive for prediction error methods because, in important cases, the derivatives up to second order are easily calculated and a suitable approximation to the Hessian naturally suggests itself, as we now show.

Take the predictor models $M(\theta), \theta \in \mathbb{R}^q$ of Section 4.2,

$$y_k = f_k(y^{k-1}; u^{k-1}, \theta) + e_k \qquad k = 0, 1, \ldots$$

in which the f_k are given functions and the e_k are zero-mean independent random variables. We assume that the f_k are twice continuously differentiable in their arguments.

We shall consider identification criteria which arise, respectively, in least squares and maximum likelihood parameter estimation, namely

$$J_1(\theta) = \operatorname{trace}\{WD(\theta)\}$$

in which W is a given positive definite matrix, and

$$J_2(\theta) = \log\det\{D(\theta)\}.$$

Here

$$D(\theta) = \frac{1}{N}\sum_k \varepsilon_k(\theta)\varepsilon_k^{\mathrm{T}}(\theta), \qquad \text{in which}$$

$$\varepsilon_k(\theta) = y_k - f_k(y^{k-1}; u^{k-1}, \theta).$$

Straightforward calculations give the first partial derivatives of J_1:

$$\frac{\partial J_1}{\partial \theta_i}(\theta) = \left(\frac{2}{N}\right)\sum_{k=1}^{N} \varepsilon_k^{\mathrm{T}}(\theta) W \frac{\partial \varepsilon_k}{\partial \theta_i}(\theta) \qquad i = 1, \ldots, q.$$

As for the first partial derivatives of J_2 we have

$$\frac{\partial J_2}{\partial \theta_i}(\theta) = \operatorname{trace}\left\{\frac{\mathrm{d}}{\mathrm{d}D}[\log\det D(\theta)]\frac{\partial D}{\partial \theta_i}(\theta)\right\}$$

$$= \operatorname{trace}\left\{D^{-1}(\theta)\frac{1}{N}\sum_k\left(\frac{\partial \varepsilon_k}{\partial \theta_i}(\theta)\varepsilon_k^{\mathrm{T}}(\theta) + \varepsilon_k(\theta)\frac{\partial \varepsilon_k^{\mathrm{T}}}{\partial \theta_i}(\theta)\right)\right\}$$

by the matrix calculus identity, Lemma D.4.4, of Appendix D,

$$= \frac{2}{N}\sum_{k=1}^{N} \varepsilon_k^{\mathrm{T}}(\theta)D^{-1}(\theta)\frac{\partial \varepsilon_k^{\mathrm{T}}}{\partial \theta_i}(\theta).$$

Also,

$$\frac{\partial^2 J_1}{\partial \theta_i \partial \theta_j} = \frac{2}{N}\left[\sum_{k=1}^{N}\frac{\partial \varepsilon_k^{\mathrm{T}}}{\partial \theta_i}(\theta)W\frac{\partial \varepsilon_k}{\partial \theta_j}(\theta) + \sum_{k=1}^{N}\varepsilon_k^{\mathrm{T}}(\theta)W\frac{\partial^2 \varepsilon_k(\theta)}{\partial \theta_i \partial \theta_j}\right]$$

and

$$\frac{\partial^2 J_2}{\partial \theta_i \partial \theta_j} = \frac{2}{N}\sum_{k=1}^{N}\frac{\partial \varepsilon_k^{\mathrm{T}}}{\partial \theta_i}(\theta)D(\theta)^{-1}\frac{\partial \varepsilon_k}{\partial \theta_j}(\theta) + \frac{2}{N}\sum_{k=1}^{N}\varepsilon_k^{\mathrm{T}}(\theta)D(\theta)^{-1}\frac{\partial^2 \varepsilon_k}{\partial \theta_i \partial \theta_j}(\theta)$$

$$-\frac{2}{N^2}\sum_{k=1}^{N}\sum_{l=1}^{N}\varepsilon_k^{\mathrm{T}}(\theta)D^{-1}(\theta)\left[\varepsilon_l(\theta)\frac{\partial \varepsilon_l^{\mathrm{T}}}{\partial \theta_i}(\theta) + \frac{\partial \varepsilon_l}{\partial \theta_i}(\theta)\varepsilon_l^{\mathrm{T}}(\theta)\right]$$

$$\cdot D^{-1}(\theta)\frac{\partial \varepsilon_k}{\partial \theta_j}(\theta).$$

We have used Lemma D.4.2. of Appendix D in calculating $\partial^2 J_2/\partial \theta_i \partial \theta_j$. It is assumed here that $D(\theta)$ is non-singular. The first and second partial derivatives of $\varepsilon_k(\theta)$ which appear in these expressions can be calculated from the formulae

$$\frac{\partial \varepsilon_k}{\partial \theta_i}(\theta) = -\frac{\partial f_k}{\partial \theta_i}(y^{k-1}; u^{k-1}, \theta)$$

and

$$\frac{\partial^2 \varepsilon_k}{\partial \theta_i \partial \theta_j}(\theta) = -\frac{\partial^2 f_k}{\partial \theta_i \partial \theta_j}(y^{k-1}; u^{k-1}, \theta).$$

Suitable approximations $J_1''(\theta)$ and $J_2''(\theta)$ to the Hessians in these cases are given by

$$\{J_1''(\theta)\}_{ij} = \frac{2}{N}\sum_{k=1}^{N}\frac{\partial \varepsilon_k^{\mathrm{T}}}{\partial \theta_i}(\theta)W\frac{\partial \varepsilon_k}{\partial \theta_j}(\theta)$$

and

$$\{J_2''(\theta)\}_{ij} = \frac{2}{N}\sum_{k=1}^{N}\frac{\partial\varepsilon_k^{\mathrm{T}}}{\partial\theta_i}(\theta)D(\theta)^{-1}\frac{\partial\varepsilon_k}{\partial\theta_j}(\theta).$$

These choices of $J_1''(\theta)$ and $J_2''(\theta)$ are positive semi-definite, and can be made positive definite by addition of a term $\alpha I, \alpha > 0$, if necessary. Notice that, in replacing the derivatives by their approximations, we have ignored the terms

$$\frac{2}{N}\sum_{k=1}^{N}\varepsilon_k^{\mathrm{T}}(\theta)W\frac{\partial^2\varepsilon_k}{\partial\theta_i\partial\theta_j}(\theta)$$

$$\frac{2}{N}\sum_{k=1}^{N}\varepsilon_k^{\mathrm{T}}(\theta)D(\theta)^{-1}\frac{\partial^2\varepsilon_k}{\partial\theta_i\partial\theta_j}(\theta)$$

and

$$\frac{2}{N^2}\sum_{k=1}^{N}\sum_{l=1}^{N}\varepsilon_k^{\mathrm{T}}(\theta)D^{-1}(\theta)\left[\varepsilon_l(\theta)\frac{\partial\varepsilon_l^{\mathrm{T}}}{\partial\theta_i}(\theta)+\frac{\partial\varepsilon_l}{\partial\theta_i}(\theta)\varepsilon_l^{\mathrm{T}}(\theta)\right]D^{-1}(\theta)\frac{\partial\varepsilon_k}{\partial\theta_j}(\theta).$$

To justify these approximations, let us suppose that the data is actually generated by the predictor model when the parameter θ takes value θ^*. Under mild assumptions on f and the noise $\{e_k\}$, it is possible to show that the omitted terms all tend to zero, as $N \to \infty$, almost surely, when $\theta = \theta^*$. It is reasonable to assume then that the terms will be small when N is large and $\theta^{(k)}$ is close to θ^*. Proof involves application of the Ergodic theorem, Theorem 1.1.15 and use of the facts that, for $k = 1, 2, \ldots, \varepsilon_k(\theta^*) = e_k$, whence for $l < k, \varepsilon_k(\theta)$ is independent of $\varepsilon_l(\theta^*)$, and for $l < k, \varepsilon_k(\theta^*)$ is independent of

$$\frac{\partial\varepsilon_l}{\partial\theta_i}(\theta^*) \quad \text{and} \quad \frac{\partial^2\varepsilon_l}{\partial\theta_i\partial\theta_j}(\theta^*).$$

4.5.2 The generalized least squares algorithm

The generalized least squares algorithm is a more specialized algorithm for identifying parameters in certain models involving correlated disturbances. The algorithm is more widely applicable than this, but for the sake of simplicity, we describe it in connection with the class of scalar models

$$A(z^{-1})y_k = B(z^{-1})u_k + \xi_k, \qquad k = 0, \ldots, N$$

in which the disturbances ξ_k are generated by the equations

$$F(z^{-1})\xi_k = e_k, \qquad k = 0, \ldots, N.$$

We take zero initial conditions ($y_k = 0, u_k = 0, \xi_k = 0$ for $k < 0$). Here the polynomials $A(\sigma), B(\sigma)$ and $F(\sigma)$ take the form

$$A(\sigma) = 1 + a_1\sigma^{-1} + \cdots + a_n\sigma^{-n}$$
$$B(\sigma) = b_1\sigma^{-1} + \cdots + b_n\sigma^{-n}$$

and

$$F(\sigma) = 1 + f_1\sigma^{-1} + \cdots + f_n\sigma^{-n}.$$

The e_k are independent random variables. The coefficients $a_1, \ldots, a_n, b_1, \ldots, b_n, f_1, \ldots, f_n$ make up the entries of the unknown vector parameter θ to be identified.

A parameter value is sought which minimizes the least squares criterion

$$\frac{1}{N}\sum_{k=1}^{N} \varepsilon_k^2(\theta) \tag{4.5.2}$$

in which the $\varepsilon_k(\theta)$ are the prediction errors corresponding to the parameter value θ.

Calculations similar to those performed in connection with the ARMAX model equations (4.4.7) give the prediction errors $\varepsilon_k(\theta)$ as

$$\varepsilon_k(\theta) = A(z^{-1})F(z^{-1})y_k - B(z^{-1})F(z^{-1})u_k. \tag{4.5.3}$$

The parameter estimation problem is therefore that of minimizing (4.5.2) over θ when $\varepsilon_k(\theta)$ is given by (4.5.3).

A minimization scheme is suggested by the observations that, if either the f_i or both a_i and b_i are fixed, then the prediction errors are linear functions of the remaining free parameter coefficients, and the minimization problem over these components can be solved in closed form.

If the f_i are fixed, the prediction errors $\varepsilon_k(\theta), k = 1, 2, \ldots$, are given by

$$\varepsilon_k(\theta) = A(z^{-1})\bar{y}_k - B(z^{-1})\bar{u}_k, \qquad k = 1, 2, \ldots, \tag{4.5.4}$$

in which the $\bar{y}_k$ and $\bar{u}_k$ are calculated from the data and the f_i according to

$$\bar{y}_k = F(z^{-1})u_k, \bar{u}_k = F(z^{-1})u_k, \qquad k = 0, 1, \ldots$$

On the other hand, if the a_i and b_i are fixed then

$$\varepsilon_k(\theta) = F(z^{-1})\bar{\varepsilon}_k \tag{4.5.5}$$

in which the $\bar{\varepsilon}_k$ are calculated from the data and the a_i and b_i according to

$$\bar{\varepsilon}_k = A(z^{-1})y_k - B(z^{-1})u_k.$$

The closed-form solutions to the problems of minimizing (4.5.2) when $\varepsilon_k(\theta)$ is defined by (4.5.4) and $\bar{y}_k, \bar{u}_k$ are known, and when $\varepsilon_k(\theta)$ is defined by (4.5.5) and the ε_k are known, are provided by the least squares theory of Section 4.4 (under the assumption that the normal equations in question have a unique solution).

It is natural then to minimize the criterion (4.5.2) alternately over the f_i and then over the a_i and b_i. This idea is the basis of the generalized least squares algorithm.

The polynomial $F_0(z^{-1})$ with leading coefficient unity, is chosen arbitrarily. (A common choice is $F_0(z^{-1}) = 1$). Sequences of polynomials $\{F_j(\sigma)\}$, $\{A_j(\sigma)\}$, $\{(B_j(\sigma)\}$ with coefficients the unknown parameters, are then generated recursively as follows. For $j = 1, 2, \ldots,$

(a) The polynomials $A_j(\sigma)$ and $B_j(\sigma)$ are chosen to have coefficients which minimize (4.5.2) where the $\varepsilon_k(\theta)$ are given by (4.5.4) and where

$$\bar{y}_k = F_{j-1}(z^{-1})y_k, \bar{u}_k = F_{j-1}(z^{-1})u_k, \qquad k \geq 0$$
$$\bar{y}_k = 0, \quad \bar{u}_k = 0 \qquad k < 0$$

and,

(b) The polynomial $F_j(\sigma)$ is chosen to have coefficients which minimize (4.5.2), where $\varepsilon_k(\theta)$ is now given by (4.5.5) and where

$$\bar{\varepsilon}_k = A_j(z^{-1})y_k - B_j(z^{-1})u_k, \qquad k \geq 0$$
$$\bar{\varepsilon}_k = 0, \qquad k < 0.$$

The recursion is terminated when the change in parameter values between iterations becomes insignificant; the coefficients of the current $A_j(\sigma)$, $B_j(\sigma)$, $F_j(\sigma)$ provide the parameter estimate.

4.6 Algorithms for on-line parameter estimation

In many applications, parameter estimates are required on-line, in the sense that we must obtain estimates based on data available at time N before new data comes in at time $N + 1$. This is the case, for example, when adaptive control schemes are implemented, because then the control strategy to be applied at a particular time depends on the parameter estimates at that time.

The parameter estimation algorithms of Section 4.5 are often infeasible for such applications because the calculations involved cannot be completed sufficiently quickly. The algorithms we now consider are devised to overcome this difficulty; they update the parameter estimates to take account of new data in a computation-saving manner. For the sake of simplicity, attention is restricted to algorithms for single input, single output systems.

4.6.1 The recursive least squares algorithm

We consider scalar ARMAX models of the form

$$y_k + a_1 y_{k-1} + \cdots + a_n y_{k-n} = b_1 u_{k-1} + \cdots + b_m u_{k-m} + e_k, \quad k = 0, 1, \ldots$$

with zero initial conditions ($y_k = 0, u_k = 0$, for $k < 0$).

Here $e_0, e_1, \ldots$ are zero-mean independent random variables. The vector of unknown parameters θ is $[a_1, \ldots, a_n, b_1, \ldots, b_m]^{\mathrm{T}}$.

Let $\hat{\theta}_N$ be the estimate based on data y^N, u^{N-1} up to time N obtained by minimizing the least squares criterion

$$V_N(\theta) = \frac{1}{N} \sum_{k=1}^{N} \varepsilon_k^2(\theta)$$

in which the $\varepsilon_k(\theta)$ are the prediction errors. These, we recall, are obtained from the equation

$$\varepsilon_k(\theta) = y_k + a_1 y_{k-1} + \cdots + a_n y_{k-n} - b_1 u_{k-1} - \cdots - b_m u_{k-m}, \quad k = 1, 2, \ldots$$

We have seen in Section 4.4 that $\hat{\theta}_N$ is given by

$$\hat{\theta}_N = (X_N^{\mathrm{T}} X_N)^{-1} X_N^{\mathrm{T}} Y_N. \tag{4.6.1}$$

Here

$$X_N = \mathrm{col}[x_1^{\mathrm{T}}, \ldots, x_N^{\mathrm{T}}], \qquad Y_N = [y_1, \ldots, y_N]^{\mathrm{T}}$$

and

$$x_k = [-y_{k-1}, -y_{k-2}, \ldots, -y_{k-n}, u_{k-1}, u_{k-2}, \ldots, u_{k-m}]^{\mathrm{T}}.$$

(It is assumed that $X_N^{\mathrm{T}} X_N$ is non-singular).

At time $N+1$, y_{N+1} and u_N (and hence x_{N+1}) become known. The least squares estimate which takes account of this new information is

$$\hat{\theta}_{N+1} = (X_{N+1}^{\mathrm{T}} X_{N+1})^{-1} X_{N+1}^{\mathrm{T}} Y_{N+1}.$$

We shall show that $\hat{\theta}_{N+1}$ can be determined from $\hat{\theta}_N$, u_N and y_{N+1} given the matrix P_N,

$$P_N = (X_N^T X_N)^{-1},$$

by means of simple calculations.

But we first take note of a matrix identity known as the *matrix inversion lemma.*

Lemma 4.6.1

Let A be an $n \times n$ matrix and let b, c be n-vectors. Suppose that A and $A + bc^T$ are non-singular and that $1 + c^T A^{-1} b \neq 0$, then

$$(A + bc^T)^{-1} = A^{-1} - (1 + c^T A^{-1} b)^{-1} A^{-1} bc^T A^{-1}.$$

PROOF We have merely to check that

$$(A^{-1} - (1 + c^T A^{-1} b)^{-1} A^{-1} bc^T A^{-1})(A + bc^T)$$

is the identity matrix. But this matrix is expressible as

$$\begin{aligned} & I + A^{-1}bc^T - (1 - c^T A^{-1} b)^{-1}(A^{-1}bc^T + A^{-1}b(c^T A^{-1} b)c^T) \\ &= I + A^{-1}bc^T - (1 + c^T A^{-1} b)^{-1}(1 + c^T A^{-1} b)A^{-1}bc^T = I \qquad \square \end{aligned}$$

Bearing in mind our assumption that $X_N^T X_N$ is non-singular, we readily confirm that the hypotheses are satisfied when we set $A = X_N^T X_N$ and $b = c = x_{n+1}$. It follows that

$$\begin{aligned} P_{N+1} &= (X_{N+1}^T X_{N+1})^{-1} = ((X_N^T X_N) + x_{N+1} x_{N+1}^T)^{-1} \\ &= [I - (1 + x_{N+1}^T P_N x_{N+1})^{-1} P_N x_{N+1} x_{N+1}^T] P_N. \end{aligned}$$

We deduce that

$$\begin{aligned} \hat{\theta}_{N+1} &= P_{N+1} X_{N+1}^T y_{N+1} = P_{N+1}(X_N^T y_N + x_{N+1} y_{N+1}) \\ &= [I - (1 + x_{N+1}^T P_N x_{N+1})^{-1} P_N x_{N+1} x_{N+1}^T] \\ &\quad \cdot P_N (X_N^T Y_N + x_{N+1} y_{N+1}) \\ &= P_N X_N^T Y_N + (1 + x_{N+1}^T P_N x_{N+1})^{-1} [(1 + x_{N+1}^T P_N x_{N+1}) \\ &\quad \cdot P_N x_{N+1} y_{N+1} - P_N x_{N+1} x_{N+1}^T P_N X_N^T Y_N \\ &\quad - P_N x_{N+1} x_{N+1}^T P_N x_{N+1} y_{N+1}] \\ &= P_N X_N^T Y_N + (1 + x_{N+1}^T P_N x_{N+1})^{-1} P_N x_{N+1} \\ &\quad \cdot [y_{N+1} - x_{N+1}^T P_N X_N^T Y_N] \\ &= \hat{\theta}_N + (1 + x_{N+1}^T P_N x_{N+1})^{-1} P_N x_{N+1} [y_{N+1} - x_{N+1}^T \hat{\theta}_N], \end{aligned}$$

by Lemma 4.6.1.

These relationships can be expressed

$$\hat{\theta}_{N+1} = \hat{\theta}_N + K_{N+1}\varepsilon_{N+1}(\hat{\theta}_N) \tag{4.6.2}$$

in which $\varepsilon_{N+1}(\hat{\theta}_N)$ is the prediction error associated with $\hat{\theta}_N$:

$$\varepsilon_{N+1}(\hat{\theta}_N) = y_{N+1} - x_{N+1}^{\mathrm{T}}\hat{\theta}_N \tag{4.6.3}$$

and

$$K_{N+1} = (1 + x_{N+1}^{\mathrm{T}}P_N x_N)^{-1}P_N x_{N+1}, \tag{4.6.4}$$

together with

$$P_{N+1} = [I - (1 + x_{N+1}^{\mathrm{T}}P_N x_{N+1})^{-1}P_N x_{N+1}x_{N+1}^{\mathrm{T}}]P_N. \tag{4.6.5}$$

Equations (4.6.2)–(4.6.5) determine thc least squares estimate $\hat{\theta}_{N+1}$, given data up to time $N+1$, as a function of $\hat{\theta}_N$ the least squares estimate given data up to time N, the new data y_{N+1}, u_N and the matrix P_N, and generate the matrix P_{N+1} in preparation for the next updating of the parameter estimate. These equations define the *recursive least squares algorithm*. The starting value for the recursion (4.6.5) is $P_0 = (X_0^{\mathrm{T}}X_0)^{-1}$. We see that P_N is updated by means of a Riccati equation.

Each step of the algorithm involves only matrix multiplication; inversion of the matrix P_{N+1}, required for the corresponding non-recursive algorithm, is avoided.

We see that the least squares procedure for ARMAX models with uncorrelated disturbances can be expressed in a manner suitable for on-line use. The on-line algorithm gives the same sequence of estimates as we would obtain by applying the procedure each time new data comes in. Off-line parameter estimation algorithms for models which permit correlated disturbances can often be adapted to give on-line algorithms too, provided certain approximations are made. A typical on-line algorithm which arises in this way has the following characteristics: the estimate $\hat{\theta}_N$ based on data up to time N is used as an initial value for one iteration of the associated off-line algorithm based on data up to time N, and approximations are introduced by means of which the gradients of the identification criterion at time $N+1$, and other useful variables, can be simply calculated from the gradients of the identification criterion at time N. It is assumed that, although the approximations will result in poorer estimates for each N, their effect will become small for large N. A number of algorithms of this kind are now described.

4.6.2 A recursive Newton–Raphson algorithm

We recall the modified Newton–Raphson algorithm of Section 4.5 as applied to scalar predictor models when a least-squares identification criterion

$$J_N(\theta) = \frac{1}{N} \sum_{i=1}^{N} \varepsilon_k^2(\theta)$$

is adopted. Here the $\varepsilon_k(\theta)$ are the prediction errors

$$\varepsilon_k(\theta) = y_k - f_k(y^{k-1}, u^{k-1}, \theta), \qquad k = 1, 2, \ldots, N \tag{4.6.6}$$

associated with the model $M(\theta)$.

Given data y^N, u^{N-1} and an estimate, θ_{old}, the algorithm supplies a revised estimate, θ_{new}, according to the rule

$$\theta_{\text{new}} = \theta_{\text{old}} - \alpha H_N(\theta_{\text{old}})^{-1} \frac{\partial J_N^{\text{T}}}{\partial \theta}(\theta_{\text{old}}). \tag{4.6.7}$$

Here

$$\frac{\partial J_N}{\partial \theta}(\theta),$$

the gradient of J_N with respect to θ, is

$$\frac{\partial J_N}{\partial \theta}(\theta) = \frac{2}{N} \sum_{i=1}^{N} \varepsilon_k(\theta)\psi_k(\theta).$$

In this expression the row vectors $\{\psi_k(\theta)\}$ are defined by the equations

$$\psi_k(\theta) = -\frac{\partial f_k}{\partial \theta}(y^{k-1}, u^{k-1}, \theta), \qquad k = 1, \ldots, N. \tag{4.6.8}$$

$H_N(\theta)$, an approximation to the Hessian of J_N at θ, is given by

$$H_N(\theta) = \frac{2}{N} \sum_{k=1}^{N} \psi_k^{\text{T}}(\theta)\psi_k(\theta)$$

and α is a suitable positive number. (It is assumed that θ_{new} given by (4.6.7) will lie in the parameter constraint set.)

Suppose that a parameter estimate θ_j has been calculated on the basis of data y_j, u_{j-1} for $j = 1, \ldots, N-1$. Further data y_N, u_{N-1} now becomes available. Formula (4.6.7) suggests that we choose a new estimate θ_N to be

$$\theta_N = \theta_{N-1} - \alpha_N H_N(\theta_{N-1})^{-1} \frac{\partial J_N^{\mathrm{T}}}{\partial \theta}(\theta_{N-1}). \tag{4.6.9}$$

Here $\{\alpha_N\}$ is a suitable sequence of positive numbers.

This expression must be approximated if it is to be of use for on-line calculations. One characteristic which makes it unsuitable is that it involves the prediction errors, the $\varepsilon_k(\theta_{N-1})$, and their gradients the $\psi_k(\theta_{N-1})$. These processes are obtained by solving the equations (4.6.6) and (4.6.8) all of which depend on θ_{N-1} and, except in special cases, knowledge of prediction errors and their gradients for previously considered parameter values does not simplify the calculations. Such simplification is achieved, however, if $\varepsilon_k(\theta_{N-1})$ and $\psi_k(\theta_{N-1})$, for $k = 1, \ldots, N$, are approximated by the vectors, written ε_k and ψ_k, which result when, for $k = 1, 2, \ldots, \theta_{N-1}$ is replaced by the currently available estimates θ_{k-1} in both of the recursive equations (4.6.6), (4.6.8). The column vectors $\{\varepsilon_k\}$ and row vectors $\{\psi_k\}$ are then defined by

$$\varepsilon_k = y_k - f_k(y^{k-1}, u^{k-1}, \theta_{k-1}), \qquad k = 1, 2, \ldots \tag{4.6.10}$$

and

$$\psi_k = -\frac{\partial f_k}{\partial \theta}(y^{k-1}, u^{k-1}, \theta_{k-1}), \qquad k = 1, 2, \ldots \tag{4.6.11}$$

Introduction of $\{\psi_k\}$ provides us with a convenient new approximation, $2R_N$, to the Hessian, where

$$R_N := \frac{1}{N} \sum_{k=1}^{N} \psi_k^{\mathrm{T}} \psi_k.$$

Notice that application of the matrix inversion lemma (Lemma 4.6.1) as in the derivation of the recursive least squares algorithm, results in recursive equations for $P_N := (NR_N)^{-1}$, namely

$$P_N = [I - (1 + \psi_N P_{N-1} \psi_N^{\mathrm{T}})^{-1} P_{N-1} \psi_N^{\mathrm{T}} \psi_N] P_{N-1}.$$

Consider now approximation of the gradient

$$\frac{\partial J_N}{\partial \theta}(\theta_{N-1}).$$

If θ_{N-1} actually minimized, J_{N-1} then we would have

$$\frac{\partial J_{N-1}}{\partial \theta}(\theta_{N-1}) = 0$$

and it would follow that

$$\frac{\partial J_N}{\partial \theta}(\theta_{N-1}) = \frac{2}{N}\sum_{k=1}^{N} \varepsilon_k(\theta_{N-1})\psi_k(\theta_{N-1})$$
$$= \frac{2}{N}\varepsilon_N(\theta_{N-1})\psi_N(\theta_{N-1}).$$

In view of this, even if θ_{N-1} is not minimizing, $(2/N)\varepsilon_N(\theta_{N-1})\psi_N(\theta_{N-1})$ is a natural choice of approximation for

$$\frac{\partial J_N}{\partial \theta}(\theta_{N-1}).$$

Furthermore $(2/N)\varepsilon_N(\theta_{N-1})\psi_N(\theta_{N-1})$ itself can be approximated by $(2/N)\varepsilon_N\psi_N$.

These approximations substituted in place of $H_N(\theta_{N-1})$ and

$$\frac{\partial J_N}{\partial \theta}(\theta_{N-1})$$

in (4.6.9) lead to the updating formula

$$\theta_N = \theta_{N-1} - (\alpha_N/N)R_N^{-1}\psi_N^{\mathrm{T}}\varepsilon_N.$$

Expressed in terms of P_N, the formula becomes

$$\theta_N = \theta_{N-1} - \alpha_N P_N \psi_N^{\mathrm{T}}\varepsilon_N \tag{4.6.12}$$

in which, we recall,

$$P_N = [I - (1 + \psi_N P_{N-1}\psi_N^{\mathrm{T}})^{-1}P_{N-1}\psi_N^{\mathrm{T}}\psi_N]P_{N-1}. \tag{4.6.13}$$

Equations (4.6.10)–(4.6.13) define the *recursive Newton–Raphson algorithm*. Solution of the recursive equations requires a suitable starting value P_0, a positive definite symmetric matrix, and an initial estimate θ_0 of the vector parameter.

For $\{\alpha_i\}$, P_0 and θ_0 appropriately chosen, and for the models considered in the derivation of the recursive least squares algorithm, the recursive Newton–Raphson algorithm is in fact the same as the recursive least squares algorithm. In this case the approximations coincide with the true values of the variables concerned.

4.6.3 The recursive generalized least squares algorithm

Consider now scalar stochastic dynamical models

$$\left.\begin{aligned} A(z^{-1})y_k &= B(z^{-1})u_k + \xi_k \\ F(z^{-1})\xi_k &= e_k \end{aligned}\right\} \qquad k = 0, 1, \ldots$$

with zero initial data ($y_k = 0$, $u_k = 0$, $\xi_k = 0$, for $k < 0$).

Here $\{e_k\}$ is a sequence of independent, zero-mean random variables. The polynomials $A(\sigma)$, $B(\sigma)$ and $F(\sigma)$ are of the form

$$A(\sigma) = 1 + a_1\sigma + \cdots + a_n\sigma^n,$$
$$B(\sigma) = b_1\sigma + \cdots + b_n\sigma^n,$$
$$F(\sigma) = 1 + f_1\sigma + \cdots + f_n\sigma^n.$$

We take $a_1,\ldots,a_n$, $b_1,\ldots,b_n$, $f_1,\ldots,f_n$ as the unknown parameters. For convenience we divide them up to form two vector parameters $\psi = (a_1,\ldots,a_n, b_1,\ldots,b_n)^T$ and $\gamma = (f_1,\ldots,f_n)^T$.

Let us recall the generalized least squares algorithm for determination of estimates given data y^N, u^{N-1} (see Section 4.5). The underlying idea is that, if either ψ or γ is fixed then the value of the other parameter, γ or ψ, which minimizes the mean square of the residuals

$$J_N(\psi,\gamma) = \frac{1}{N}\sum_{i=1}^{N} \varepsilon^2(\psi,\gamma)$$

can be obtained by solution of the normal equations for a simple least squares problem involving uncorrelated disturbances; after j iterations of the algorithm, when estimates ψ_j, γ_j have been determined, the next iteration yields estimates ψ_{j+1}, γ_{j+1} where ψ_{j+1} minimizes $\psi \to J_N(\psi,\gamma_j)$ and γ_{j+1} minimizes $\gamma \to J_N(\psi_{j+1},\gamma)$.

One way in which the algorithm can be modified for on-line use is to couple it with the recursive least squares algorithm and to introduce certain approximations. Suppose that estimates ψ_N, γ_N based on data y^N, u^{N-1} are available and new data y_{N+1}, u_N comes in. One iteration of the generalized least squares algorithm, applied with ψ_N, γ_N as initial values, gives estimates $\bar{\psi}$, $\bar{\gamma}$ which minimize $\psi \to J_{N+1}(\psi,\gamma_N)$, $\gamma \to J_{N+1}(\bar{\psi},\gamma)$. In order to apply the recursive least squares algorithm to determine $\bar{\psi}$, $\bar{\gamma}$ we require the solution of two Riccati equations. These equations, determined at each step by the most recent estimates γ_N, $\bar{\psi}$ of the parameters γ and ψ must be solved over the time interval 1 to $N+1$. The computational burden of updating the parameters is reduced if we take as new estimates ψ_{N+1}, γ_{N+1}, approximations to $\bar{\psi}$, $\bar{\gamma}$, instead of $\bar{\psi}$, $\bar{\gamma}$ themselves, calculated from solutions to two approximating Riccati equations, determined at each time step j, not by the most recent parameter estimates γ_N, ψ_{N+1}, but by the estimates γ_j, ψ_{j+1} available at time j; in order to calculate the parameter estimates ψ_{N+1}, γ_{N+1} we need only advance the solutions to the approximating Riccati equations by one step, since the solutions at time N are available from calculation of ψ_N, γ_N.

The recursive generalized least squares algorithm updates estimates in this way. A more detailed description of the algorithm is as follows: Vectors $\hat{\psi}_0$, $\hat{\gamma}_0$ and positive definite matrices P_0, Q_0 are supplied as starting values. $\hat{\psi}_N$, $\hat{\gamma}_N$, P_N, Q_N, $N = 1, 2, \ldots$ are then calculated by the formulae

$$\begin{aligned}\psi_{N+1} &= \hat{\psi}_N + K_{N+1}(y_{N+1}(\hat{\gamma}_N) - x_{N+1}^{\mathrm{T}}\hat{\psi}_N)\\ K_{N+1} &= (1 + x_{N+1}^{\mathrm{T}}P_N x_{N+1})^{-1}P_N x_{N+1}\\ P_{N+1} &= [I - (1 + x_{N+1}^{\mathrm{T}}P_N x_{N+1})^{-1}P_N x_{N+1}x_{N+1}^{\mathrm{T}}]P_N\end{aligned}$$

in which

$$\begin{aligned}y_k(\hat{\gamma}_N) &= F_N(z^{-1})y_k, \qquad k = N - n + 1, \ldots, N + 1\\ u_k(\hat{\gamma}_N) &= F_N(z^{-1})u_k, \qquad k = N - n + 1, \ldots, N + 1\\ x_{N+1} &= [-y_N(\hat{\gamma}_N), -y_{N-1}(\hat{\gamma}_N), \ldots, -y_{N-n+1}(\hat{\gamma}_N),\\ &\qquad u_N(\hat{\gamma}_N), \ldots, u_{N-n+1}(\hat{\gamma}_N)]^{\mathrm{T}}\end{aligned}$$

and $F_N(\sigma)$ is the polynomial with coefficients entries of $\hat{\gamma}_N$ together with

$$\begin{aligned}\hat{\gamma}_{N+1} &= \hat{\gamma}_N + L_{N+1}(\eta_{N+1}\hat{\psi}_{N+1}) - \xi_{N+1}^{\mathrm{T}}\hat{\gamma}_N)\\ L_{N+1} &= (1 + \xi_{N+1}^{\mathrm{T}}Q_N\xi_{N+1})^{-1}Q_N\xi_{N+1}\\ Q_{N+1} &= [I - (1 + \xi_{N+1}^{\mathrm{T}}Q_N\xi_{N+1})^{-1}Q_N\xi_{N+1}\xi_{N+1}^{\mathrm{T}}]Q_N\end{aligned}$$

in which

$$\begin{aligned}\eta_k(\hat{\psi}_{N+1}) &= A_{N+1}(z^{-1})y_k - B_{N+1}(z^{-1})u_k, \quad k = N - n + 1, \ldots, N + 1\\ \xi_{N+1} &= [-\eta_N(\hat{\psi}_{N+1}), \ldots, -\eta_{N-n+1}(\hat{\psi}_{N+1})]^{\mathrm{T}}\end{aligned}$$

and $A_{N+1}(\sigma)$, $B_{N+1}(\sigma)$ are the polynomials with coefficients entries of $\hat{\psi}_{N+1}$.

4.6.4 The extended matrix algorithm

This is another algorithm which permits 'correlated disturbances'. We describe the form it takes for scalar ARMAX models:

$$A(z^{-1})y_k = B(z^{-1})u_k + C(z^{-1})e_k, \qquad k = 0, 1, \ldots$$

with zero initial data ($y_k = 0$, $u_k = 0$, $e_k = 0$, for $k < 0$).

Here e_k is a sequence of zero-mean, independent random variables. The polynomials $A(\sigma)$, $B(\sigma)$ and $C(\sigma)$ are of the form:

$$A(\sigma) = 1 + a_1\sigma + \cdots + a_n\sigma^n, \qquad B(\sigma) = b_1\sigma + \cdots + b_n\sigma^n,$$
$$C(\sigma) = 1 + c_1\sigma + \cdots + c_n\sigma^n.$$

and the coefficients $a_1, \ldots, a_n, b_1, \ldots, b_n, c_1, \ldots, c_n$ make up the vector θ of unknown parameters.

Suppose that at time k we had knowledge of past disturbances $e_j, j < k$. Then we could write the system equations

$$y_k = x_k^{\mathrm{T}} \theta + e_k \tag{4.6.14}$$

in which

$$x_k := (-y_{k-1}, \ldots, -y_{k-n}, u_{k-1}, \ldots, u_{k-n}, e_{k-1}, \ldots, e_{k-n})^{\mathrm{T}},$$

was a known vector. Past disturbances merely have the role of additional inputs to the system in this hypothetical situation, and the least squares parameter estimation problem takes a form to which the recursive least squares algorithm is applicable.

Of course past disturbances are not known and it is therefore necessary to estimate them. An estimate $\hat{\varepsilon}_k$ of the disturbance e_k based on past parameter estimates $\hat{\theta}_j$, $j < k$, and data available up to time k together with an estimate $\hat{x}_k$ of the regression vector x_k are easily obtained by solving the recursive equations

$$\hat{\varepsilon}_j = y_j - \hat{x}_j^{\mathrm{T}} \hat{\theta}_{j-1}$$

and

$$\hat{x}_j = (-y_{j-1}, \ldots, -y_{j-n}, u_{j-1}, \ldots, u_{j-n}, \hat{\varepsilon}_{j-1}, \ldots, \hat{\varepsilon}_{j-n})^{\mathrm{T}} \qquad j = 1, 2, \ldots$$

The extended matrix algorithm is the algorithm which results from applying the recursive least squares algorithm for the model equation (4.6.14) when $\hat{x}_k$ replaces x_k.

A vector $\hat{\theta}_0$ and a positive definite, symmetric matrix P_0 is supplied. $\hat{\theta}_N$, P_N, $N = 1, 2, \ldots$ are then calculated according to the formulae

$$\begin{aligned}
\hat{\theta}_{N+1} &= \hat{\theta}_N + K_{N+1} \hat{\varepsilon}_{N+1} \\
K_{N+1} &= (1 + x_{N+1}^{\mathrm{T}} P_N x_{N+1})^{-1} P_N x_{N+1} \\
P_{N+1} &= [I - (1 + \hat{x}_{N+1}^{\mathrm{T}} P_N \hat{x}_{N+1})^{-1} P_N \hat{x}_{N+1} \hat{x}_{N+1}^{\mathrm{T}}] P_N
\end{aligned}$$

in which

$$\hat{\varepsilon}_k = y_k - \hat{x}_k^{\mathrm{T}} \hat{\theta}_{N+1}, \qquad k = N - n + 1, \ldots, N + 1$$

and

$$\hat{x}_{N+1} = (-y_N, \ldots, -y_{N-n+1}, u_N, \ldots, u_{N-n+1}, \hat{\varepsilon}_N, \ldots, \hat{\varepsilon}_{N-n+1})^{\mathrm{T}}.$$

4.7 Bias arising from correlated disturbances

It is a straightforward matter to apply maximum likelihood estimation methods (for Gaussian disturbances) or least squares methods when the models considered are described by vector difference equations:

$$A(z^{-1})y_k = B(z^{-1})u_{k-1} + w_k,$$

when the coefficient matrices of the polynomials $A(\sigma)$ and $B(\sigma)$ to be estimated are unconstrained and when the disturbance sequence $\{w_k\}$ is zero-mean uncorrelated; as we have seen, the identification problem reduces to minimization of a quadratic functional, and it can therefore be solved in closed form. When the disturbances $\{w_k\}$ are not uncorrelated but are modelled, say, by

$$w_k = C(z^{-1})e_k$$

in which $\{e_k\}$ is a sequence of uncorrelated random variables and the coefficient matrices of the polynomial $C(\sigma)$ are to be estimated, then the identification procedures give rise to minimization problems involving non-quadratic objective functionals, and for this reason their implementation is a much more formidable task.

In the circumstances one might be tempted to estimate parameters under the hypothesis that the disturbances were uncorrelated, even if there were reason to doubt the hypothesis. However, this is not advisable because we can expect that disturbance correlation will give rise to biased estimates. The following simple example illustrates the point.

Suppose that a dynamical system with scalar input and output is described by

$$y_k = ay_{k-1} + w_k, \qquad k \in \mathbb{Z} \tag{4.7.1}$$

and that the disturbances w_k are generated by

$$w_k = e_k + ce_{k-1}, \qquad k \in \mathbb{Z}. \tag{4.7.2}$$

Here $\{e_k\}$ is a sequence of zero-mean uncorrelated random variables with uniformly bounded fourth-order moments. a and c are given real numbers and $|a| < 1$. We assume that the variance of e_k, which we write $R_{ee}(0)$, is positive and does not depend on k.

The least squares estimate $\hat{a}_N$ of a, given data for times $k = 1, \ldots, N$ and calculated without regard to the correlation of the disturbances is:

$$\hat{a}_N = \left(\frac{1}{N}\sum_{k=1}^{N} y_{k-1}^2\right)^{-1}\left(\frac{1}{N}\sum_{k=1}^{N} y_k y_{k-1}\right).$$

Now it is not difficult to show that, under our assumptions on $\{e_k\}$, there exist constants $c>0$, $\lambda\in(0,1)$ such that for d_k taken as either y_{k-1}^2 or $y_k y_{k-1}$, $k=1,2,\ldots$, we have

$$\operatorname{cov}\{d_t, d_{t+s}\} \le c\lambda^s \qquad \text{for } t,s\ge 0.$$

It follows from the ergodic theorem (Theorem 1.1.15) that

$$\frac{1}{N}\sum_{k=1}^{N} y_{k-1}^2 \to R_{yy}(0) \qquad \text{a.s.}$$

and

$$\frac{1}{N}\sum_{k=1}^{N} y_k y_{k-1} \to R_{yy}(1) \qquad \text{a.s.}$$

where

$$R_{yy}(0)=Ey_{k-1}^2 \quad \text{and} \quad R_{yy}(1)=Ey_k y_{k-1}.$$

The number $R_{yy}(0)$ is positive. We deduce that, a.s., $\hat{a}_N$ is defined for all N sufficiently large and

$$\hat{a}_N \to \frac{R_{yy}(1)}{R_{yy}(0)} \qquad \text{as } N\to\infty, \qquad \text{a.s.} \tag{4.7.3}$$

The asymptotic value of the estimate of a is now compared with the true value. From (4.7.1) and (4.7.2),

$$y_k y_{k-1} = a y_{k-1}^2 + (e_k + c e_{k-1}) y_{k-1}, \qquad k\in\mathbb{Z}$$

and

$$y_k e_k = a y_{k-1} e_k + (e_k + c e_{k-1}) e_k, \qquad k\in\mathbb{Z}.$$

Taking expectations and noting that e_k is uncorrelated with e_{k-1} and y_{k-1} we conclude that

$$R_{yy}(1) = aR_{yy}(0) + cR_{ye}(0)$$

and $R_{ye}(0)=R_{ee}(0)$ where $R_{ye}(0)=E\{y_k e_k\}$. It follows from these equations and (4.7.3) that

$$\hat{a}_N \to a + c\frac{R_{ee}(0)}{R_{yy}(1)}, \qquad \text{as } N\to\infty \qquad \text{a.s.}$$

We see that an asymptotic bias is present of $cR_{ee}(0)/R_{yy}(1)$. This will be zero only if c is zero, that is, only if the disturbances are uncorrelated.

One situation in which we can disregard correlation of the disturbances and still obtain unbiased estimates is when the system and model equations involve no autoregressive terms, or in other words take the form

$$y_k = B(z^{-1})u_k + w_k$$

Least squares estimates of the matrix coefficients of $B(\sigma)$ will be unbiased even if $\{w_k\}$ is a correlated sequence. Indeed, the analysis of least squares estimation for static models provided in Section 4.3 is applicable (see Example 4.3.1) and this establishes that the estimates are unbiased even with a finite number N of data points (see Proposition 4.3.2). We can expect though that the estimates (for each N) will have larger variance when correlation of the disturbances is disregarded than when allowance is made for it in the estimation scheme.

4.8 Three-stage least squares and order determination for scalar ARMAX models

This section concerns parameter estimation for scalar ARMAX systems

$$A(z^{-1})y_k = B(z^{-1})u_k + C(z^{-1})e_k \tag{4.8.1}$$

in which $\{e_k\}$ is a normal white noise sequence with variance σ^2. We have seen above that maximum likelihood estimation of the parameters A, B, C, σ^2 is in general a nonlinear minimization problem but that if $C(z^{-1}) = 1$ then it reduces to least squares estimation of A and B which, computationally, is a very much simpler task. A similar reduction applies if $C(z^{-1})$ is any known stable polynomial. To see this, write (4.8.1) as

$$\frac{A(z^{-1})}{C(z^{-1})}y_k = \frac{B(z^{-1})}{C(z^{-1})}u_k + e_k \tag{4.8.2}$$

and define filtered sequences $\bar{y}_k$, $\bar{u}_k$ as follows

$$\begin{aligned} C(z^{-1})\bar{y}_k &= y_k \\ C(z^{-1})\bar{u}_k &= u_k. \end{aligned} \tag{4.8.3}$$

Then $\bar{y}_k$, $\bar{u}_k$ satisfy

$$A(z^{-1})\bar{y}_k = B(z^{-1})\bar{u}_k + e_k$$

and we can estimate A, B by least squares, as before but using the filtered data $(\bar{y}_k, \bar{u}_k)$ in place of the original data (y_k, u_k).

When $C(z^{-1})$ is unknown it is natural to consider replacing the nonlinear maximization of the likelihood by a sequence of least squares operations in which the data is filtered as in (4.8.3) but with the 'true' C replaced by some estimate. This, indeed, is the idea behind the Generalized Least Squares algorithm described in Section 4.5. Another algorithm along the same lines is the so-called three-stage least squares algorithm, described as follows. It is assumed that a data sequence $\{y_k, u_k, k = 1, 2, \ldots, N\}$ is given. The degrees of the polynomials $A(z^{-1})$, $B(z^{-1})$, $C(z^{-1})$ and an integer p ($p = 10$ is a typical value in applications) are pre-specified.

Three-stage least squares algorithm (Mayne and Firoozan (1982))

(a) Estimate the parameters in the model

$$\mathscr{A}(z^{-1})y_k = \mathscr{B}(z^{-1})u_k$$

by least squares. Here

$$\mathscr{A}(z^{-1}) = 1 + \alpha_1 z^{-1} + \cdots + \alpha_p z^{-p}$$
$$\mathscr{B}(z^{-1}) = \beta_0 + \beta_1 z^{-1} + \cdots + \beta_p z^{-p}.$$

(b) Form the residual sequence

$$\hat{\varepsilon}_k = \hat{\mathscr{A}}(z^{-1})y_k - \hat{\mathscr{B}}(z^{-1})u_k \tag{4.8.4}$$

where $\hat{\mathscr{A}}$, $\hat{\mathscr{B}}$ are formed from the parameter estimates of part (a).

(c) Estimate the parameters in the model

$$A(z^{-1})y_k = B(z^{-1})u_k + C(z^{-1})\hat{\varepsilon}_k \tag{4.8.5}$$

by least squares. Denote the estimates $\hat{A}_1$, $\hat{B}_1$, $\hat{C}_1$.

(d) Filter the data through $\hat{C}_1(z^{-1})$, giving filtered data $\bar{y}_k$, $\bar{u}_k$, $\bar{\varepsilon}_k$:

$$\hat{C}_1(z^{-1})\bar{y}_k = y_k$$
$$\hat{C}_1(z^{-1})\bar{u}_k = u_k$$
$$\hat{C}_1(z^{-1})\bar{\varepsilon}_k = \hat{\varepsilon}_k.$$

(e) Re-estimate the parameters in (4.8.5) replacing y_k, u_k, $\hat{\varepsilon}_k$ by $\bar{y}_k, \bar{u}_k, \bar{\varepsilon}_k$. This gives the final estimates $\hat{A}_2$, $\hat{B}_2$, $\hat{C}_2$.

As its name suggests, the algorithm involves just three least squares estimations, in contrast to the generalized least squares algorithm where filtering and least squares estimation are repeated until some criterion is satisfied.

Steps (a)–(c) of the algorithm are clearly motivated as follows: denote $\mathscr{A}_0 = A/C$ and $\mathscr{B}_0 = B/C$. Then (4.8.2) becomes

$$\mathscr{A}_0(z^{-1})y_1 = \mathscr{B}_0(z^{-1})u_k + e_k.$$

This gets rid of the unwanted $C(z^{-1})$, but $\mathscr{A}_0$, $\mathscr{B}_0$ are now infinite-degree polynomials. We therefore truncate them to polynomials $\mathscr{A}$, $\mathscr{B}$ of pth degree where p is 'large'. The residuals $\hat{\varepsilon}_k$ given by (4.8.4) then approximate the noise sequence e_k, enabling us to estimate A, B and C by least squares. For the final steps (d) and (e) we behave as if $\hat{C}_1$ were the 'true' value, filtering the data to remove noise correlation.

The three stage least squares algorithm is justified, apart from the above motivation, by its large sample behaviour in case the data is actually generated by an ARMAX system (4.8.1) with known order and constant but unknown parameters A_0, B_0, C_0.

Consider first the no-input case in which the data $\{y_1, \ldots, y_N\}$ is generated by the ARMA system

$$A_0(z^{-1})y_k = C_0(z^{-1})e_k. \tag{4.8.6}$$

It is assumed that A_0 and C_0 have no common factors and that the zeros of $\sigma \to A_0(\sigma)$ and $\sigma \to C_0(\sigma)$ all lie outside the closed unit disc. The model set is the set of ARMA models

$$A(z^{-1})y_k = C(z^{-1})e_k,$$

where

$$A(z^{-1}) = 1 + a_1 z^{-1} + \cdots + a_n z^{-n}$$
$$C(z^{-1}) = 1 + c_1 z^{-1} + \cdots + c_l z^{-l},$$

with parameter vector $\theta^{\mathrm{T}} = (a_1, \ldots, a_n, c_1, \ldots, c_l)$. The parameter vector corresponding to the true system A_0, C_0 is denoted θ_0.

For a given data set the parameter estimate $\hat{\theta}$ given by the 3-stage least squares algorithm (i.e. containing the coefficients of the estimated polynomials $\hat{A}_2, \hat{C}_2$) depends both on the length N of the data sequence and the degree p of the polynomial $\mathscr{A}(z^{-1})$ used in step (a). We write it $\hat{\theta}(N, p)$. Mayne and Firoozan (1982) demonstrate the following large sample properties.

Proposition 4.8.1

Under the conditions stated above:

(a) For each p there is a vector $\theta(p)$ such that

$$\lim_{N \to \infty} \hat{\theta}(N, p) = \theta(p) \qquad \text{a.s.}$$

(b) $\lim_{p \to \infty} \theta(p) = \theta_0$.

The result says that for any fixed p the parameter estimates may be asymptotically biased ($\theta(p) \neq \theta_0$) but the bias $\theta(p) - \theta_0$ can be made arbitrarily small by choosing p sufficiently large. The proof of Proposition 4.8.1 is complicated and we must refer the reader to the original paper for this. The paper also contains important results relating to the large-sample distribution of $\hat{\theta}(N, p)$ which show that this estimator is asymptotically efficient, i.e. it has properties similar to those of the classical maximum likelihood estimator described in Proposition 4.1.3.

Proposition 4.8.1 also holds for ARMAX systems if the input sequence $\{u_k\}$ satisfies a condition of 'persistent excitation'. This condition is discussed below in Chapter 5.

Since the bias $\theta(p) - \theta_0$ disappears as $p \to \infty$ it seems that if p is allowed to increase with N then a sequence of estimators converging to the true value might be obtained. Thus we take $p = p(N)$ for some increasing function $p(\cdot)$ and ask whether $p(\cdot)$ can be chosen so that $\hat{\theta}(N, p(N)) \to \theta_0$ a.s. This question has been investigated by Hannan and Kavalieris (1983), who show that indeed $\hat{\theta}(N, p(N)) \to \theta_0$ a.s. if

$$\limsup_{N \to \infty} p(N) \left(\frac{\log N}{N} \right)^{1/2} > 0.$$

Thus in particular the choice $p(N) = (N/\log N)^{1/2}$ would suffice. With this choice, $p(100) = 5$, $p(10^4) = 33$. For a given data set of fixed length the appropriate choice of p depends on the positions of the zeros of $C_0(z^{-1})$ (which are, of course, not known in advance). Since the polynomial $\mathscr{A}(z^{-1})$ in step (a) of the algorithm is intended to approximate $A_0(z^{-1})/C_0(z^{-1})$ it is clear that a relatively small value of p will suffice if the zeros of $C_0(z^{-1})$ are well inside the unit circle. In practice a value of p in the range 10–15 seems adequate in many applications, but the method may be expected to run into trouble if $C_0(z^{-1})$ is only marginally stable.

Order determination

In the three-stage least squares algorithm as described above the orders l, n of C_0 and A_0 are assumed known. It has recently been shown by Hannan and Rissanen (1982) that a modification of the algorithm will supply consistent estimates of l and n *and* of the parameter vector θ. Before describing this, we discuss model order determination in somewhat more general terms.

Model order selection for the static least-squares problem was discussed in Section 4.3. Models of successively higher order are fitted and the correct order is identified by a statistical test based on the rate of decrease with model order n of the residual sum of squares function $S_N(n, \theta) = \varepsilon^T(\theta)\varepsilon(\theta)$ (see Proposition 4.3.8 *et seq.*). It was pointed out that this test could be interpreted as selecting those values of n, θ which give the absolute minimum of a function $A_N(n, \theta)$ defined by

$$A_N(n, \theta) = \log S_N(n, \theta) + \tilde{\kappa} n \tag{4.8.7}$$

for some constant $\tilde{\kappa}$. The statistical testing theory is only valid for static models, but a similar procedure is often used for determining the order of dynamical models such as the ARMA model (4.8.6).

Criteria of the form (4.8.7) were introduced by Akaike (1969), and are called AIC criteria (IC for 'information criterion'). They represent a quantitative formulation of the so-called 'principle of parsimony' in model-building, namely that, other things being equal, the model with the smallest number of parameters should be preferred. In (4.8.7), $\log S_N(n, \theta)$ decreases with increasing n but the second term $\tilde{\kappa} n$ imposes a penalty for introducing more parameters. By choosing (n, θ) to minimize $A_N(n, \theta)$ we achieve a trade-off between accurate model-fitting (small $S_N(n, \theta)$) and parsimony of parametrization (small n). The relative weights are controlled by the constant $\tilde{\kappa}$.

A slightly different approach to order determination starts from the maximum likelihood method. For concreteness we discuss this in the context of the ARMA model (4.8.6) although the ideas apply more generally. If the disturbance sequence $\{e_k\}$ is normal with mean 0 and unknown variance σ^2 then the likelihood function is

$$L_N(l, n, \theta, \sigma^2) = \frac{1}{(2\pi\sigma^2)^{N/2}} \exp\left(-\frac{1}{2\sigma^2} \sum_{k=1}^{N} \varepsilon_k^2(\theta) \right).$$

As indicated, it depends on the orders n, l of $A(z^{-1})$ and $C(z^{-1})$, on the parameters $\theta^T = (a_1, \ldots, a_n, c_1, \ldots, c_l)$, on the noise variance σ^2 and on the number N of data points. As before we denote

$$S_N(\theta) = \sum_{k=1}^{N} \varepsilon_k^2(\theta).$$

Parameter estimates are obtained by maximizing L_N over the range of allowable parameter values. We cannot, however, regard l and n as parameters and estimate them in this way: clearly $\min_\theta S_N(\theta)$ decreases as n and l increase (since the minimum is being taken over a larger set) and thus the 'maximum likelihood' estimates of n, l will be whatever largest value we regard as allowable. Several authors, including Akaike (1977), Rissanen (1978) and Schwarz (1978) have examined this situation and concluded, independently and by widely differing arguments, that the appropriate quantity to be maximized is

$$BIC_N(l, n, \theta, \sigma^2) = \log L_N - \tfrac{1}{2}(l+n)\log N.$$

As in the case of the AIC criterion, the BIC criterion introduces a linear penalty for increasing the number of parameters. The weighting of this penalty is however dependent on the number N of data points. Let us denote by $\hat{\theta}_N$ and $\hat{\sigma}_N^2$ the maximum likelihood estimates of θ and σ^2 (for fixed n, l, N). Then it is easily checked that

$$\hat{\sigma}_N^2 = \frac{1}{N} S_N(\hat{\theta}_N)$$

and that

$$\log L_N(l, n, \hat{\theta}_N, \hat{\sigma}_N^2) = -\frac{N}{2}(\log \hat{\sigma}_N^2 + \log 2\pi - 1).$$

Thus maximizing BIC_N is equivalent to minimizing

$$\log \hat{\sigma}_N^2 + (l+n)\frac{\log N}{N}. \tag{4.8.8}$$

The exact arguments advanced in favour of this procedure need not detain us here, particularly since these arguments do not directly imply any optimality properties of the estimates $\hat{l}, \hat{n}$ produced. Instead we introduce a family of criteria

$$H_c(l, n, \theta, \sigma^2) = \log \hat{\sigma}_N^2 + (l+n)\frac{c(N)}{N}$$

where $c(N)$ is an increasing function of N (thus (4.8.8) is the special case with $c(N) = \log N$). We estimate l, n, θ, σ^2 by minimizing H_c, where n, l are limited by $n \leq \bar{n}(N)$, $l \leq \bar{l}(N)$. Here $\bar{n}$, $\bar{l}$ are *a priori* upper bounds, possibly depending on N. We now ask how the function $c(N)$ should be chosen to obtain various desirable properties for the corresponding estimates. Properties which might be required are: (a) high probability of selecting the correct model order for finite data sets, or (b) consistency, i.e. asymptotically correct choice of l, n, θ, σ^2 as $N \to \infty$. No theory is available for (a), but Hannan (1980) shows that consistency holds for certain choices of $c(N)$ including the choice $c(N) = \log N$ corresponding to the BIC criterion.

Let us now return to the three-stage least squares parameter estimation algorithm. Since this is intended as an approximation to the maximum likelihood method, it is natural to suppose that it might be combined with the order determination methods outlined above to yield consistent estimates of model order and parameters. Such a result has been demonstrated by Hannan and Rissanen (1982), for ARMA models (see also Hannan and Kavalieris (1983)). The three-stage least squares algorithm is as stated before (with $u_k \equiv 0$), except that step (c) is replaced by (c′).

(c′) For each (n, l) calculate

$$\tilde{\sigma}_{n,l}^2 = \inf_{\theta} \frac{1}{N} \sum_{k=1}^{N} (A(z^{-1})y_k - C(z^{-1})\hat{\varepsilon}_k)^2$$

where $\theta^{\mathrm{T}} = (a_1, \ldots, a_n, c_1, \ldots, c_l)$. Choose $\hat{n}, \hat{l}$ in the range $0 \le n \le \bar{n}(N)$, $0 \le l \le \bar{l}(N)$ to minimize

$$\log \tilde{\sigma}_{n,l}^2 + (l + n)\frac{c(N)}{N}.$$

Denote by $A_1(z^{-1})$, $C_1(z^{-1})$ the least squares estimates of $A(z^{-1})$, $C(z^{-1})$ with orders $\hat{n}, \hat{l}$ respectively.

Hannan and Rissanen (1982) show that consistent estimates of n, l, θ are obtained under the same conditions as before if we take

$$c(N) = (\log N)^{1+\delta}, \qquad \bar{n}(N) = \bar{l}(N) = (\log N)^{\beta}. \qquad (4.8.9)$$

Here δ, β are arbitrary strictly positive constants. This is a satisfying result because it means that the entire identification procedure, including model order determination, can be carried out by a simple combination of least squares estimators. Many computer packages incorporating least squares estimation are available. Undoubtedly the results stated above apply to ARMAX models if the input is persistently exciting.

Computational experience of this method is reported by Hannan and Rissanen (1982) and by Kountzeris (1984) using simulated data.

The method works well except when $A_0(z^{-1})$ and $C_0(z^{-1})$ almost contain common factors (in which case any identification method would have difficulty in deciding whether a cancellation had taken place or not). The order p of the AR used in step (a) seems not to be critical. Kountzeris (1984) reports that for data sets of length N between 500 and 1500 a value of δ between 1.5 and 2.5 maximizes the frequency of correct order selection (upwards of 80% in straightforward cases). In applications, the upper bounds $\bar{n}$, $\bar{l}$ would normally be set at an *a priori* fixed value rather than being calculated from some formula as they are in (4.8.9).

Application of the above procedures to real data has so far not been investigated in any detail. Here we must drop the assumption that the data is generated by a 'true' ARMA model and regard the problem as that of selecting (n, l, θ) to give a 'best' model according to some criterion such as minimizing prediction error taking into account errors of model estimation. The same procedure may be used, but possibly some different function $c(N)$ might be appropriate. Exactly how this function should be chosen in a 'prediction error' context remains a subject for future research.

Notes

System identification is a field with a multidisciplinary base which has been in a state of active development for twenty years. It is not surprising then that it has generated an extensive literature. For an overview of the field, and a source of references, we refer the reader to the book by Goodwin and Payne (1977), the survey by Åström and Eykhoff (1971) and tutorial papers in a special issue of *Automatica* (1981). A comprehensive account, including treatment of non-stationary models and many practical details of data analysis, is given by Box and Jenkins (1976). Further material on important topics in systems identification not entered into in this book can be found in Goodwin and Payne (1977) (experiment design, procedures for estimating time-varying parameters and other topics), in Söderström and Stoica (1980) (the instrumental variables technique) and Gustavson *et al.* (1977), Clark (1976) and Hannan *et al.* (1980) (unique parametrization and model class selection).

Section 4.1 Detailed coverage of point estimation theory is provided in a number of books on statistics (Kendall and Stuart, 1979, for example).

Section 4.3 For refinements and extensions of the theory of least squares and of maximum likelihood parameter estimation see (Kendall and Stuart (1979) and Rao (1965). Note that the normal equations of least squares theory are often ill-conditioned. Robust procedures for their solution are described in Golub (1965). Our approach in this section to estimation of model order is a classical one (Lehman, 1959).

Sections 4.2 *and* 4.4 The idea of formulating stochastic models as predictor models and of interpreting least squares and maximum likelihood procedures for dynamical systems as prediction error methods, which provides the framework for these sections, has been emphasized by Ljung (1978), and Caines (1976), though it is implicit in earlier literature. Proof of results on the asymptotic distributions of parameter estimates described in Section 4.4.4 are to be found in Ljung and Caines (1979).

Section 4.5 The modified Newton–Raphson algorithm was proposed by Åström and Bohlin (1965). The generalized least squares algorithm (in a slightly different form) was devised by Clarke (1967).

Section 4.6 It is known that direct implementation of the recursive least squares algorithm can give rise to numerical instability. For modifications of the algorithm which are robust see Hanson and Lawson (1969). The recursive generalized least squares algorithm is due to Hastings-James and Sage (1969). The extended matrix method, first described in Panushka (1968), was proposed independently by a number of authors and a variety of names have been given to it, including Panushka's method and the approximate maximum likelihood method. There is evidence that the method can give estimates which are not consistent (Ljung *et al.*, 1975). For a full treatment of recursive identification algorithms and their implementation we refer to the recent book of Ljung and Söderström (1983).

Section 4.8 A procedure involving the first two stages of the three-stage least squares algorithm was introduced by Durbin (1960); the algorithm as given is due to Mayne and Firoozan (1982).

The development of order determination methods is outlined with references in the main body of this section. The asymptotic distribution of order estimates given by the AIC criterion has been calculated by Shibata (1976).

The algorithms we have given are for off-line identification. Recursive counterparts of these algorithms, suitable for on-line use, are given by Mayne, Åström and Clark (1984) and by Hannan and Rissanen (1982).

References

Akaike, H. (1969) Fitting autoregressive models for prediction. *Ann. Inst. Statist. Math.* **21**, 243–247.

Akaike, H. (1977) On entropy maximization principle. In *Applications of Statistics* (ed. P. R. Krishnaiah) North Holland, Amsterdam.

Åström, K. J. and Bohlin, T. (1965) Numerical identification of linear dynamical system operating data. From *Theory of Self-adaptive Control Systems* (ed. P. Hammond) Plenum Press, New York.

Åström, K. J. and Eykhoff, P. (1971) System identification – a survey. *Automatica*, **7**, 123–162.

Automatica (1981) Special Issue on Identification and System Parameter Estimation, **17(1)**.

Box, G. E. P. and Jenkins, G. M. (1976) *Time Series Analysis Forecasting and Control* (2nd edn) Holden-Day, San Francisco.

Caines, P. E. (1976) Prediction error identification methods for stationary stochastic processes. *IEEE Trans. Automatic Control*, **AC-21**, 500–506.

Clark, J. M. C. (1976) The consistent selection of parametrizations in system identification. *Proc. JACC.*

Clarke, D. W. (1967) Generalized least-squares estimation of the parameters of a dynamic model. *Proc. IFAC Symp. Identification and Automatic Control Systems*, Prague, Czechoslovakia.

Durbin, J. (1960) The fitting of time-series models. *Int. Statist. Rev.*, **28**, 233–244.

Golub, C. (1965) Numerical method for solving linear least squares problems. *Numerische Mathematik*, **7(3)**, 206–216.

Goodwin, C. G. and Payne, R. L. (1977) *Dynamic System Identification: Experiment Design and Data Analysis*, Academic Press, New York.

Gustavson, I., Ljung, L. and Söderström, T. (1977) Identification of processes in closed loop – identification and accuracy aspects. *Automatica*, **13**, 59–75.

Hannan, E. J. (1980) The estimation of the order of an ARMA process. *Ann. Statist.* **8**, 1071–1081.

Hannan, E. J., Dunsmuir, W. and Deistler, M. (1980) Estimation of vector ARMAX models. *J. Multivariate Anal.*, **10**, 275–295.

Hannan, E. J. and Kavalieris, L. (1983) Linear estimation of ARMA processes. *Automatica*, **19**, 447–448.

Hannan, E. J. and Rissanen, J. (1982) Recursive estimation of mixed autoregressive-moving average order. *Biometrika*, **69**, 81–94.

Hanson, R. J. and Lawson, C. L. (1969) Extensions and applications of the Householder algorithms, for solving linear squares problems. *Math. Comput.*, **23**, 787–812.

Hastings-James, R. and Sage, M. E. (1969) Recursive generalized least squares procedure for on-line identification of process parameters. *Proc. IEE*, **116**, 12.

Kendall, M. G. and Stuart, A. (1979) *The Advanced Theory of Statistics*, vol. 2, (4th edn), Griffin, London.

Kountzeris, A. (1984) Model order selection in identification, MSc Thesis, Imperial College, London University.

Lehman, E. L. (1959) *Testing Statistical Hypotheses*, Wiley, New York.

Ljung, L. (1978) Convergence analysis of parametric identification methods. *IEEE Trans. Automatic Control*, **AC-23: 3**, 770–783.

Ljung, L. and Caines, P. E. (1979) Asymptotic normality of prediction error estimators for approximate system models. *Stochastics*, **3**, 29–46.

Ljung, L. and Söderström, T. (1983) *Theory and Practice of Recursive Identification*. M.I.T. Press, Cambridge, Mass.

Ljung, L., Söderström, T. and Gustavson, I. (1975) Counterexamples to general convergence of a commonly used recursive identification method. *IEEE Trans. Automatic Control*, **AC-20: 5**, 643–653.

Mayne, D. Q. and Firoozan, F. (1982) Linear identification of ARMA processes. *Automatica*, **18**, 461–466.

Mayne, D. Q., Åström, K. J. and Clark, J. M. C. (1984) A new algorithm for recursive estimation of parameters in controlled ARMA processes. *Automatica*, to appear.

Panushka, V. (1968) An adaptive recursive least squares identification procedure. *Proc. JACC.*

Rao, C. R. (1965) *Linear Statistical Inference and its Applications*, Wiley, New York.

Rissanen, J. (1978) Modelling by shortest data description. *Automatica*, **14**, 467–471.

Schwarz, G. (1978) Estimating the dimension of a model. *Ann. Statist.*, **6**, 461–465.

Shibata, R. (1976) Selecting the order of an autoregressive model by Akaike's information criterion. *Biometrika*, **63**, 117–126.

Söderström, T. and Stoica, P. G. (1980) *Instrumental Variable Methods for System Identification*, Springer, Berlin.

CHAPTER 5

Asymptotic analysis of prediction error identification methods

Chapter 4 provided for the most part merely a description of identification methods for dynamical systems. It is true that if we limit attention to simple moving-average models with uncorrelated disturbances and if we assume that the system is describable within the model set, then the models can be reformulated as static models to which the analysis of Section 4.3 is applicable and we can deduce certain properties of the estimates. However, the question remains open of how good are the estimates when more complicated models are considered, or when the model set does not contain a description of the system. The analysis which follows is centred on this question.

Ideally, we want precise information about the quality of an estimate which results from applying an identification method to data up to termination time N. Except in very restrictive circumstances, analysis of estimates based on a data record of finite length is quite intractable. However, it is not so difficult to investigate asymptotic properties of the estimates in the limit as $N \to \infty$; this is our more modest objective. The results of the analysis suggest what estimates would be obtained from an application of an off-line identification algorithm, or from application of an on-line algorithm which approximates the off-line one, as $N \to \infty$. Of course, a major shortcoming of our theory is that, by the nature of asymptotic analysis, it does not tell us how large N must be for the theory to give a reasonable picture of the parameter estimates based on a data record of length N.

The chapter is organized with the interests partly in mind of readers who wish to understand the results without following all details of the proofs. A central role is played in the analysis by a general convergence theorem, Theorem 5.2.1. While the significance of Theorem 5.2.1 is easily grasped, its proof is rather intricate. In the body of the chapter we explore the implications of Theorem 5.2.1. A proof of the theorem is provided in Appendix A.

5.1 Preliminary concepts and definitions

A number of definitions associated with a general formulation of prediction error identification methods will now be given. These definitions will be convenient when we come to state the hypotheses under which the analysis applies.

There are basically three ingredients in the description of a prediction error identification method: the system which generates the data, the model set, and finally the identification criterion which governs selection of the model. It is helpful at the outset to consider these individually.

5.1.1 The system

The system is the source of two stochastic processes, the r-vector output process $\{y_k\}$ and the m-vector input process $\{u_k\}$. Realization of these processes (up to time N in the case of $\{y_k\}$ and up to time $N-1$ in the case of $\{u_k\}$) constitute the data at time N.

Our aim is selection of a model whose response is a good approximation to that of the system, as a result of analysis of the data. What kind of assumptions need to be made about the system for this to be possible? We think of the system as defined by a family of recursive equations driven by a sequence of independent random variables, the disturbances. These equations, the system equations, incorporate feedback relationships which generate the input. They supply the input and the output at time k as a function of the disturbances which enter the system after an arbitrary, earlier time l, and of the state at time l, which summarizes the effect on the subsequent response of disturbances occuring at, or before, time l. Now we can expect that analysis of the data will supply a good model only when the state at time l, which is not observed, does not have a predominant effect on the response at times much later than k. Insensitivity of the subsequent response to the state at time l is a characteristic of stable systems. It is natural therefore in the analysis of identification methods to assume at the very least that the system is stable.

A notion of stability, suitable in the context of identification, is suggested by properties of linear models in which a 'stable' transfer function relates the disturbances $\{e_k\}$ and the outputs $\{y_k\}$:

$$y_k = g(z^{-1})^{-1}T(z^{-1})e_k, \qquad k \in \mathbb{Z}.$$

Here the e_k are uniformly bounded in fourth moment. $T(\sigma)$ is a matrix of polynomials in σ and $g(\sigma)$ is a polynomial in σ such that the zeros of $\sigma \to g(\sigma)$ lie outside the closed unit disc. If $y_{k,l}$, $k = l, l+1, \ldots$, is the output when disturbances e_j, $j = l, l-1, \ldots$, are ignored (we take this to mean that they are set to zero) then we know from Proposition 2.1.2 that there exist constants $c > 0$, $\lambda \in (0, 1)$ such that

$$E\|y_k - y_{k,l}\|^4 \leq c\lambda^{k-l}, \qquad k \geq l. \tag{5.1.1}$$

If the e_k are independent, then $y_{k,l}$ is a function of $e_{l+1}, \ldots, e_k$ but $y_{k,l}$ is independent of e_k, $k \leq l$.

The inequality (5.1.1), augmented by an analogous inequality for the input, is taken as defining system stability.

Definition 5.1.1

The system which generates the data is said to be *stable* if there exist independent random variables $\{v_k\}_{k \in \mathbb{Z}}$ and constants $c \in (0, \infty)$, $\lambda \in (0, 1)$ with the following properties: y_k, u_k are functions of $v_k, v_{k-1}, \ldots$ for $k = 0, 1, \ldots$ and given integers k, l, with $k \geq l \geq 1$, random variables $y_{k,l}, u_{k,l}$ can be found which are functions of $v_{l+1}, v_{l+2}, \ldots, v_k$ and are such that

$$E\|y_k - y_{k,l}\|^4 \leq c\lambda^{k-l}$$
$$E\|u_k - u_{k,l}\|^4 \leq c\lambda^{k-l}.$$

(It is understood that $y_{k,k} = 0, u_{k,k} = 0$ so that the inequalities imply, in particular, that $E\|y_k\|^4 \leq c, E\|u_k\|^4 \leq c$).

Generally speaking, the random variables $\{v_k\}$ are interpreted as disturbances entering the system. The random variables $y_{k,l}$ and $u_{k,l}$ are in such circumstances usually chosen to be the output and input generated by the system equations when we set the disturbances to zero for time $k \leq l$.

Since our definition of stability was motivated by properties of linear models we would expect that, at the very least, systems describable by linear models of Chapter 2, which are stable in the customary sense, would also be stable in the sense of Definition 5.1.1. This is the case, as we now show.

Consider a linear dynamical system with feedback, in ARMAX form:

$$\begin{aligned} A(z^{-1})y_k &= B(z^{-1})u_{k-1} + C(z^{-1})e_k \\ D(z^{-1})u_k &= E(z^{-1})y_k + w_k, \qquad k\in\mathbb{Z}. \end{aligned} \tag{5.1.2}$$

or having the state-space description

$$\begin{aligned} x_{k+1} &= Ax_k + Bu_k + Ke_k, \\ y_k &= Hx_k + e_k, \\ u_k &= Mx_k + w_k, \qquad k\in\mathbb{Z}. \end{aligned} \tag{5.1.3}$$

In (5.1.2), $A(\sigma)$, $B(\sigma)$, $C(\sigma)$, $D(\sigma)$, $E(\sigma)$ are polynomials with coefficients $r\times r$, $r\times m$, $r\times r$, $r\times m$, $r\times r$ matrices respectively. In (5.1.3), A, B, K, H, M are $n\times n, n\times m, n\times r, r\times n, m\times n$ matrices respectively.

In either case we assume that the disturbances, $e_k, w_j, k, j\in\mathbb{Z}$ are independent and there exists a constant c such that

$$E\|e_k\|^4 \le c, E\|w_k\|^4 \le c, \qquad \text{all } k\in\mathbb{Z}. \tag{5.1.4}$$

Proposition 5.1.2

Suppose (5.1.4) is satisfied the system $\{y_k, u_k\}$ is stable if either

(a) $\{y_k\}, \{u_k\}$ are generated by ARMAX equations (5.1.2), $\det A(0) \neq 0$, $\det D(0) \neq 0$ and the zeros of $\sigma \to \det[A(\sigma) - \sigma B(\sigma)D^{-1}(\sigma)E(\sigma)]$ and $\sigma \to \det D(\sigma)$ lie outside the closed unit disc; or,

(b) $\{y_k\}$, $\{u_k\}$ are generated by state-space equations (5.1.3), and the eigenvalues of $A + BM$ are contained in the open unit disc.

We mention that the hypotheses in Proposition 5.1.2 assure stability of the closed-loop transfer functions through which the inputs and outputs are expressed in terms of the disturbances. In view of the theory of Section 2.1 then, we may take equations (5.1.2), or (5.1.3) to define the outputs and inputs as fourth-order random variables, under the hypotheses.

PROOF Let us take the independent random variables $\{v_k\}$ required for verification of stability to be

$$v_k = \begin{pmatrix} w_k \\ e_k \end{pmatrix}, \qquad k\in\mathbb{Z}.$$

Formal manipulation of the transfer functions associated with system

(5.1.2), or with system (5.1.3), gives the following expression for y_k and u_k in terms of the composite disturbance vectors v_k:

$$y_k = S(z^{-1})v_k; \qquad u_k = T(z^{-1})v_k.$$

In the case of ARMAX description,

$$S(\sigma) = \bar{A}^{-1}(\sigma)[\sigma B(\sigma)D^{-1}(\sigma) \,\vdots\, C(\sigma)]$$

and

$$T(\sigma) = D^{-1}(\sigma)[E(\sigma)S(\sigma) + (I \,\vdots\, 0)]$$

in which

$$\bar{A}(\sigma) = (A(\sigma) - \sigma B(\sigma)D^{-1}(\sigma)E(\sigma)).$$

In the case of a state-space description,

$$S(\sigma) = [HW^{-1}(\sigma)B \,\vdots\, HW^{-1}(\sigma)K + I]$$

and

$$T(\sigma) = [MW^{-1}(\sigma)B + I \,\vdots\, MW^{-1}(\sigma)B]$$

in which

$$W(\sigma) = [\sigma I - (A + BM)].$$

It is not difficult to deduce from the hypotheses that, in either case, $S(\sigma)$ and $T(\sigma)$ are expressible as

$$S(\sigma) = s(\sigma)^{-1}\tilde{S}(\sigma) \quad \text{and} \quad T(\sigma) = t(\sigma)^{-1}\tilde{T}(\sigma)$$

in which $\tilde{S}(\sigma)$, $\tilde{T}(\sigma)$ are matrices with entries polynomials in σ, and $s(\sigma)$, $t(\sigma)$ are polynomials with the properties that the zeros of $\sigma \to s(\sigma)$ and $\sigma \to t(\sigma)$ lie outside the closed unit disc.

By the theory of Section 2.1, equations (5.1.2), or equations (5.1.3), define the outputs and the inputs as fourth-order random variables.

For given $l \geq 1$ define $y_{k,l}$ and $u_{k,l}$ by

$$y_{k,l} = S(z^{-1})v_{k,l} \quad \text{and} \quad u_{k,l} = T(z^{-1})v_{k,l}$$

in which

$$v_{k,l} = \begin{cases} v_k & \text{for} \quad k > l \\ 0 & \text{for} \quad k \leq l \end{cases}.$$

Notice that $y_{k,l}$ and $u_{k,l}$ are functions of $v_{l+1}, v_{l+2}, \ldots, v_k$

It follows from Proposition 2.1.2 that there exist constants $\lambda\in(0,1), c_1>0$ such that

$$E\|y_k - y_{k,l}\|^4 + E\|u_k - u_{k,l}\|^4 \le c_1 \sum_{j=k-l}^{\infty} \lambda^j E\|v_{k-j}\|^4$$
$$\le c_2\lambda^{k-l}, \qquad k\ge l$$

where $c_2 = c_1 c(1-\lambda)^{-1}$. Here c is the constant of hypothesis (5.1.4).

We deduce from these properties of the random variables $y_{k,l}, u_{k,l}$ that the system is stable. □

5.1.2 The model set

A family of models is supplied, members of which are specified by a vector parameter θ which ranges over a subset D of $\mathbb{R}^q$. The model corresponding to the parameter value θ is written $\mathcal{M}(\theta)$.

We shall assume that the models $\mathcal{M}(\theta)$ can be formulated as predictor models of the type considered in Section 4.2:

$$y_k = f_k(\theta; y^{k-1}, u^{k-1}) + e_k, \qquad k = 1, 2, \ldots$$

Here, as previously, y^{k-1} denotes $\{y_{k-1}, y_{k-2}, \ldots, y_0\}$ and u^{k-1} denotes $\{u_{k-1}, u_{k-2}, \ldots, u_0\}$. $\{e_k\}$ is a sequence of independent zero-mean random variables. For $k=1,2,\ldots, f_k(\cdot;\cdot,\cdot)$: $\mathbb{R}^q \times \mathbb{R}^{rk} \times \mathbb{R}^{mk} \to \mathbb{R}^r$ is a deterministic function.

Selection of a predictor model amounts to selection of the predictors, the $f_k(\theta;\cdot,\cdot)$. We shall consider parameter estimation schemes which involve minimization of a function of the predictors evaluated at the data. It is natural to limit attention to models for which data in the distant past has little effect on the current prediction; this will mean that all data will make a significant contribution to choice of the model and not just the early data. The precise conditions we shall impose on the models are suggested by those which the linear models of Chapter 2 can be expected to satisfy. These conditions are embodied in the following definition.

Definition 5.1.3

The predictors associated with the family of models $\{\mathcal{M}(\theta): \theta\in D\}$ are *uniformly stable* if D is compact, and there exist constants $c>0$, $\lambda\in(0,1)$ and an open neighbourhood $\mathcal{D}$ of D with the following properties:

(a) $\theta \to f_k(\theta; \alpha^{k-1}, \beta^{k-1})$ is continuously differentiable on $\mathscr{D}$ for arbitrary vectors $\alpha^{k-1}, \beta^{k-1}$ and for $k = 1, 2, \ldots$
(b) $\| f_k(\theta; 0^{k-1}, 0^{k-1}) \| \leq c$ for all $\theta \in \mathscr{D}, k = 1, 2, \ldots$ (here 0^{k-1} denotes zero vectors of appropriate dimensions) and
(c) For q_k taken to be the function f_k, and also the function $(\partial/\partial\theta) f_k$, we have

$$\| q_k(\theta; \alpha^{k-1}, \beta^{k-1}) - q_k(\theta; \bar{\alpha}^{k-1}, \bar{\beta}^{k-1}) \|$$
$$\leq c \sum_{s=0}^{k-1} \lambda^{k-s} [\| \alpha_s - \bar{\alpha}_s \| + \| \beta_s - \bar{\beta}_s \|]$$

for arbitrary vectors α^{k-1} $(= \alpha_{k-1}, \ldots, \alpha_0)), \beta^{k-1}, \bar{\alpha}^{k-1}, \bar{\beta}^{k-1}$, $k = 1, 2, \ldots$

Built in to the definition of a uniformly stable family of predictors is the requirement that the parameter constraint set D be compact. The constraints on the parameter θ which define D may be seen as safeguards introduced into the identification algorithm under consideration which restrict the size of the estimates and steer them away from values for which the predictors are barely stable. These safeguards may be purely notional; we can expect that if the system is stable and if the system can be closely approximated by a model, then unconstrained minimization of the identification criteria will yield estimates which are confined to some set D with properties as described in Definition 5.1.3.

Let us examine conditions under which model sets comprising stochastic dynamical models of Chapter 2 yield uniformly stable predictors. Consider predictors $f_k(\theta; \cdot, \cdot)$ associated with a family of ARMAX models:

$$\begin{aligned} A_\theta(z^{-1}) y_k &= B_\theta(z^{-1}) u_{k-1} + C_\theta(z^{-1}) e_k \qquad && k \geq 0 \\ y_k &= 0, u_k = 0, e_k = 0, && k < 0 \end{aligned} \tag{5.15}$$

or associated with a family of state-space models:

$$\begin{aligned} x_{k+1} &= A_\theta x_k + B_\theta u_k + K_\theta e_k, \qquad && k \geq 0 \\ y_k &= H_\theta x_k + e_k, && k \geq 0 \\ x_0 &= 0. \end{aligned} \tag{5.1.6}$$

In either case, $\{e_k\}$ is a sequence of independent, zero-mean random variables, and θ ranges over an open set $\tilde{\mathscr{D}}$ which contains the compact set D of permitted parameter values.

In (5.1.5), $A_\theta(\sigma), B_\theta(\sigma), C_\theta(\sigma)$ are polynomials in σ with matrix coefficients whose entries are continuously differentiable functions in θ on $\tilde{\mathscr{D}}$. We assume that

$$C_\theta(0) = A_\theta(0) = I.$$

In (5.1.6), $A_\theta, B_\theta, K_\theta, H_\theta$ are matrices whose entries are continuously differentiable functions in θ on $\tilde{\mathscr{D}}$.

The predictors associated with the models (5.1.5) are

$$f_k(\theta; \alpha^{k-1}, \beta^{k-1}) = \hat{y}_k, \qquad k = 1, 2, \ldots$$

where the $\hat{y}_k$ are obtained from the recursive equations

$$C_\theta(z^{-1})\hat{y}_k = z[C_\theta(z^{-1}) - A_\theta(z^{-1})]\alpha_{k-1} + B_\theta(z^{-1})\beta_{k-1}, \qquad k = 0, 1, \ldots$$

$$\hat{y}_k = 0, \alpha_k = 0, \beta_k = 0, \qquad k < 0.$$

The predictors associated with the models (5.1.6) are

$$f_k(\theta; \alpha^{k-1}, \beta^{k-1}) = H_\theta \bar{x}_k, \qquad k = 1, 2, \ldots$$

in which the $\bar{x}_k$ are obtained from

$$\bar{x}_{k+1} = A_\theta \bar{x}_k + B_\theta \beta_k + K_\theta(\alpha_k - H_\theta \bar{x}_k), \qquad k = 0, 1, \ldots$$
$$\bar{x}_0 = 0.$$

(These formulae were derived in Section 2.6.)

Proposition 5.1.4

The predictors associated with the family of models $\{\mathscr{M}(\theta): \theta \in D\}$ are uniformly stable if either:

(a) The models are the ARMAX models (5.1.5) and, for each $\theta \in D$, the zeros of $\sigma \to \det C_\theta(\sigma)$ lie outside the closed unit disc; or,
(b) The models are the state-space models (5.1.6) and, for each $\theta \in D$, the eigenvalues of $A_\theta - K_\theta H_\theta$ lie in the open unit disc.

PROOF (a) Here we consider ARMAX models (5.1.5). By hypothesis, the zeros of $\sigma \to \det C_\theta(\sigma)$ lie outside the closed unit disc. for all $\theta \in D$. The coefficients of the polynomial $C_\theta(\sigma)$ are continuous in θ on the open set $\tilde{\mathscr{D}}$ which contains D, and the determinant function is continuous. We can deduce therefore from the compactness of D that there exist a bounded open set $\mathscr{D}$ which satisfies

$$D \subset \mathscr{D} \subset \tilde{\mathscr{D}},$$

and $\varepsilon > 0$ such that the zeros of $\sigma \to \det C_\theta(\sigma)$ lie in the set $\{\sigma \in \mathbb{C}: |\sigma| > 1+\varepsilon\}$ for all $\theta \in \mathscr{D}$. We make this choice of $\mathscr{D}$.

The first two conditions in Definition 5.1.3 are obviously satisfied, and we attend only to the third.

Let α^{k-1} $(=(\alpha_{k-1},\ldots,\alpha_0))$, β^{k-1} and $\bar{\alpha}^{k-1}, \bar{\beta}^{k-1}$ be two pairs of vectors at which $f_k(\theta;\cdot,\cdot)$ is evaluated, and let θ_j be an arbitrary component of the parameter θ.

It is easy to show that Δf_k defined by

$$\Delta f_k = \begin{bmatrix} f_k(\theta;\alpha^{k-1},\beta^{k-1}) - f_k(\theta;\bar{\alpha}^{k-1},\bar{\beta}^{k-1}) \\ \dfrac{\partial}{\partial\theta_j} f_k(\theta;\alpha^{k-1},\beta^{k-1}) - \dfrac{\partial}{\partial\theta_j} f_k(\theta;\bar{\alpha}^{k-1},\bar{\beta}^{k-1}) \end{bmatrix},$$

is generated by the recursive equations

$$\begin{aligned} \tilde{C}_\theta(z^{-1})\Delta f_i &= \tilde{A}_\theta(z^{-1})(\alpha_{i-1} - \bar{\alpha}_{i-1}) + \tilde{B}_\theta(z^{-1})(\beta_{i-1} - \bar{\beta}_{i-1}) \\ &\qquad\qquad i = 0, 1, \ldots \\ \Delta f_i &= 0, \alpha_i = \bar{\alpha}_i = 0, \beta_i = \bar{\beta}_i = 0, \qquad i < 0 \end{aligned} \tag{5.1.7}$$

in which

$$\tilde{C}_\theta(\sigma) = \left[\begin{array}{c|c} C_\theta(\sigma) & 0 \\ \hline \dfrac{\partial}{\partial\theta} C_\theta(\sigma) & C_\theta(\sigma) \end{array}\right]$$

$$\tilde{A}_\theta(\sigma) = \sigma^{-1} \begin{bmatrix} C_\theta(\sigma) - A_\theta(\sigma) \\ \dfrac{\partial}{\partial\theta_j}[C_\theta(\sigma) - A_\theta(\sigma)] \end{bmatrix}$$

and

$$\tilde{B}_\theta(\sigma) = \begin{bmatrix} B_\theta(\sigma) \\ \dfrac{\partial}{\partial\theta_j} B_\theta(\sigma) \end{bmatrix}$$

By (5.1.7) and in view of the special structure of $\tilde{C}_\theta$, Δf_k can be expressed in terms of the composite vectors γ_i:

$$\begin{aligned} &\gamma_i = \operatorname{col}(\alpha_i - \bar{\alpha}_i, \beta_i - \bar{\beta}_i), \qquad i = 0, \ldots, k-1 \\ &\gamma_i = 0, \qquad i < 0 \end{aligned}$$

as

$$\Delta f_k = [g_\theta(z^{-1})]^{-1} G_\theta(z^{-1}) \gamma_{k-1}.$$

Here

$$g_\theta(\sigma) = (\det C_\theta(\sigma))^2$$

and $G_\theta(\sigma)$ is a polynomial in σ whose coefficients depend continuously on θ.

By construction, the zeros of $\sigma \to g_\theta(\sigma)$ are contained in $\{\sigma \in \mathbb{C}: |\sigma| > 1 + \varepsilon\}$ for all $\theta \in \mathscr{D}$. Since $\mathscr{D}$ is bounded, the coefficients of G_θ remain in bounded sets as θ ranges in $\mathscr{D}$. We now apply Proposition 2.1.2 when we identify $\{e_k\}$ with the deterministic sequence $\{\cdots 0, 0, \gamma_0, \gamma_1, \ldots\}$ and set $d = 1$. We conclude that there exist constants $\bar{c} > 0$ and $\lambda \in (0, 1)$, which do not depend on $\theta \in \mathscr{D}$ or k, such that

$$\| \Delta f_k \| \leq \bar{c} \sum_{i=0}^{k-1} \lambda^{k-i} \| \gamma_i \|. \tag{5.1.8}$$

We can assume that, in these inequalities, the norms are so chosen that $\|\gamma_i\| = \|\alpha_i\| + \|\beta_i\|$. Bearing in mind the definition of Δf_k and γ_i, we deduce from (5.1.8) that the third condition of Definition 5.1.3 is satisfied for our chosen λ and some $c > 0$.

(b) The case of state-space models can be treated in exactly the same way, following reformulation of the state-space equations as ARMAX equations. This is possible since the hypotheses on the state-space models imply that the associated ARMAX models satisfy the conditions of part (a). □

5.1.3 The identification criterion

Prediction error formulation of an identification method requires specification of a sequence of functions $l_k(\cdot,\cdot), k = 1, 2, \ldots,$ from the space $\mathbb{R}^q \times \mathbb{R}^r$ to the space of $d \times d$ matrices, and a real-valued function $h(\cdot)$ with domain the space of $d \times d$ matrices.

Let y^N, u^{N-1} be data at time N. Let $\varepsilon_k(\theta)$, $k = 1, 2, \ldots$ be the prediction errors associated with model $\mathscr{M}(\theta)$:

$$\varepsilon_k(\theta) = y_k - f_k(\theta; y^{k-1}, u^{k-1}), \qquad k = 1, 2, \ldots$$

We seek a parameter value which minimizes the identification criterion

$$\theta \to V_N(\theta; y^N, u^{N-1})$$

in which

$$V_N(\theta; y^N, u^{N-1}) = h(Q_N(\theta; y^N, u^{N-1}))$$

and

$$Q_N(\theta; y^N, u^{N-1}) = \frac{1}{N} \sum_{k=1}^{N} l_k(\theta, \varepsilon_k(\theta)).$$

It is convenient to collect together under the following definition those properties shared by the commonly used identification criteria.

Definition 5.1.5

The identification criterion is said to be *quadratically bounded* if $h(\cdot)$ is a continuous function and there exist an open neighbourhood $\mathscr{D}$ of the parameter constraint set D and a constant $c > 0$ with the properties:

(a) The functions $l_k(\cdot, \cdot)$, $k = 1, 2, \ldots$ are continuously differentiable on $\mathscr{D} \times \mathbb{R}^r$

(b) $\|l_k(\theta, 0)\| \leq c$, for all $\theta \in \mathscr{D}$, $k = 1, 2, \ldots$

(c) $\left\| \frac{\partial}{\partial \varepsilon} l_k(\theta, \varepsilon) \right\| \leq c\|\varepsilon\|$, for all $\theta \in \mathscr{D}$, $\varepsilon \in \mathbb{R}^r$, $k = 1, 2, \ldots$ and;

(d) $\left\| \frac{\partial}{\partial \theta} l_k(\theta, \varepsilon) \right\| \leq c\|\varepsilon\|^2$, for all $\theta \in \mathscr{D}$, $\varepsilon \in \mathbb{R}^r$, $k = 1, 2, \ldots$

Examples of quadratically bounded identification criteria are the least squares criterion

$$V_N(\theta, y^N, u^{N-1}) = \frac{1}{N} \sum_{k=1}^{N} \varepsilon_k^{\mathrm{T}}(\theta) W_k \varepsilon_k(\theta)$$

in which the weighting matrices W_k are uniformly bounded; and the criterion

$$V_N(\theta, y^N, u^{N-1}) = \det \frac{1}{N} \sum_{k=1}^{N} \varepsilon_k(\theta) \varepsilon_k^{\mathrm{T}}(\theta)$$

which, we recall, arises in connection with maximum likelihood identification methods when the disturbances are assumed to be gaussian (see Section 4.4).

5.2 Asymptotic properties of the parameter estimates

Let θ_N minimize the identification criterion

$$\theta \to V_N(\theta; y^N, u^{N-1})$$

associated with data y^N, u^{N-1} at time N. θ_N then is a parameter estimate obtained under ideal circumstances when exact criterion minimization is possible.

We consider the problem of characterizing the set into which the estimate θ_N converges as the length N of the data record tends to infinity.

In the event that the limit

$$\lim_{N\to\infty} EV_N(\theta; y^N, u^{N-1}) \tag{5.2.1}$$

exists for all $\theta \in D$, we might expect that, as $N \to \infty$, θ_N converges into the set

$$\{\theta: \lim_{N\to\infty} EV_N(\theta; y^N, u^{N-1}) = \min_{\psi\in D} \lim_{N\to\infty} EV_N(\psi; y^N, u^{N-1})\} \tag{5.2.2}$$

almost surely. This would mean that the identification method supplies a parameter value which minimizes the expected value of the identification criterion in the limit as $N \to \infty$.

The description we now give of the asymptotic properties of θ_N involves a limiting set which has the general features of the set (5.2.2), but differs from it in two respects. Firstly, $EV_N(\theta; y^N, u^{N-1})$, which can be written $Eh(Q_N(\theta; y^N, u^{N-1}))$, is replaced by $h(EQ_N(\theta; y^N, u^{N-1}))$. This means that, in the absence of further assumptions, we sacrifice the interpretation of the limiting set as a set of parameter values which minimize, in the limit, the expected value of the original identification criterion. Secondly, we make allowance for possible non-existence of the limit

$$\lim_{N\to\infty} h(EQ_N(\theta; y^N, u^{N-1}))$$

by introduction of the 'lim inf' operation.

Theorem 5.2.1

Suppose that the system which generates the data is stable, the predictors associated with the model set are uniformly stable, and the identification criterion is quadratically bounded. Then

$$\theta_N \to D_I, \qquad \text{a.s.,} \qquad \text{as } N \to \infty,$$

where

$$D_I = \{\theta: \max_{\psi\in D} \liminf_{N\to\infty} [h(EQ_N(\theta)) - h(EQ_N(\psi))] = 0\}. \tag{5.2.3}$$

Convergence is understood in the sense that, almost surely, for every $\varepsilon > 0$ there exists $\bar{N}$ such that, for all $N > \bar{N}$, the set $\{\theta : |\theta - \theta_N| \leq \varepsilon\} \cap D_I$ is non-empty.

PROOF See Appendix A. □

We refer to Definitions 5.1.1, 5.1.3 and 5.1.5 for explanation of the terms 'stable' system, 'uniformly stable' predictors and 'quadratically bounded' criterion.

The theorem is important because of the ease with which it can be applied to give more explicit information about the limit set in special situations of interest. A number of such applications will shortly be given.

To illustrate the connection between the set D_I and the set (5.2.2), let us suppose that the function $h(\cdot)$ is linear and that the limit (5.2.1) exists, for any $\theta \in D$. Under these assumptions the expectation operator and the action of h commute and we have, for given $\theta \in D$,

$$\max_{\psi \in D} \liminf_{N \to \infty} [h(EQ_N(\theta; y^N, u^{N-1})) - h(EQ_N(\psi; y^N, u^{N-1}))]$$

$$= \lim_{N \to \infty} EV_N(\theta; y^N, u^{N-1}) - \min_{\psi \in D} \lim_{N \to \infty} EV_N(\psi; y^N, u^{N-1}).$$

It is clear from this equation that the two sets D_I and (5.2.2) coincide.

5.3 Consistency

An indication that an identification scheme will perform satisfactorily is provided by the property that, if the data is assumed to be generated by a system which, asymptotically, is indistinguishable from a model in the model set considered, then the parameter estimate θ_N will converge into the set of parameter values associated with such models. This is the property of consistency.

We shall interpret the notion that the data is generated by a system asymptotically indistinguishable from one of the models as meaning that the set D_T, defined by

$$D_T = \left\{ \theta \in D : \lim_{N \to \infty} \frac{1}{N} \sum_{k=1}^{N} E\| \hat{y}_k - \hat{y}_k(\theta) \|^2 = 0 \right\}, \tag{5.3.1}$$

is non-empty. Here $\hat{y}_k$ is the conditional expectation of y_k given $y_{k-1}, y_{k-2}, \ldots$ and $u_{k-1}, u_{k-2}, \ldots$, and $\hat{y}_k(\theta)$ is the (one-step-ahead) predictor associated with model $\mathscr{M}(\theta)$, evaluated at the data.

According to this interpretation, we assess a model by the quality of its predictors and view $\hat{y}_k$ as the best predictor of y_k given knowledge of the true system. The model $\mathcal{M}(\theta)$ and the system are taken to be asymptotically indistinguishable if the mean square difference between the predictors supplied by $\mathcal{M}(\theta)$ and those available if the system were known is zero in the limit as the number of data points tends to infinity.

For the sake of simplicity we now limit attention to identification criteria which arise in least squares estimation,

$$V_N(\theta) = \operatorname{trace}\left[\frac{1}{N}\sum_{k=1}^{N} \varepsilon_k(\theta)\varepsilon_k^{\mathrm{T}}(\theta)\right],$$

and in maximum likelihood estimation based on assumptions of gaussian disturbances, namely,

$$V_N(\theta) = \det\left[\frac{1}{N}\sum_{k=1}^{N} \varepsilon_k(\theta)\varepsilon_k^{\mathrm{T}}(\theta)\right]$$

(see Section 4.4).

When such criteria are adopted, and when the system is asymptotically indistinguishable from a member of the model set, a more refined description of the limiting set can be given than that in Theorem 5.2.1.

Proposition 5.3.1

Suppose that the data is generated by a stable system and the predictors associated with the model set are uniformly stable. Suppose also that D_T (given by (5.3.1)) is non-empty, that

$$l_k(\theta, \varepsilon) = \varepsilon\varepsilon^{\mathrm{T}}, \qquad \text{for } k = 1, 2, \ldots, \theta \in \mathbb{R}^q, \quad \varepsilon \in \mathbb{R}^p,$$

and either

$$h(\cdot) = \operatorname{trace}(\cdot) \tag{5.3.2a}$$

or

$$h(\cdot) = \det(\cdot), \tag{5.3.2b}$$

and there exists $\delta > 0$ such that

$$E[(y_k - \hat{y}_k)(y_k - \hat{y}_k)^{\mathrm{T}}] > \delta I, \qquad \text{for } k = 1, 2, \ldots \tag{5.3.3}$$

Then

$$\theta_N \to \left\{\theta : \liminf_{N\to\infty} \frac{1}{N}\sum_{k=1}^{N} E\|\hat{y}_k - \hat{y}_k(\theta)\|^2 = 0\right\}, \quad \text{a.s.} \qquad \square$$

Proof of the proposition requires the following estimate on the determinant of the sum of two symmetric matrices.

Lemma 5.3.2

Let A, B be symmetric $n \times n$ matrices. Suppose that A is positive definite and that B is non-negative definite. Then

$$\det(A+B) \geq \det(A) + \frac{\det(A)}{n\lambda_{\max}(A)}\operatorname{trace}\{B\}.$$

Here $\lambda_{\max}(A)$ denotes the maximum eigenvalue of A.

PROOF Let $A^{1/2}$ be a positive definite square root of A (see Appendix D.1). By the properties of the determinant function

$$\begin{aligned}\det(A+B) &= \det(A^{1/2}(I + A^{-1/2}BA^{-1/2})A^{1/2}) \\ &= \det(A^{1/2})\det(I + A^{-1/2}BA^{-1/2})\det(A^{1/2}) \\ &= \det(A)\det(I + A^{-1/2}BA^{-1/2}) \\ &= \det(A)\prod_{i=1}^{n}(1+d_i). \qquad (5.3.4)\end{aligned}$$

Here the d_i are the eigenvalues of $A^{-1/2}BA^{-1/2}$.
Since the trace of B is equal to the sum of the eigenvalues of B,

$$\lambda_{\max}(B) \geq \operatorname{trace}\{B\}/n. \qquad (5.3.5)$$

Let y be an eigenvector of B corresponding to $\lambda_{\max}(B)$ and choose x such that $y = A^{-1/2}x$. Then, since for a symmetric matrix S

$$\lambda_{\max}(S) = \max_{z \neq 0}\frac{z^{\mathrm{T}}Sz}{\|z\|^2}$$

and the maximum is achieved at any eigenvector corrresponding to $\lambda_{\max}(S)$, we have

$$\begin{aligned}\lambda_{\max}(A^{-1/2}BA^{-1/2}) &= \max_{z \neq 0}\frac{z^{\mathrm{T}}A^{-1/2}BA^{-1/2}z}{\|z\|^2} \\ &\geq \frac{x^{\mathrm{T}}A^{-1/2}BA^{-1/2}x}{\|x\|^2} \\ &= \frac{y^{\mathrm{T}}By}{\|y\|^2}\frac{\|y\|^2}{y^{\mathrm{T}}Ay} \\ &\geq \lambda_{\max}(B)\Big/\max_{z \neq 0}\frac{z^{\mathrm{T}}Az}{\|z\|^2} \\ &\geq \frac{\lambda_{\max}(B)}{\lambda_{\max}(A)}.\end{aligned}$$

But the eigenvalues d_i, $i = 1,\dots,n$, are non-negative; it follows that

$$\prod_i (1 + d_i) \geq 1 + \frac{\lambda_{\max}(B)}{\lambda_{\max}(A)}.$$

We deduce now from this inequality, (5.3.4) and (5.3.5) that

$$\det(A + B) \geq \det(A) + \frac{\det(A)}{n\lambda_{\max}(A)} \operatorname{trace}\{B\}. \qquad \square$$

PROOF OF PROPOSITION 5.3.1 The identification criterion is (in either case (5.3.2a) or (5.3.2b)) quadratically bounded, the system is stable and the predictors are uniformly stable. It follows from Theorem 5.2.1 then that θ_N converges, a.s., into the set D_I (see (5.2.3)). Take $\theta \in D_I$. We shall show that

$$\liminf_{N \to \infty} \left\{ \frac{1}{N} \sum_{k=1}^{N} \| \hat{y}_k - \hat{y}_k(\theta) \|^2 \right\} = 0 \tag{5.3.6}$$

and thereby prove the proposition.

We define v_k, $k = 1, 2, \dots$ by

$$v_k = y_k - \hat{y}_k.$$

Then, for $k = 1, 2, \dots,$ and $\psi \in D$

$$\begin{aligned} E\varepsilon_k(\psi)\varepsilon_k^{\mathrm{T}}(\psi) &= E[y_k - \hat{y}_k(\psi)][y_k - \hat{y}_k(\psi)]^{\mathrm{T}} \\ &= E[\hat{y}_k - \hat{y}(\psi) + v_k][\hat{y}_k - \hat{y}_k(\psi) + v_k]^{\mathrm{T}} \end{aligned}$$

by definition of v_k

$$= E[\hat{y}_k - \hat{y}_k(\psi)][\hat{y}_k - \hat{y}_k(\psi)]^{\mathrm{T}} + E v_k v_k^{\mathrm{T}} \tag{5.3.7}$$

since v_k has zero mean and is uncorrelated with y^{k-1}, u^{k-1}.

Suppose first that (5.3.2a) is true. Take ψ to be any point in D. Then since $\theta \in D_I$,

$$\begin{aligned} 0 \geq \liminf_{N \to \infty} &\left\{ \operatorname{trace}\left(E \frac{1}{N} \sum_{k=1}^{N} \varepsilon_k(\theta)\varepsilon_k^{\mathrm{T}}(\theta) \right) \right. \\ &\left. - \operatorname{trace}\left(E \frac{1}{N} \sum_{k=1}^{N} \varepsilon_k(\psi)\varepsilon_k^{\mathrm{T}}(\psi) \right) \right\} \\ = \liminf_{N \to \infty} &\left\{ \operatorname{trace}\left(\frac{1}{N} \sum_{k=1}^{N} E[\hat{y}_k - \hat{y}_k(\theta)][\hat{y}_k - \hat{y}_k(\theta)]^{\mathrm{T}} \right) \right. \\ &\left. - \operatorname{trace}\left(\frac{1}{N} \sum_{k=1}^{N} E[\hat{y}_k - \hat{y}_k(\psi)][\hat{y}_k - \hat{y}_k(\psi)]^{\mathrm{T}} \right) \right\} \end{aligned}$$

by (5.3.7)

$$= \liminf_{N\to\infty} \left\{ \frac{1}{N} \sum_{k=1}^{N} E\|\hat{y}_k - \hat{y}_k(\theta)\|^2 - \frac{1}{N} \sum_{k=1}^{N} E\|\hat{y}_k - \hat{y}_k(\psi)\|^2 \right\}. \tag{5.3.8}$$

Since D_T is non-empty, by assumption, we may choose $\psi \in D_T$. Then

$$\lim_{N\to\infty} \frac{1}{N} \sum_{k=1}^{N} E\|\hat{y}_k - \hat{y}_k(\psi)\|^2$$

exists and is zero. It follows from (5.3.8) that

$$0 \geq \liminf_{N\to\infty} \left\{ \frac{1}{N} \sum_{k=1}^{N} E\|\hat{y}_k - \hat{y}_k(\theta)\|^2 \right\}.$$

We have shown (5.3.6), as required.

Now suppose that (5.3.2b) is true. Again take ψ an arbitrary vector in D. In this case we deduce from the hypothesis $\theta \in D_I$, and (5.3.7), that

$$0 \geq \liminf_{N\to\infty} \left\{ \det E \frac{1}{N} \sum_{k=1}^{N} \varepsilon_k(\theta)\varepsilon_k^{\mathrm{T}}(\theta) - \det E \frac{1}{N} \sum_{k=1}^{N} \varepsilon_k(\psi)\varepsilon_k^{\mathrm{T}}(\psi) \right\}$$

$$= \liminf_{N\to\infty} \left\{ \det(S_N(\theta) + P_N) - \det(S_N(\psi) + P_N) \right\}$$

where

$$S_N(\eta) = \frac{1}{N} \sum_{k=1}^{N} E(\hat{y}_k - \hat{y}_k(\eta))(\hat{y}_k - \hat{y}_k(\eta))^{\mathrm{T}}$$

and

$$P_N = \frac{1}{N} \sum_{k=1}^{N} E v_k v_k^{\mathrm{T}}.$$

This inequality can be written

$$\liminf_{N\to\infty} \{\det(S_N(\theta) + P_N) - \det P_N - \Delta_N(\psi)\} \leq 0 \tag{5.3.9}$$

in which $\Delta_N(\psi) = \det(S_N(\psi) + P_N) - \det(P_N)$.

We now choose $\psi \in D_T$. It is easy to show that $S_N(\psi) \to 0$ as $N \to \infty$. Since $\{P_N\}$ is a bounded sequence (the v_k have uniformly bounded second order moments, remember) and the function $\det(\cdot)$ is continuous, we have that the functions $L \to \det(L + P_N)$ are continuous at zero, uniformly in N. It follows that

$$\lim_{N\to\infty} \Delta_N(\psi)$$

exists and is zero. Equation (5.3.9) therefore implies

$$\liminf_{N\to\infty} \{\det(S_N(\theta) + P_N) - \det P_N\} \leq 0. \tag{5.3.10}$$

However, in view of hypothesis (5.3.3) and the uniform boundedness of the second moments of the v_k, there exist $\alpha, \bar{\alpha} > 0$ such that $\det(P_k) < \alpha$ and $\lambda_{\max}(P_k) > \bar{\alpha}$, $k = 1, 2, \ldots$ We deduce from Lemma 5.3.2 that there exists also some $c > 0$ such that

$$\det\{S_N(\theta) + P_N\} - \det P_N \geq c \operatorname{trace}\{S_N(\theta)\}$$

$$= c\left[\frac{1}{N}\sum_{k=1}^{N} \|\hat{y}_k - \hat{y}_k(\theta)\|^2\right] \qquad \text{for } N = 1, 2, \ldots$$

From this inequality and (5.3.10) it follows that

$$\liminf_{N\to\infty}\left\{\frac{1}{N}\sum_{k=1}^{N} \|\hat{y} - \hat{y}_k(\theta)\|^2\right\} \leq 0.$$

We have shown that (5.3.6) is true. □

Consistency properties of identification schemes which involve a variety of model sets can be deduced from Proposition 5.3.1, when the system is describable by a model in the model set. We find, typically, that a scheme is consistent provided the closed-loop system which generates the data is stable, the models supply uniformly stable predictors and that an additive term in the input is 'persistently exciting'.

A persistently exciting input $\{w_k, k \in \mathbb{Z}\}$ is one which, loosely speaking, is sufficiently varied that the resulting data provides as much information about the input/output characteristics of stable linear systems on which it acts as consideration of *all* possible inputs. The precise form that our definition of a persistently exciting input takes is suggested by consideration of least squares identification of the parameters $a_1, \ldots, a_M$ in the system described by the equations

$$y_k = a_1 w_{k-1} + \cdots + a_M w_{k-M} + e_k, \qquad k = 1, 2, \ldots, \tag{5.3.11}$$

from observations of $\{y_k\}$ and knowledge of the inputs, here written

$\{w_k\}$. $\{e_k\}$ is a sequence of zero-mean, uncorrelated, random variables with common variance $\sigma^2(\sigma^2 > 0)$. The least squares estimate $\hat{a}_{(N+1)}$ of the unknown parameters a with components $a_1, \ldots, a_M$, based on $N+1$ data points, is unbiased and has covariance

$$\Sigma_N = \sigma^2 \left[\sum_{k=1}^{N} w_k(M) w_k^{\mathrm{T}}(M) \right]^{-1}$$

where

$$w_k(M) = \mathrm{col}[w_k, w_{k-1}, \ldots, w_{k-M}]. \tag{5.3.12}$$

(These properties are deduced from the results of Section 4.3.)

A sufficient condition that the estimates $\hat{a}_{(N)}$ converge to the true parameter values in mean square, and therefore that the input/output characteristics of the system are fully determined in the limit as $N \to \infty$, is that there exists $\delta > 0$ such that

$$\frac{1}{N} \sum_{i=1}^{N} w_k(M) w_k^{\mathrm{T}}(M) \geq \delta I \tag{5.3.13}$$

(in the sense of the usual ordering of symmetric matrices), for all N sufficiently large. For then, given arbitrary $\xi \in \mathbb{R}^M$, the variance $\xi^{\mathrm{T}} \Sigma_N \xi$ of the estimate $\xi^{\mathrm{T}} \hat{a}_{(N)}$ of $\xi^{\mathrm{T}} a$ is bounded by $\sigma^2 \| \xi \|^2 / \delta N$, and this last number tends to zero as $N \to \infty$.

Existence of a number $\delta > 0$ such that (5.3.13) holds is a suitable defining property for a persistently exciting input relevant to simple models with the description (5.3.11); for more complex systems it is often necessary to modify the definition and require that (5.3.13) holds for arbitrary M.

Definition 5.3.3

Let w_k, $k \in \mathbb{Z}$, be a sequence of vector random variables with uniformly bounded second moments. For any integer $M \geq 1$, we define $w_k(M)$ by (5.3.12). The sequence $\{w_k\}$ is said to be a *persistently exciting sequence of order* M if there exist $\delta > 0$ and $N_0 > 0$ such that

$$\frac{1}{N} \sum_{k=1}^{N} E w_k(M) w_k^{\mathrm{T}}(M) \geq \delta I, \qquad \text{for } N \geq N_0. \tag{5.3.14}$$

It is said to be *persistently exciting of infinite order* if there exists $\delta > 0$ with the following property: corresponding to any integer $M \geq 1$, an integer N_0 can be chosen such that (5.3.14) is satisfied.

Notice that we have defined sequences of *random variables* which are persistently exciting; this is with a view to considering inputs which involve, possibly, a random component. Of course the definition subsumes that of persistently exciting deterministic sequences.

We see that any sequences $\{w_k\}$ of zero-mean, independent vector random variables with uniformly bounded second-order moments is a persistently exciting (stochastic) sequence of infinite order, provided there exists $\delta > 0$ such that $Ew_k w_k^{\mathrm{T}} \geq \delta I$, for all k sufficiently large. If we assume in addition that the sequence $\{w_k\}$ has uniformly bounded fourth-order moments, then realizations of $\{w_k\}$ define persistently exciting deterministic sequences of infinite order, almost surely (this last assertion can be deduced from Theorem 1.1.15 and the fact that a countable intersection of probability-one events is a probability-one event). Recursive procedures which generate persistently exciting deterministic sequences are also available.

The following lemma provides a direct connection between the notion of a persistently exciting input as we have defined it and the property that the input uniquely determines the input/output characteristics of stable systems on which it acts.

Lemma 5.3.4

Let $L(\sigma)$ be an $r \times m$ matrix of rational functions in σ. It is assumed the zeros of the denominator of each entry of $L(\sigma)$ lie outside the closed unit disc. Let $\{w_k\}_{k\in\mathbb{Z}}$ be a sequence of m-vector random variables which have uniformly bounded second-order moments, and suppose that

$$\liminf_{N\to\infty} \frac{1}{N} \sum_{k=1}^{N} E \| L(z^{-1}) w_k \|^2 = 0. \tag{5.3.15}$$

Then $L(\sigma)$ is identically zero if either of the following conditions is satisfied:

(a) $L(\sigma)$ can be expressed as a polynomial in σ:

$$L(\sigma) = L_0 + L_1\sigma + \cdots + L_M\sigma^M$$

of degree M and $\{w_k\}$ is a persistently exciting sequence of order M.

(b) $\{w_k\}$ is a persistently exciting sequence of infinite order.

PROOF (a) Suppose that $L(\sigma) = L_0 + L_1\sigma + \cdots + L_M\sigma^M$ and $\{w_k\}$ is persistently exciting of order M. In this case

$$L(z^{-1})w_k = L(M)w_k(M)$$

in which $L(M) = [L_0 \vdots L_1 \vdots \ldots \vdots L_M]$ and $w_k(M) = \mathrm{col}[w_k, w_{k-1}, \ldots, w_{k-M}]$. For $k = 1, 2, \ldots,$

$$\|L(z^{-1})w_k\|^2 = \|L(M)w_k(M)\|^2 = \mathrm{trace}\, L(M)w_k(M)w_k^{\mathrm{T}}(M)L^{\mathrm{T}}(M)$$

$$= \sum_{i=1}^{r} l_i^{\mathrm{T}} w_k(M)w_k^{\mathrm{T}}(M)l_i.$$

In this last expression, l_i denotes the ith column of $L^{\mathrm{T}}(M)$. It follows that

$$\frac{1}{N}\sum_{k=1}^{N} E\|L(z^{-1})w_k\|^2 = \sum_{i=1}^{r} l_i^{\mathrm{T}}\left[\frac{1}{N}\sum_{k=1}^{N} Ew_k(M)w_k^{\mathrm{T}}(M)\right]l_i.$$

By the persistent excitation hypothesis, however, there exist $\delta > 0$ and N_0 such that

$$\frac{1}{N}\sum_{k=1}^{N} Ew_k(M)w_k^{\mathrm{T}}(M) \geq \delta I, \qquad \text{for } N \geq N_0.$$

For $N \geq N_0$, then,

$$\frac{1}{N}\sum_{k=1}^{N} E\|L(z^{-1})w_k\|^2 \geq \delta \sum_{i=1}^{r} l_i^{\mathrm{T}} l_i = \delta \sum_{i=0}^{M} \|L_i\|^2. \tag{5.3.16}$$

Here, $\|L_i\|$ denotes the trace norm $(\mathrm{trace}[L_iL_i^{\mathrm{T}}])^{1/2}$. By (5.3.15) and (5.3.16)

$$0 = \liminf_{N\to\infty} \frac{1}{N}\sum_{k=1}^{N} E\|L(z^{-1})w_k\|^2 \geq \delta \sum_{i=0}^{M} \|L_i\|^2$$

which implies that $L(\sigma) = L_0 + L_1\sigma + \cdots + L_M\sigma^M = 0$.

(b) Now suppose that $\{w_k\}$ is a persistently exciting sequence of infinite order. Let $\{L_i\}$ be the coefficients in the formal expansion of $L(\sigma)$ about $\sigma = 0$:

$$L(\sigma) = \sum_{i=0}^{\infty} L_i\sigma^i.$$

Bearing in mind that the w_k have uniformly bounded second-order moments, we can deduce from Proposition 2.1.2 that there exist constants $c > 0$ and $\lambda \in (0, 1)$, such that for arbitrary k, J, K,

$$E\left\|\sum_{i=J}^{K} L_i w_{k-i}\right\|^2 \leq c\lambda^J. \tag{5.3.17}$$

Given an integer $M > 0$, $L(z^{-1})w_k$ can be written as the sum of two terms

$$L(z^{-1})w_k = \sum_{i=0}^{M} L_i w_{k-i} + \sum_{i=M+1}^{\infty} L_i w_{k-i}.$$

Writing $\sum_{i=0}^{M} L_i w_{k-i}$ simply as $\sum$, we have

$$\begin{aligned} E\|L(z^{-1})w_k\|^2 &= E\left\|\sum_{i=0}^{M}\right\|^2 + 2E\left[\left(\sum_{i=0}^{M}\right)^{\mathrm{T}}\left(\sum_{i=M+1}^{\infty}\right)\right] + E\left\|\sum_{i=M+1}^{\infty}\right\|^2 \\ &\geq E\left\|\sum_{i=0}^{M}\right\|^2 - 2\left(E\left\|\sum_{i=0}^{M}\right\|^2\right)^{1/2}\left(E\left\|\sum_{i=M+1}^{\infty}\right\|^2\right)^{1/2} \end{aligned}$$

by Schwarz's inequality

$$\geq E\left\|\sum_{i=0}^{M} L_i w_{k-i}\right\|^2 - 2c\lambda^{(M+1)/2}.$$

by (5.3.17). It follows that, for $N = 1, 2, \ldots,$

$$\frac{1}{N}\sum_{k=1}^{N} E\|L(z^{-1})w_k\|^2 \geq \frac{1}{N}\sum_{k=1}^{N} E\left\|\sum_{i=0}^{M} L_i w_{k-i}\right\|^2 - 2c\lambda^{(M+1)'2}. \quad (5.3.18)$$

$\{w_k\}$ is a persistently exciting sequence of infinite order and so is certainly of order M; in consequence there exist $\delta > 0$ (δ does not depend on M) and $N_0 > 0$ such that (5.3.14) is true. We deduce, using the arguments of part (a), that

$$\frac{1}{N}\sum_{k=1}^{N} E\left\|\sum_{i=0}^{M} L_i w_{k-i}\right\|^2 \geq \delta \sum_{i=0}^{M}\|L_i\|^2 \quad \text{for } N \geq N_0 \quad (5.3.19)$$

It follows from (5.3.15), (5.3.18) and (5.3.19) that

$$0 = \liminf_{N\to\infty}\frac{1}{N}\sum_{k=1}^{N} E\|L(z^{-1})w_k\|^2 \geq \delta\sum_{i=0}^{M}\|L_i\|^2 - 2c^2\lambda^{M+1}.$$

Taking the limit $M \to \infty$ and remembering that δ does not depend on M, we deduce that

$$\sum_{i=0}^{\infty}\|L_i\|^2 = 0.$$

It follows that $L(\sigma) = \sum_{i=0}^{\infty} L_i \sigma^i = 0$. □

It is clear from this lemma that, if $L_1(\sigma)$ and $L_2(\sigma)$ are stable transfer functions, if w_k is a persistently exciting sequence of infinite order, and if $L_1(z^{-1})w_k$ and $L_2(z^{-1})w_k$ are sufficiently close in the sense that

$$\liminf_{N\to\infty} \frac{1}{N} \sum_{i=0}^{\infty} E\|L_1(z^{-1})w_k - L_2(z^{-1})w_k\|^2 = 0,$$

then $L_1 = L_2$.

Finally we prove a representative theorem on consistency. Here ARMAX models are considered:

$$A_\theta(z^{-1})y_k = B_\theta(z^{-1})u_{k-1} + C_\theta(z^{-1})e_k.$$

Associated with each model are transfer functions $A_\theta(\cdot)^{-1}B_\theta(\cdot)$ and $A_\theta^{-1}(\cdot)C_\theta(\cdot)$ which relate the outputs to the inputs and disturbances. The theorem asserts that the estimates provide, in the limit, the correct transfer functions. (We view two transfer functions as the same if their values coincide wherever they are both defined). This is clearly the best that can be achieved from a procedure which involves processing input/output data.

Theorem 5.3.5

Consider a system described by the ARMAX equations

$$A_0(z^{-1})y_k = B_0(z^{-1})u_{k-1} + C_0(z^{-1})e_k, \qquad k\in\mathbb{Z}$$

in which $A_0(\sigma)$, $B_0(\sigma)$, $C_0(\sigma)$ are polynomials with matrix coefficients. The e_k are independent, zero-mean random variables which have uniformly bounded fourth-order moments and are such that

$$Ee_k e_k^{\mathrm{T}} \geq \alpha I, \qquad \text{for } k\in\mathbb{Z}$$

for some $\alpha > 0$.

Suppose that the control u_k is generated by linear feedback acting on y^k and u^{k-1} with an additive disturbance, w_k, independent of e_j, $j\in\mathbb{Z}$:

$$H(z^{-1})u_k = F(z^{-1})y_k + w_k \tag{5.3.20}$$

and that the w_k have uniformly bounded fourth-order moments. Here, $H(\sigma)$ and $F(\sigma)$ are polynomials with matrix coefficients.

Let the predictors be calculated on the basis of models $\{\mathcal{M}(\theta): \theta\in D\}$:

$$A_\theta(z^{-1})y_k = B_\theta(z^{-1})u_{k-1} + C_\theta(z^{-1})e_k, \qquad k = 0, 1, \ldots$$

$$y_k = 0, \qquad u_k = 0, \qquad e_k = 0, \qquad k < 0$$

in which $A(\sigma)$, $B(\sigma)$ and $C(\sigma)$ are polynomials with matrix coefficients continuously differentiable in θ on some open set $\mathscr{D}$ containing D, a compact set.

We assume the following.

Conditions on the identification criterion:

$$V_N(\theta) = \frac{1}{N} \sum_{k=1}^{N} \|\varepsilon_k(\theta)\|^2$$

or

$$V_N(\theta) = \det \frac{1}{N}\left[\sum_{k=1}^{N} \varepsilon_k(\theta)\varepsilon_k^{\mathrm{T}}(\theta)\right]$$

in which $\varepsilon_k(\theta)$ is the prediction error associated with $\mathscr{M}(\theta)$.

Stability of the closed loop system:

The zeros of $\sigma \to \det(A_0(\sigma) - \sigma B_0(\sigma)H^{-1}(\sigma)F(\sigma))$ and of $\sigma \to \det H(\sigma)$ lie outside the closed unit disc.

Uniform stability of the predictors:

$C_\theta(0) = I$, $A_\theta(0) = I$ and the zeros of $\sigma \to \det C_\theta(\sigma)$ lie outside the closed unit disc for all $\theta \in D$.

The true system can be represented within the model set:

$$A_{\theta^*}^{-1}(\cdot)B_{\theta^*}(\cdot) = A_0^{-1}(\cdot)B_0(\cdot) \quad \text{and} \quad A_{\theta^*}^{-1}(\cdot)C_{\theta^*}(\cdot) = A_0^{-1}(\cdot)C_0(\cdot)$$

for some $\theta^* \in D$.

Persistent excitation:

The disturbance w_k is a persistently exciting sequence of infinite order.

Then

$$\begin{aligned}\theta_N \to \{\theta : A_\theta^{-1}(\cdot)B_\theta(\cdot) &= A_0^{-1}(\cdot)B_0(\cdot)\\ \text{and} \quad A_\theta^{-1}(\cdot)C_\theta(\cdot) &= A_0^{-1}(\cdot)C_0(\cdot)\}, \qquad \text{a.s.}\end{aligned}$$

as $N \to \infty$.

PROOF We first check that asymptotic analysis of θ_N is covered by Proposition 5.3.1. In view of Propositions 5.1.2 and 5.1.4 the system which generates the data is stable, and the predictors are uniformly stable. The hypotheses on the identification criterion imposed in Proposition 5.3.1 are also true since $e_k = y_k - \hat{y}_k$ and, by assumption, $e_k e_k^{\mathrm{T}} \geq \alpha I$, $k \in \mathbb{Z}$, for some $\alpha > 0$.

Proposition 5.3.1 will be applicable then if we can show that

$$\lim_{N\to\infty} \frac{1}{N} \sum_{k=1}^{N} E\|\hat{y}_k - \hat{y}_k(\theta^*)\|^2 = 0 \tag{5.3.21}$$

for this will mean that the set D_T defined by (5.3.1) is non-empty.

Now $\hat{y}_k$ is given by

$$\hat{y}_k = (I - C_0^{-1}A_0)y_k + z^{-1}C_0^{-1}B_0u_k \tag{5.3.22}$$

(for simplicity we have written A_0 in place of $A_0(z^{-1})$, etc.). The predictors $\hat{y}_k(\theta)$ supplied by $\mathcal{M}(\theta)$, $\theta \in D$, are

$$\hat{y}_k(\theta) = (I - C_\theta^{-1}A_\theta)(y_k - \xi_k) + z^{-1}C_\theta^{-1}B_\theta(u_k - \eta_k) \tag{5.3.23}$$

in which

$$\xi_k = \begin{cases} 0 & k \geq 0 \\ y_k & k < 0 \end{cases} \quad \text{and} \quad \eta_k = \begin{cases} 0 & k \geq 0 \\ u_k & k < 0 \end{cases}.$$

The presence of the terms ξ_k, η_k in this last equation is due to our choice of zero data values prior to time $k = 0$ for the purpose of calculating the predictors.

By hypothesis, $A_{\theta^*}^{-1}B_{\theta^*} = A_0^{-1}B_0$, $A_{\theta^*}^{-1}C_{\theta^*} = A_0^{-1}C_0$. Since $C_\theta(\sigma)$ is invertible for some σ ($C_\theta(0) = I$, remember), it follows that $C_{\theta^*}^{-1}A_{\theta^*} = C_0^{-1}A_0$. But then $C_{\theta^*}^{-1}B_{\theta^*} = C_0^{-1}B_0$ since

$$C_{\theta^*}^{-1}B_{\theta^*} = C_{\theta^*}^{-1}A_{\theta^*}A_{\theta^*}^{-1}B_{\theta^*} = C_0^{-1}A_0A_0^{-1}B_0 = C_0^{-1}B_0.$$

Subtracting equation (5.3.23) from equation (5.3.22) therefore gives

$$\hat{y}_k - \hat{y}_k(\theta^*) = (I - C_{\theta^*}^{-1}A_{\theta^*})\xi_k + z^{-1}C_{\theta^*}^{-1}B_{\theta^*}\eta_k. \tag{5.3.24}$$

Now the y_k are uniformly bounded in second moment since the closed-loop system is stable. Because the roots of $\sigma \to \det H(\sigma)$ lie outside the closed unit disc and the w_k are uniformly bounded in second moment, it follows from Proposition 2.1.2 that the u_k given by the feedback equation (5.3.20) are also bounded in second moment. Since the η_k and ξ_k are zero for $k \geq 0$, and since the zeros of $\sigma \to \det C_{\theta^*}(\sigma)$ lie outside the closed unit disc, it follows from

Proposition 2.1.1 applied to equation (5.3.24) that

$$E\|\hat{y}_k - y_k(\theta^*)\|^2 \to 0 \qquad \text{as } k \to \infty. \tag{5.3.25}$$

The property (5.3.21) is then true and we can apply Proposition 5.3.1. This gives

$$\theta_N \to \tilde{D}_I$$

where

$$\tilde{D}_I = \left\{\theta : \liminf_{N\to\infty} \frac{1}{N}\sum_{k=1}^{N} E\|\hat{y}_k - \hat{y}_k(\theta)\|^2 = 0\right\}.$$

Take θ an arbitrary point in $\tilde{D}_I$. We shall complete the proof by showing that $A_\theta^{-1}B_\theta = A_0^{-1}B_0$ and $A_\theta^{-1}C_\theta = A_0^{-1}C_0$.

Substitution of the feedback control equation into the system equation gives

$$y_k = Me_k + \tilde{M}w_{k-1} \tag{5.3.26}$$

where $M = (A_0 - z^{-1}B_0H^{-1}F)^{-1}C_0$, $\tilde{M} = (A_0 - z^{-1}B_0H^{-1}F)^{-1}B_0H^{-1}$ and

$$u_k = H^{-1}FMe_k + H^{-1}(I + z^{-1}F\tilde{M})w_k. \tag{5.3.27}$$

We also have from (5.3.22) and (5.3.23) that

$$\hat{y}_k - \hat{y}_k(\theta) = K_\theta y_k - L_\theta u_{k-1} + d_k \tag{5.3.28}$$

where $K_\theta = C_\theta^{-1}A_\theta - C_0^{-1}A_0$, $L_\theta = C_\theta^{-1}B_\theta - C_0^{-1}B_0$ and

$$d_k = (I - C_\theta^{-1}A_\theta)\xi_k + C_\theta^{-1}B_\theta\eta_{k-1}.$$

The reasoning that led to (5.3.25) gives

$$E\|d_k\|^2 \to 0 \qquad \text{as } k \to \infty. \tag{5.3.29}$$

Now

$$2\|K_\theta y_k - L_\theta u_{k-1} + d_k\|^2 \geq \|K_\theta y_k - L_\theta u_{k-1}\|^2 - 2\|d_k\|^2$$

by properties of the Euclidean norm. It follows from (5.3.28) that

$$\begin{aligned} 2E\|\hat{y}_k - \hat{y}_k(\theta)\|^2 &\geq E\|K_\theta y_k - L_\theta u_{k-1}\|^2 - 2E\|d_k\|^2 \\ &= E\|(K_\theta - z^{-1}L_\theta H^{-1}F)Me_k\|^2 \\ &\quad + E\|[(K_\theta - z^{-1}L_\theta H^{-1}F)\tilde{M} - L_\theta H^{-1}]w_{k-1}\|^2 \\ &\quad - 2E\|d_k\|^2 \end{aligned} \tag{5.3.30}$$

(we have used equations (5.3.26) and (5.3.27) and also the fact that $\{e_k\}$ is a zero-mean sequence, independent of $\{w_k\}$).

Since $\theta \in \tilde{D}_I$,

$$\liminf_{k\to\infty} E\|\hat{y}_k - \hat{y}_k(\theta)\|^2 = 0.$$

It follows from (5.3.29) and (5.3.30) that

$$\liminf_{k\to\infty} E\|(K_\theta - z^{-1}L_\theta H^{-1}F)Me_k\|^2 = 0 \tag{5.3.31}$$

and

$$\liminf_{k\to\infty} E\|[K_\theta - z^{-1}L_\theta H^{-1}F)\tilde{M} - L_\theta H^{-1}]w_{k-1}\|^2 = 0. \tag{5.3.32}$$

The e_k are independent, zero-mean random variables such that $\text{cov}\{e_k\} > \alpha I$, $k \in \mathbb{Z}$, for some $\alpha > 0$, and, in consequence, they define a persistently exciting sequence of infinite order. We deduce from (5.3.31) and Lemma 5.3.4 that

$$(K_\theta - z^{-1}L_\theta H^{-1}F)M = 0.$$

However, $M(\sigma)$ is invertible for some σ (in fact $M(0) = I$). It follows that

$$K_\theta - z^{-1}L_\theta H^{-1}F = 0. \tag{5.3.33}$$

But then, by (5.3.32)

$$\liminf_{k\to\infty} E\|L_\theta H^{-1}w_{k-1}\|^2 = 0.$$

Since $\{w_k\}$ is a persistently exciting sequence of infinite order we can call upon Lemma 5.3.4 again, and deduce that $L_\theta H^{-1} = 0$. However $H(\sigma)$ is invertible for some σ, and we conclude that $L_\theta = 0$. Equation (5.3.33) now gives $K_\theta = 0$. Recalling the definition of L_θ and K_θ, we see that $C_\theta^{-1}A_\theta = C_0^{-1}A_0$ and $C_\theta^{-1}B_\theta = C_0^{-1}B_0$. But $A_\theta(\sigma)$ and $C_\theta(\sigma)$ are invertible for some σ; it follows that

$$A_\theta^{-1}C_\theta = (C_\theta^{-1}A_\theta)^{-1} = A_0^{-1}C_0$$

and

$$A_\theta^{-1}B_\theta = (A_0^{-1}C_0C_\theta^{-1})B_\theta = A_0^{-1}C_0C_0^{-1}B_0 = A_0^{-1}B_0. \qquad \square$$

Analogous properties of identification schemes which involve models in state space form can be derived from Proposition 5.3.1 by reformulation of the state-space models as ARMAX models; the conditions on the matrices in the state-space models which must be imposed in order that the corresponding ARMAX models satisfy the hypotheses of Proposition 5.3.1 are essentially those appearing in Propositions 5.1.2 and 5.1.4.

5.4 Interpretation of identification in terms of system approximation

The basis of consistency analysis is the hypothesis that the system is describable within the model set. Yet, strictly speaking, imposition of this hypothesis is seldom justified. In typical applications the system has a very complicated structure. One can expect of a model no more than that it reproduces, sufficiently accurately, certain significant features of the system. (This is not to dismiss consistency analysis: a method which cannot select a correct model when the model set contains the true system description is unlikely to be a good one, and consistency analysis therefore gives us grounds for ruling out certain methods, at the very least.)

An important issue then is behaviour of identification methods when the model set does not include the system description. We might hope that, in these circumstances, the model selected is a best approximation, in some sense, to the system. This is the gist of the following proposition. It gives conditions under which, in the limit, the identification method supplies a model which best approximates the system, for the given input, in the sense that the mean square difference between predictors, $\hat{y}_k$, based on knowledge of the true system and predictors, $\hat{y}_k(\theta)$, based on the models, is a minimum.

Proposition 5.4.1

Suppose that the system which generates the data is stable and that the models provide uniformly stable predictors. Suppose also that the identification criterion V_N is the least squares criterion

$$V_N(\theta) = \frac{1}{N}\sum_{k=1}^{N} \varepsilon_k^{\mathrm{T}}(\theta)\varepsilon_k(\theta), \qquad N = 1, 2, \ldots,$$

and the limit

$$W(\theta) = \lim_{N\to\infty}\frac{1}{N}\sum_{k=1}^{N} E\|\hat{y}_k - \hat{y}_k(\theta)\|^2, \tag{5.4.1}$$

exists for every $\theta\in D$. Then

$$\theta_N \to \{\theta\colon W(\theta) = \min_{\psi\in D} W(\psi)\} \qquad \text{a.s.}$$

PROOF By Theorem 5.2.1, θ_N converges, almost surely, into the set

$$D_I = \left\{\theta\colon \max_{\psi\in D}\left(\liminf_{N\to\infty}\frac{1}{N}\sum_{k=1}^{N}(E\varepsilon_k^{\mathrm{T}}(\theta)\varepsilon_k(\theta) - E\varepsilon_k^{\mathrm{T}}(\psi)\varepsilon_k(\psi))\right) = 0\right\}.$$

But for any $\theta \in D$,

$$\varepsilon_k(\theta) = \hat{y}_k - \hat{y}_k(\theta) + v_k$$

where $v_k = y_k - \hat{y}_k$ and

$$E\|\varepsilon_k^{\mathrm{T}}(\theta)\varepsilon_k(\theta)\|^2 = E\|\hat{y}_k - \hat{y}_k(\theta)\|^2 + E\|v_k\|^2,$$

since v_k has zero mean, and is uncorrelated with $\hat{y}_k - \hat{y}_k(\theta)$. From this equation and from existence of the limit (5.4.1), for arbitrary $\theta \in D$, we deduce that the inclusion $\theta \in D_I$ can be equivalently stated:

$$\lim_{N\to\infty} \frac{1}{N}\sum_{k=1}^{N} \|\hat{y}_k - \hat{y}_k(\theta)\|^2 \leq \lim_{N\to\infty} \frac{1}{N}\sum_{k=1}^{N} E\|\hat{y}_k - \hat{y}_k(\psi)\|^2,$$

for all $\psi \in D$. θ_N behaves then as claimed. □

The hypothesis that the limit (5.4.1) exists, for each $\theta \in D$, is a reasonable one when the system is time-invariant and driven by a disturbance which is a stationary process, and when the input takes the form of time-varying linear feedback (defined through parameters which have limits as time tends to infinity) together with an additive disturbance which is stationary.

We stress that the proposition states that the identification methods, under appropriate conditions, select a model $\mathscr{M}(\theta^*)$ which is a best approximation to the system, in a certain natural sense, *merely for the particular input considered.* It does not claim that the model $\mathscr{M}(\theta^*)$ is a good approximation for arbitrary inputs. Indeed, the model $\mathscr{M}(\theta^*)$ and the system could be ill-matched for inputs differing from that of the identification experiment and in consequence $\mathscr{M}(\theta^*)$ could be quite unsuitable for the application intended. This point is illustrated by the phenomenon of self-tuning (see Chapter 7), which we now discuss.

The problem considered is that of identifying parameters to select a model, with a view to designing a feedback controller which, when applied to the system, results in as small an output variance as possible.† We suppose that the system is describable within the model set to the following extent:

(a) The model set contains a model which correctly describes the deterministic part of the system; but;
(b) The models are driven by disturbances which are not correlated, even though the disturbances driving the system are correlated.

†Mimimum variance control is studied in Sections 7.1 and 7.3.

Two approaches to tackling the problem suggest themselves. The first is to separate the tasks of identification and control implementation. Here, a model $\mathcal{M}(\theta^*)$ is selected on the basis of an identification experiment in which data is obtained by applying an input suitable for identification purposes, say a persistently exciting input, to the system. A controller is then chosen, which is a minimum-variance controller with respect to $\mathcal{M}(\theta^*)$, and applied to the system. The second approach is to combine identification and control implementation. Stated more precisely, the approach is to select, at time k, a model $\mathcal{M}(\theta_k)$ on the basis of data y^k, u^{k-1}, to calculate a minimum-variance controller for $\mathcal{M}(\theta_k)$, to implement this control to obtain y^{k+1} and u_k, and so on.

In each of these two approaches two 'crimes' are committed. We implement a control which, in the first place, is designed for a model in which the disturbances are uncorrelated when in fact the disturbances in the system are correlated, and which, in the second place, is designed on the basis of biased estimates of the parameters in the deterministic part of the system. (The bias here results from a disregard of the disturbance correlation in the system; see Section 4.7.)

It is not surprising then that the first approach, separate identification and control implementation, typically gives rise to biased estimates and to a control with poor properties. The second approach, integrated identification and control, also gives rise to biased estimates. What is, at first sight, remarkable is that the second approach can supply, in the limit as $k \to \infty$, a control which is optimal, in the sense that the corresponding output of the system has minimum variance. This behaviour, cancellation of errors from two different sources, is the self-tuning phenomenon.

Some light is shed on the phenomenon by Proposition 5.4.1. Suppose that the parameter estimates θ_k obtained from the second approach converge (we write the limit $\bar{\theta}$). If the control is chosen according to a minimum variance strategy then, in the limit, $\hat{y}_k(\bar{\theta})$ is zero. Proposition 5.4.1 indicates that $E\|\hat{y}_k - \hat{y}_k(\bar{\theta})\|^2$ will be as small as possible. But

$$E\|\hat{y}_k - \hat{y}_k(\bar{\theta})\|^2 = E\|\hat{y}_k\|^2 = E\|y_k\|^2 - E\|v_k\|^2.$$

Here the v_k's are the prediction errors based on knowledge of the true system and do not depend on the control. Consequently the model selected in the limit gives rise to a control strategy which minimizes $E\|y_k\|^2$.

On the other hand, Proposition 5.4.1 has little bearing on the first approach, where the identification and control implementation phases are separated. Here, according to Proposition 5.4.1, the predictors associated with the model $\mathscr{M}(\theta^*)$ selected will be close to those associated with the true system for the input used in the identification experiment. However, if a control law is now selected to make the predictors for $\mathscr{M}(\theta^*)$ zero, and if the resulting input is different from that of the identification experiment, we can come to no conclusions about the variance of the output for this input.

Notes

The asymptotic analysis of this chapter is limited to investigation of subsets in parameter space into which estimates obtained from off-line prediction error identification schemes converge in the limit as the number of data points tends to infinity. For purposes of estimating confidence intervals it is desirable also to have information about the asymptotic distributions of the estimates: a proof of asymptotic normality of the estimates, essentially the framework of this chapter, is given in Ljung and Caines (1979), as discussed in Section 4.4.4. Study of the asymptotic properties of estimates supplied by on-line identification algorithms, which we do not enter into in this book, is the subject of much recent research (see, e.g. Kushner and Clarke (1978), Ljung (1977) and Solo (1979)).

The development of this chapter follows that in Ljung's important paper (1978). The consistency results of Section 5.3, which apply when the model set contains a true description of the system, have antecedents in a list of papers extending over many years on the consistency of the maximum likelihood method, in the contexts of independent samples, of time series analysis and of stochastic dynamical systems. Some references to this earlier literature are Caines (1976), Dunsmuir and Hannan (1976), Hannan (1973), Ljung (1976), Wald (1949), Walker (1964) and Whittle (1961).

References

Caines, P. E. (1976) Prediction error methods for stationary stochastic processes. *IEEE Trans. Automatic Control*, **AC-21**, 500–506.

Dunsmuir, W. and Hannan, E. J. (1976) Vector linear time series models. *Adv. Applied Probability*, **8**, 339–364. (see also corrections and extensions provided by M. Deistler, W. Dunsmuir and E. J. Hannan, *ibid.*, **10** (1978), 360–372).

Hannan, E. J. (1973) The asymptotic theory of linear time series models. *J. Applied Probability*, **10**, 130–145.

Kushner, H. J. and Clarke, D. S. (1978) *Stochastic Approximation Methods for Constrained and Unconstrained Systems*, Springer, New York.

Ljung, L. (1976) On the consistency of prediction error identification methods. In *System Identification: Advances and Case Studies* (eds R. K. Mehra and D. G. Lainiotis) Academic Press, New York.

——(1977) Analysis of recursive algorithms. *IEEE Trans. Automatic Control*, **AC-22**(4), 551–575.

——(1978) Convergence analysis of parametric identification methods. *IEEE Trans. Automatic Control*, **AC-23**(5), 770–783.

Ljung, L. and Caines, P. E. (1979) Asymptotic normality of prediction error estimators for approximate system models. *Stochastics*, **3**(1) 29–46.

Solo, V. (1979) The convergence of AML. *IEEE Trans. Automatic Control*, **AC-24**(6), 958–962.

Wald, A. (1949) Note on the consistency of the maximum likelihood estimate. *Ann. Math. Statistics*, **20**, 595–601.

Walker, A. M. (1964) Asymptotic properties of least estimates of parameters of the spectrum of a stationary non-deterministic time series. *J. Australian Math. Soc.*, **4**, 363–388.

Whittle, P. (1961) Gaussian estimation in stationary time series. *Bull. Inst. Int. Statist.*, **33**, 1–26.

CHAPTER 6

Optimal control for state-space models

This chapter concerns optimal control problems for the state-space models discussed in Chapters 2 and 3. The state and observation processes x_k and y_k are given respectively by the equations

$$x_{k+1} = A(k)x_k + B(k)u_k + C(k)w_k \tag{6.0.1}$$

$$y_k = H(k)x_k + G(k)w_k \tag{6.0.2}$$

where w_k is a white-noise sequence. We now wish to choose the control sequence u_k so that the system behaves in some desirable way. We have to settle two questions at the outset, namely what sort of controls are to be allowed (or, are *admissible*) and what the control objective is.

The simplest class of controls is that of *open-loop* controls which are just deterministic sequences $u_0, u_1, \ldots,$ chosen *a priori*. In this case the observation equation (6.0.2) is irrelevant since the system dynamics are entirely determined by the state equation (6.0.1). As we shall see in Section 6.1, open-loop controls are in some sense adequate for non-stochastic problems ($w_k \equiv 0$). Generally, however, it is better to use some form of *feedback control*. Such a control selects a value of u_k on the basis of measurements or observations of the system. We have *complete observations* if the state vector x_k can be measured directly, and, since the future evolution of the system depends only on its current state and future controls and noise, the natural form of control is then *state feedback*: $u_k = u_k(x_k)$. The functions $u_1(\cdot), u_2(\cdot), \ldots$ are sometimes described as a *control policy* since they constitute a decision rule: *if* the state at time k is x, *then* the control applied will be $u = u_k(x)$. Again, the observations y_k are irrelevant in this situation. In the case of *noisy measurements* or *partial observations*, however, x_k cannot be measured directly and only the sequence $y_0, y_1, \ldots, y_k$ is available. Feedback control now means that u_k is determined on the basis of the available measurements: $u_k = u_k(y_0, y_1, \ldots, y_k)$. In this case, since y_k is not the state of the system, one generally does better by allowing dependence on all past observations, not just on the current observation y_k. Finally, we shall assume throughout that the control *values* are unconstrained. It would be perhaps more realistic to restrict the values of the controls by introducing constraints of the form $|u_k| \leq 1$. While this causes no theoretical difficulties, it would make the calculation of explicit control policies substantially more difficult.

We now turn to the control objective. In classical control system design the objectives are qualitative in nature: one specifies certain stability and transient response characteristics, and any design which meets the specification will be regarded as satisfactory. The 'pole shifting' controllers considered in Chapter 7 follow this general philosophy. Here, however, our formulation is in terms of *optimal control*. The idea is as follows: the class of admissible controls is specified precisely and a scalar performance criterion or cost function $C(u)$ is associated with each control. We can then ask which control achieves the minimum cost; this control is *optimal*. Once the three ingredients (system dynamics, admissible controls and cost criterion) are specified, determination of the optimal control is in principle a purely mathematical problem involving no 'engineering judgement'. Indeed, optimal control theory has often been criticized precisely on these grounds. It may well be that a control which is theoretically optimal is subjectively quite unsatisfactory. If it is, this will be because the system model is inadequate or because the cost criterion fails to take account of all the relevant features of the problem. On the other hand, a more realistic model or a criterion which *did* include all the relevant features might well lead to an impossibly complicated optimization problem. As usual, the true situation is a trade-off between realistic modelling and mathematical tractability, and this is where the engineering judgement comes in.

In this chapter we shall study *linear regulator* problems, where the cost criterion is given by

$$C_N(u) = E\left[\sum_{k=0}^{N-1} \|Dx_k + Fu_k\|^2 + x_N^{\mathrm{T}} Q x_N\right]. \tag{6.0.3}$$

The number N of stages in the problem is called the *time horizon* and we shall consider both the finite-horizon ($N < \infty$) and infinite-horizon ($N = \infty$) cases. Further discussion of the cost function $C_N(u)$

will be found in Section 6.1. It implies a general control objective of regulating the state x_k to 0 while not using too much control energy as measured by the quantity $u_k^T F^T F u_k$. Note that the quantity in square brackets in (6.0.3) is a random variable and we obtain a scalar cost function (as required for optimization) by taking its expected value, which is practical terms means that we are looking for a control policy which gives the minimum average cost over a long sequence of trials.

The optimization problem represented by equations (6.0.1)–(6.0.3) is known as the LQG problem since it involves a *linear* system (6.0.1), (6.0.2), a *quadratic* cost criterion (6.0.3) and *gaussian* or normal white-noise disturbances in the state-space model. (For reasons explained below, $\{w_k\}$ is assumed here to be a sequence of independent normal random variables rather than a 'wide-sense' white noise as generally considered in previous chapters.) It is sufficiently general to be applicable in a wide variety of cases and the optimal control is obtained in an easily implemented form. It also has, as we shall see, close relations with the Kalman filter.

In addition to the standard linear regulator as defined above we shall study the same problem with *discounted costs*:

$$C^{\rho}(u) = E\left[\sum_{k=1}^{N-1} \rho^k \| Dx_k + Fu_k \|^2 + \rho^N x_N^T Q x_N\right]$$

where ρ is a number, $0 < \rho < 1$. There are important technical reasons for introducing the *discount factor* ρ, but there is also a financial aspect to it. Suppose that money can be invested at a constant interest rate $r\%$ per annum and one has to pay bills of £a_0, £$a_1, \ldots$ each year starting at the present time. What capital is needed to finance these bills entirely out of investment income? Since £1 now is worth £$(1 + 0.01\,r)^k$ in k years' time, the amount required is $\sum_k a_k \rho^k$ where $\rho = (1 + 0.01\,r)^{-1}$ and this is one's total debt *capitalized at its present value*. In particular, a constant debt of £a/year in perpetuity can be financed with a capital of

$$£ \sum_{k=0}^{\infty} a\rho^k = £a/(1 - \rho).$$

An important feature of this result is that while the total amount of debt is certainly infinite, it nevertheless has a finite capital value. Similarly, in the control problems, the discount factor enables us to attach a finite cost (and therefore consider optimization) in cases where without discounting the cost would be $+\infty$ for all control policies. Of course it is not realistic to assume that interest rates will remain constant for all time, and a more subjective interpretation of $C^{\rho}(u)$ is simply to say that it attaches small importance to costs which have to be paid at some time in the distant future.

In the three sections of this chapter we discuss the linear regulator problem in three stages. First, in Section 6.1 we consider the deterministic case when $w_k = 0$. Many of the 'structural features' of the LQG problem are already present in this case, and the optimal control turns out to be linear feedback: $u_k = -M(k)x_k$ for a precomputable sequence of matrices $M(k)$. This same control is shown in Section 6.2 to be optimal also in the stochastic case with complete observations, the effect of the noise being simply to increase the cost. Finally we consider the 'full' LQG problem in Section 6.3 and show that the optimal control is now $-M(k)\hat{x}_{k|k-1}$ where $\hat{x}_{k|k-1}$ is the best estimate of the state given the observations, generated by the Kalman filter. This results demonstrates the so-called 'certainty-equivalence' principle: if the state cannot be observed directly, estimate it and use the estimate as if it were the true state. We also discuss an idea of somewhat wider applicability known as the 'separation principle'.

6.1 The deterministic linear regulator

6.1.1 Finite time horizon

In this section we consider control of the linear system

$$x_{k+1} = A(k)x_k + B(k)u_k \tag{6.1.1}$$

for $k = 0, 1, \ldots, N$ with a given initial condition x_0. We wish to choose a control sequence $u = (u_0, u_1, \ldots, u_{N-1})$ so as to minimize the cost[†]

$$J_N(u) = \sum_{k=0}^{N-1} \|D(k)x_k + F(k)u_k\|^2 + x_N^{\mathrm{T}} Q x_N. \tag{6.1.2}$$

Here $D(k)$, $F(k)$ are matrices of dimensions $p \times n$, $p \times m$ respectively and Q is a non-negative definite symmetric $n \times n$ matrix. It will be assumed throughout that the $m \times m$ matrices $F^{\mathrm{T}}(k)F(k)$ are strictly positive definite, which implies in particular that we must have $p \geq m$.

We shall also study various infinite-time problems related to (6.1.1)–(6.1.2), i.e. consider what happens as $N \to \infty$.

[†] We denote the cost by J_N in the deterministic case, reserving C_N for the average cost in the stochastic problem.

The cost function $J_N(u)$ is somewhat different from that conventionally employed in treatments of this subject. The more usual form of cost function is

$$\tilde{J}_N(u) = \sum_{k=0}^{N-1} (x_k^{\mathrm{T}} Q(k) x_k + u_k^{\mathrm{T}} R(k) u_k) + x_N^{\mathrm{T}} Q x_N$$

where $Q(k)$, $R(k)$ are symmetric non-negative definite matrices (strictly positive definite in the case of $R(k)$). This has more intuitive appeal since the terms involving x_k penalize deviation of x_k from 0 while $\Sigma u_k^{\mathrm{T}} R(k) u_k$ is a measure of control energy. Thus the control problem is to steer x_k to zero as quickly as possible without expending too much control energy; energy expenditure can be penalized more or less heavily by appropriate specification of the matrices $R(k)$. This cost function is, however, a special case of (6.1.2): take $p = n + m$ and

$$D(k) = \left[\begin{array}{c} Q^{1/2}(k) \\ \hline 0 \end{array}\right] \qquad F(k) = \left[\begin{array}{c} 0 \\ \hline R^{1/2}(k) \end{array}\right]$$

where $Q^{1/2}(k)$, $R^{1/2}(k)$ are any 'square roots' of $Q(k)$, $R(k)$, i.e. satisfy $(Q^{1/2}(k))^{\mathrm{T}} Q^{1/2}(k) = Q(k)$ (and similarly for $R^{1/2}(k)$). Such square roots always exist for non-negative definite symmetric matrices, as shown in Appendix D, Proposition D.1.3.

We prefer the cost function (6.1.2) because of its extra generality, but more importantly because it connects up naturally with the formulation of the Kalman filter given in Chapter 3. This will become apparent below.

The control problem (6.1.1)–(6.1.2) can in principle be regarded as an unconstrained minimization problem. For a given sequence $u = (u_0, u_1, \ldots, u_{N-1})$ and initial condition x_0, the corresponding x_k sequence can be computed from the state equations (6.1.1):

$$\begin{aligned} x_1 &= A(0)x_0 + B(0)u_0 \\ x_2 &= A(1)x_1 + B(1)u_1 \\ &= A(1)A(0)x_0 + A(1)B(0)u_0 + B(1)u_1, \quad \text{etc.} \end{aligned}$$

Substituting in (6.1.2), we obtain $J_N(u)$ explicitly as a function of the mN-vector $u = \mathrm{col}\{u_0, u_1, \ldots, u_{N-1}\}$ and one could now use 'standard' hill-climbing techniques to find the vector u^* which minimizes $J_N(u)$. This would, however, be a very unsatisfactory way of solving the problem. Not only is the dimension mN very large even for innocuous-looking problems, but also we have thrown away an essential feature of the problem, namely its dynamic structure, and therefore calculation of the optimal u^* would give us very little insight into what is really happening in the optimization process.

A solution method which uses in an essential way the dynamic nature of the problem is R. Bellman's technique of dynamic programming. Introduced by Bellman in the mid-1950s, dynamic programming has been the subject of extensive research over the years and the associated literature is now enormous. We propose to discuss it here only to the extent necessary to solve the problem at hand. The basic idea is, like many good ideas, remarkably simple, and is known as Bellman's *principle of optimality*. Suppose that u^* is an optimal control for the linear regulator problem (6.1.1)–(6.1.2), that is to say,

$$J_N(u^*) \le J_N(u)$$

for all other controls $u = (u_0, u_1, \ldots, u_{N-1})$. Let $x_0^* = x_0, x_1^*, \ldots, x_N^*$ be the corresponding state trajectory given by (6.1.1) with $u_k = u_k^*$. Now fix an integer j, $0 \le j < N$, and consider the 'intermediate' problem of minimizing

$$J_{N,j}(u^{(j)}) = \sum_{k=j}^{N-1} \| D(k)x_k + F(k)u_k \|^2 + x_N^{\mathrm{T}} Q x_N$$

over controls $u^{(j)} = (u_j, u_{j+1}, \ldots, u_{N-1})$, subject to the dynamics (6.1.1) as before with the 'initial condition'

$$x_j = x_j^*.$$

The intermediate problem is thus to optimize the performance of the system over the last $N-j$ stages, starting at a point x_j^* which is on the optimal trajectory for the *overall* optimization problem. The principle of optimality states that *the control* $u^{*(j)} = (u_j^*, u_{j+1}^*, \ldots, u_{N-1}^*)$ *is optimal for the intermediate problem.* Put another way, if u^* is optimal for the overall problem then $u^{*(j)}$ is optimal over the last $N-j$ stages starting at x_j^*. The reason for this is fairly clear: if $u^{*(j)}$ were *not* optimal for the intermediate problem then there would be some sequence $\tilde{u}^{(j)} = (\tilde{u}_j, \tilde{u}_{j+1}, \ldots, \tilde{u}_{N-1})$ such that

$$J_{N,j}(\tilde{u}^{(j)}) < J_{N,j}(u^{*(j)}).$$

Now consider the control u^0 defined as follows:

$$u_k^0 = \begin{cases} u_k^* & k < j \\ \tilde{u}_k & k \ge j \end{cases}$$

and let x_k^0 be the corresponding trajectory. Then $x_k^0 = x_k^*$ for $k \leq j$ and hence

$$\begin{aligned} J_N(u^0) &= \sum_{k=0}^{j-1} \|D(k)x_k^* + F(k)u_k^*\|^2 + J_{N,j}(\tilde{u}^{(j)}) \\ &< \sum_{k=0}^{j-1} \|D(k)x_k^* + F(k)u_k^*\|^2 + J_{N,j}(u^{*(j)}) \\ &= J_N(u^*). \end{aligned} \tag{6.1.3}$$

But this contradicts the supposition that u^* is optimal. Thus $u^{*(j)}$ must be optimal for the intermediate problem, as claimed.

In the preceding argument, the system started in a fixed but arbitrary state x_0. However, there is nothing special about the initial time zero: the same argument implies that if $\{x_k^*, u_k^*, k \geq j\}$ is an optimal control-trajectory sequence for the intermediate problem starting at $x_j = x$ (arbitrary) then $\{x_k^*, u_k^*, k \geq j'\}$ is optimal for the further intermediate problem starting at $x_j = x_{j'}^*$ for any j' between j and $N-1$.

The principle of optimality is turned into a practical solution technique as follows. Let $V_j(x)$ be the minimum cost for the intermediate problem starting at $x_j = x$. This is known as the *value function* at time j. Then taking $j' = j+1$, the above argument indicates that V_j ought to satisfy

$$V_j(x) = \min_v \left[\|D(j)x + F(j)v\|^2 + V_{j+1}(A(j)x + B(j)v) \right] \tag{6.1.4}$$

the minimum being taken over all m-vectors v. Essentially, this comes from calculations similar to (6.1.3) above. If $x_j = x$ and control $u_j = v$ is applied, then:

(a) The cost paid at time j is $\|D(j)x + F(j)v\|^2$.
(b) The next state is $x_{j+1} = A(j)x + B(j)v$.

Thus $V_{j+1}(A(j)x + B(j)v)$ is the minimal cost for the rest of the problem if control value v is applied at stage j. So certainly

$$V_j(x) \leq \|D(j)x + F(j)v\|^2 + V_{j+1}(A(j)x + B(j)v) \tag{6.1.5}$$

and this holds for any value of v. On the other hand, if $\{x_k^*, u_k^*\}$ is optimal over the last $N-j$ stages starting at $x_j^* = x$, then the principle of optimality indicates that

$$V_l(x_l^*) = \sum_{k=l}^{N-1} \|D(k)x_k^* + F(k)u_k^*\|^2 + x_N^{*\mathrm{T}} Q x_N^*$$

where l is either j or $j+1$, and this shows since $x_j^* = x$ that

$$V_j(x) = \|D(j)x + F(j)u_j^*\|^2 + V_{j+1}(A(j)x + B(j)u_j^*). \qquad (6.1.6)$$

Now (6.1.5) and (6.1.6) together imply that (6.1.4) holds.

Equation (6.1.4) is known as the *Bellman equation* and is the basic entity in discrete-time dynamic programming since it enables the optimal control u^* to be determined. Note that *at the terminal time N* the value function is

$$V_N(x) = x^{\mathrm{T}}Qx, \qquad (6.1.7)$$

since no further control is possible and one has no choice but to pay the terminal cost of $x^{\mathrm{T}}Qx$. Applying (6.1.4) with $j = N-1$ gives

$$V_{N-1}(x) = \min_v [\|D(N-1)x + F(N-1)v\|^2$$
$$+ (A(N-1)x + B(N-1)v)^{\mathrm{T}}Q(A(N-1)x + B(N-1)v)]$$

and hence determines $V_{N-1}(x)$. Now using (6.1.4) again we can calculate $V_{N-2}, V_{N-3}, \ldots, V_0$. By definition, $V_0(x_0)$ is then the minimal cost for the overall problem starting at state x_0. From (6.1.5) and (6.1.6), the optimal control u_j^* is just the value of v that achieves the minimum in (6.1.4) with $x = x_j^*$.

Before proceeding any further let us consolidate the discussion so far. We have used the principle of optimality to obtain the Bellman equation (6.1.4) and this suggests the procedure outlined above for obtaining an optimal control. Having arrived at this procedure, however, we can verify that it is correct by a simple and self-contained argument; this will be given below. Thus the principle of optimality is actually only a heuristic device which tells us why we would expect the Bellman equation to take the form it does; it does not appear in the final formulation of any results. One could present the theory without mentioning the principle of optimality at all, but this would involve pulling the Bellman equation out of the hat, and readers would be left wondering – at least, we *hope* they would be left wondering – where it came from.

Theorem 6.1.1 (Verification theorem)

Suppose $V_{N-1}(x), V_{N-2}(x), \ldots, V_0(x)$ satisfy the Bellman equation (6.1.4) with terminal condition (6.1.7). Suppose that the minimum in (6.1.4) is achieved at $v = u_j^0(x)$, i.e.

$$\begin{aligned}&\|D(j)x + F(j)u_j^0(x)\|^2 + V_{j+1}(A(j)x + B(j)u_j^0(x))\\ &\leq \|D(j)x + F(j)v\|^2 + V_{j+1}(A(j)x + B(j)v)\end{aligned}$$

for all m-vectors v. Now define (x_k^*, u_k^*) recursively as follows:

$$x_0^* = x_0 \tag{6.1.8}$$

$$\left.\begin{aligned} u_k^* &= u_k^0(x_k^*) \\ x_{k+1}^* &= Ax_k^* + Bu_k^* \end{aligned}\right\} \quad k = 0, 1, \ldots, N-1. \tag{6.1.9}$$

Then $u^* = (u_0^*, \ldots, u_{N-1}^*)$ is an optimal control and the minimum cost is $V_0(x_0)$.

PROOF Let $u = (u_0, \ldots, u_{N-1})$ be *any* control and $x_0, \ldots, x_N$ the corresponding trajectory, always with the same initial point x_0. Then from (6.1.4) we have

$$V_j(x_k) \leq \|D(j)x_j + F(j)u_j\|^2 + V_{j+1}(x_{j+1}). \tag{6.1.10}$$

Hence

$$\begin{aligned} V_N(x_N) - V_0(x_0) &= \sum_{k=0}^{N-1} (V_{k+1}(x_{k+1}) - V_k(x_k)) \\ &\geq -\sum_{k=0}^{N-1} \|D(j)x_j + F(j)u_j\|^2. \end{aligned} \tag{6.1.11}$$

Since $V_N(x_N) = x_N^{\mathrm{T}} Q x_N$ this shows that

$$V_0(x_0) \leq J_N(u). \tag{6.1.12}$$

On the other hand, by definition, equality holds in (6.1.10) and hence in (6.1.11) when $x_j = x_j^*$, $u_j = u_j^*$, so that

$$V_0(x_0) = J_N(u^*). \tag{6.1.13}$$

Now (6.1.12), (6.1.13) say that u^* is optimal and that the minimal cost is $V_0(x_0)$. □

Two remarks are in order at this point:

1. Note that the optimal control is obtained in *feedback form*, i.e. x_k^* is generated by

$$x_{k+1}^* = A(k)x_k^* + B(k)u_k^0(x_k^*)$$

where $u_k^0(\cdot)$ is a pre-determined function. (See Fig. 6.1(a).) One could in principle obtain the same cost $V_0(x_0)$ by calculating the u_k^* sequence

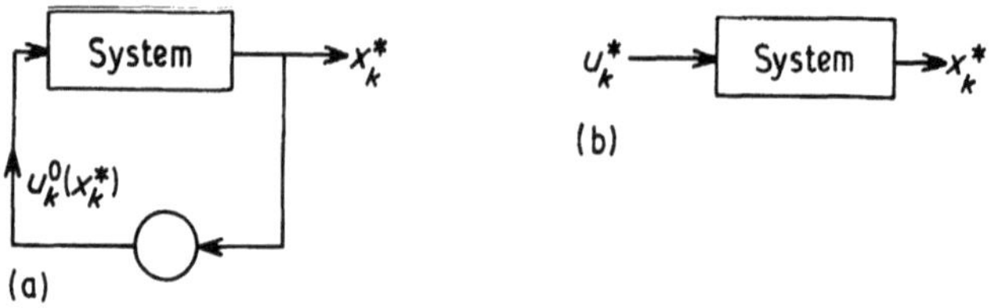

Fig. 6.1 *(a) Feedback control; (b) Open loop control.*

explicitly and applying it in open loop (Fig. 6.1(b)) but such a procedure has serious disadvantages. Using the dynamic programming approach, we have in fact not only solved the original overall control problem but have solved all the intermediate problems as well: an argument identical to that given above shows that the control u_k^* generated by (6.1.9) with any initial condition $x_j^* = x$ is optimal for the control problem over the last $N - j$ stages starting at $x_j = x$. Thus if for some reason the system gets 'off course' the feedback controller continues to act optimally for the remaining stages of control. On the other hand, the values u_k^* calculated for the open-loop control of Fig. 6.1(b) are based on a specific starting point x_0 and if this is erroneous or if an error occurs at some intermediate point then the u_k^* sequence will no longer be optimal.

2. Nothing so far depends on the quadratic nature of the cost function (6.1.2). Similar results would be obtained for any scalar cost function of the form

$$J'_N(u) = \sum_{k=0}^{N-1} l(k, x_k, u_k) + g(x_N). \tag{6.1.14}$$

We have seen above that the basic step in solving the optimal control problem is to calculate the value functions $V_{N-1}(x), \ldots, V_0(x)$. With general cost functions $J'(u)$ as in (6.1.14) this involves an immense amount of work since the whole function $V_k(\cdot)$ has to be calculated and not just the value $V_k(x)$ at some specific point x. The advantage of the quadratic cost (6.1.2) is that the value functions take a simple parametric form and can be computed in an efficient way. Indeed, the value functions are themselves quadratic forms, as the following result shows.

Theorem 6.1.2

The solution of the Bellman equation (6.1.4), (6.1.7) for the linear regular problem (6.1.1), (6.1.2) is given by

$$V_k(x) = x^{\mathrm{T}} S(k) x \qquad k = 0, 1, \ldots, N \tag{6.1.15}$$

where $S(0), \ldots, S(N)$ are symmetric non-negative definite matrices defined by (6.1.20) below. The optimal feedback control is

$$u_j^1(x) = -M(j)x$$

where

$$M(j) = [B^{\mathrm{T}}(j)S(j+1)B(j) + F^{\mathrm{T}}(j)F(j)]^{-1} \cdot [B^{\mathrm{T}}(j)S(j+1)A(j) + F^{\mathrm{T}}(j)D(j)]. \tag{6.1.16}$$

We see that the optimal controller has a very simple structure, namely *linear feedback* of the state variables. The notation u_j^1 for optimal control is used for consistency with the discounted cost case to be discussed below.

PROOF Note that the result is certainly true at $k = N$ since $V_N(x) = x^{\mathrm{T}} Q x$. To show that it holds for $k < N$ we use backwards induction: supposing (6.1.15) holds for $k = j + 1$ we show that it holds for $k = j$. Taking $V_{j+1}(x) = x^{\mathrm{T}} S(j+1) x$, the Bellman equation (6.1.4) becomes

$$V_j(x) = \min_v [\,\|D(j)x + F(j)v\|^2 + (x^{\mathrm{T}} A^{\mathrm{T}}(j) + v^{\mathrm{T}} B^{\mathrm{T}}(j)) \cdot S(j+1)(A(j)x + B(j)v)]. \tag{6.1.17}$$

The quantity in square brackets on the right-hand side is equal to

$$v^{\mathrm{T}}(B^{\mathrm{T}} S(j+1) B + F^{\mathrm{T}} F)v + 2x^{\mathrm{T}}(A^{\mathrm{T}} S(j+1) B + D^{\mathrm{T}} F)v + x^{\mathrm{T}}(A^{\mathrm{T}} S(j+1) A + D^{\mathrm{T}} D)x \tag{6.1.18}$$

where we temporarily write $B(j) = B$, etc. Now if R is a symmetric positive definite matrix and a an m-vector then

$$(v + a)^{\mathrm{T}} R (v + a) = v^{\mathrm{T}} R v + 2a^{\mathrm{T}} R v + a^{\mathrm{T}} R a$$

i.e.

$$v^{\mathrm{T}} R v + 2a^{\mathrm{T}} R v = (v + a)^{\mathrm{T}} R (v + a) - a^{\mathrm{T}} R a.$$

Clearly this expression is minimized over v at $v = -a$ and the minimum value is $-a^{\mathrm{T}} R a$. In order to identify this with the first two terms in (6.1.18) we require

$$R = B^{\mathrm{T}}S(j+1)B + F^{\mathrm{T}}F$$
$$Ra = (B^{\mathrm{T}}S(j+1)A + F^{\mathrm{T}}D)x.$$

Now by assumption $F^{\mathrm{T}}F$, and hence R, is strictly positive definite, and therefore a is specified by

$$a = R^{-1}(B^{\mathrm{T}}S(j+1)A + F^{\mathrm{T}}D)x.$$

Thus the right-hand side of (6.1.17) is equal to

$$x^{\mathrm{T}}[A^{\mathrm{T}}S(j+1)A + D^{\mathrm{T}}D - (A^{\mathrm{T}}S(j+1)B + D^{\mathrm{T}}F)$$
$$R^{-1}(B^{\mathrm{T}}S(j+1)A + F^{\mathrm{T}}D)]x. \qquad (6.1.19)$$

Hence $V_j(x) = x^{\mathrm{T}}S(j)x$ where $S(j)$ is given by the expression in the square brackets in (6.1.19) and $S(j) \geq 0$ by (6.1.17). Thus $V_k(x)$ is a quadratic form, as in (6.1.15), for all $k = 0, 1, \ldots, N$. Note from the above analysis (specifically from (6.1.19)) that the matrices $S(k)$ can be computed recursively backwards in time starting with $S(N) = Q$. In fact, writing out (6.1.19) in full we see that the $S(k)$ are generated by

$$\begin{aligned}
S(N) &= Q \\
S(j) &= A^{\mathrm{T}}(j)S(j+1)A(j) + D^{\mathrm{T}}(j)D(j) - (A^{\mathrm{T}}(j)S(j+1)B(j) \\
&\quad + D^{\mathrm{T}}(j)F(j))(B^{\mathrm{T}}(j)S(j+1)B(j) + F^{\mathrm{T}}(j)F(j))^{-1} \\
&\quad \cdot(B^{\mathrm{T}}(j)S(j+1)A(j) + F^{\mathrm{T}}(j)D(j)) \\
j &= N-1, N-2, \ldots, 0. \qquad (6.1.20)
\end{aligned}$$

Applying the dynamic programming results, the optimal feedback control is the value of v that achieves the minimum in (6.1.16), and this is equal to $-a$, so that

$$\begin{aligned}
u_j^1(x) = &-[B^{\mathrm{T}}(j)S(j+1)B(j) + F^{\mathrm{T}}(j)F(j)]^{-1} \\
&\cdot[B^{\mathrm{T}}(j)S(j+1)A(j) + F^{\mathrm{T}}(j)D(j)]x.
\end{aligned}$$

This completes the proof. □

Filtering/control duality

A very important feature of the above result is its close connection to the Kalman filter discussed in Section 3.3. Equation (6.1.20) is a Riccati equation of exactly the same type as that appearing in the Kalman filter equations, with the distinction that (6.1.20) evolves backwards from a terminal condition at time N whereas the filtering Riccati equation (3.3.6) for the estimation error covariance $P(j)$ evolves forward from an initial condition at $j=0$. The Kalman gain $K(j)$ is related to $P(j)$ in exactly the same way that the control gain $M(j)$ is related to $S(j)$, except for transposition. Specifically, the correspondence between the two problems is as shown in Table 6.1.

Table 6.1

Filtering	*Control*
(time) j	$N-j$
$A(j)$	$A^{\mathrm{T}}(j)$
$H(j)$	$B^{\mathrm{T}}(j)$
$C(j)$	$D^{\mathrm{T}}(j)$
$G(j)$	$F^{\mathrm{T}}(j)$
$P(j)$	$S(j)$
$K(j)$	$M^{\mathrm{T}}(j)$

This means that if we take the filtering Riccati equation (3.3.6), make the time substitution $j \to N-j$ and relabel A, H, C, G as $A^{\mathrm{T}}, B^{\mathrm{T}}, D^{\mathrm{T}}, F^{\mathrm{T}}$ respectively, then we get precisely (6.1.20). The same relabelling applied to the expression (3.3.5) for $K(j)$ produces $M^{\mathrm{T}}(j)$. Thus the Riccati equations (6.1.20) and (3.3.6) are the same in all but notation. This will be very important when we come to consider various properties of the Riccati equation, since its solution can be regarded interchangeably as the value function for a control problem or the error covariance for a filtering problem, and various facts can be deduced from one or other of these interpretations.

Discounted costs

Let us now specialize to the time-invariant system

$$x_{k+1} = Ax_k + Bu_k \tag{6.1.21}$$

(i.e. $A(k)=A$, $B(k)=B$ for all k) and consider minimizing a discounted cost of the form

$$J_N^{\rho}(u) = \sum_{k=0}^{N-1} \rho^k \|Dx_k + Fu_k\|^2 + \rho^N x_N^{\mathrm{T}} Q x_N \tag{6.1.22}$$

where D, F, Q are fixed matrices and ρ is the discount factor $(0<\rho\leq 1)$. This is actually a special case of the preceding problem (take

$D(k) = \rho^{k/2}D$, $F(k) = \rho^{k/2}F$ and replace Q by $\rho^N Q$); but there is another way of looking at it which provides a little more insight. Write

$$\begin{aligned} J_N^\rho(u) &= \sum_{k=0}^{N-1} \rho^k \|Dx_k + Fu_k\|^2 + \rho^N x_N^T Q x_N \\ &= \sum_{k=0}^{N-1} \|D\rho^{k/2}x_k + F\rho^{k/2}u_k\|^2 + \rho^N x_N^T Q x_N \\ &= \sum_{k=0}^{N-1} \|Dx_k^\rho + Fu_k^\rho\|^2 + x_N^{\rho T} Q x_N^\rho \end{aligned} \tag{6.1.23}$$

where we have defined

$$\begin{aligned} x_k^\rho &:= \rho^{k/2}x_k \\ u_k^\rho &:= \rho^{k/2}u_k. \end{aligned} \tag{6.1.24}$$

Multiplying (6.1.21) by $\rho^{(k+1)/2}$ gives

$$\rho^{(k+1)/2}x_{k+1} = \rho^{1/2}A\rho^{k/2}x_k + \rho^{1/2}B\rho^{k/2}u_k$$

i.e.

$$x_{k+1}^\rho = A^\rho x_k^\rho + B^\rho u_k^\rho \tag{6.1.25}$$

where $A^\rho := \rho^{1/2}A$, $B^\rho := \rho^{1/2}B$. But (6.1.23)–(6.1.25) constitute a time-invariant linear regulator problem in standard non-discounted form. The optimal control is therefore

$$\begin{aligned} u_k^\rho &= -(B^{\rho T}S^\rho(k+1)B^\rho + F^T F)^{-1}(B^{\rho T}S^\rho(k+1)A^\rho + F^T D)x_k^\rho \\ &=: -M^\rho(k)x_k^\rho \end{aligned}$$

where $S^\rho(k)$ is the solution of (6.1.20) with A replaced by $\rho^{1/2}A$ and B replaced by $\rho^{1/2}B$. In view of (6.1.24) the optimal control u_k is expressed in terms of the 'real' state x_k by

$$u_k = -M^\rho(k)x_k.$$

Thus the discounted cost problem is solved simply by taking the *undiscounted* problem and making the substitutions $A \to \rho^{1/2}A$, $B \to \rho^{1/2}B$.

6.1.2 Infinite-time problems

In this section we will continue to assume that the system and costs are time-invariant, i.e. the matrices A, B, D, F do not depend on the time, k.

In many control problems no specific terminal time N is involved and one wishes the system to have good 'long-run' performance. This suggests replacing (6.1.2) by a cost

$$J_\infty(u) = \sum_{k=0}^{\infty} \|Dx_k + Fu_k\|^2. \tag{6.1.26}$$

It is not obvious that the problem of minimizing $J_\infty(u)$ subject to the dynamics (6.1.1) makes sense: it might be the case that $J_\infty(u) = +\infty$ for all controls u. Note, however, that the problem *does* make sense as long as there is at least one control u such that $J_\infty(u) < \infty$. A simple sufficient condition for this is that the pair (A, B) be *stabilizable*, i.e. there exists an $m \times n$ matrix M such that $A - BM$ is stable. Taking for u the feedback control $\bar{u}_k = -Mx_k$, the system dynamics become

$$x_{k+1} = (A - BM)x_k.$$

Now since $A - BM$ is stable, it follows from Proposition D.3.1, Appendix D, that there exist constants $c > 0$ and $a \in (0, 1)$ such that

$$\|x_k\| \leq ca^k \|x_0\|$$

Since $\|(D - FM)x\| \leq K\|x\|$ for some constant K, the cost using control $\bar{u}$ is

$$\begin{aligned} J_\infty(\bar{u}) &= \sum_{k=0}^{\infty} \|(D - FM)x_k\|^2 \\ &\leq K^2 \sum_{k=0}^{\infty} \|x_k\|^2 \\ &\leq c^2K^2 \|x_0\|^2 \sum_{k=0}^{\infty} a^{2k} \\ &= c^2K^2 \|x_0\|^2/(1 - a^2). \end{aligned}$$

Thus with any stabilizing control, the norm of x_k decays sufficiently fast to give a finite total cost. We will therefore assume henceforth that the pair (A, B) is stabilizable.

If $V_k(x)$ is the value function at time k for the infinite-time problem then it seems likely that V_k does not actually depend on k, since, there being no 'time horizon' and the coefficients being time-invariant, the problem facing the controller is the same at time k as at time zero, except for some change in the initial state. Recalling the Bellman equation (6.1.4), this suggests that the value function $V \equiv V_k$ should satisfy

$$V(x) = \min_v [\|Dx + Fv\|^2 + V(Ax + Bv)]. \tag{6.1.27}$$

Note this is no longer a recursion but is an implicit equation which may or may not be satisfied by a particular function V.

Proposition 6.1.3

Suppose that V is a solution of (6.1.27) such that V is continuous and $V(0) = 0$,[†] and that $u^1(x)$ achieves the minimum on the right, i.e. for all vectors v,

$$\| Dx + Fu^1(x) \|^2 + V(Ax + Bu^1(x)) \leq \| Dx + Fv \|^2 + V(Ax + Bv).$$

Suppose also that u^1 is a stabilizing control in the sense that $\| x_k \| \to 0$ as $k \to \infty$, where x_k is the trajectory corresponding to u^1, i.e.

$$x_{k+1} = Ax_k + Bu^1(x_k).$$

Then $u^1(x)$ is optimal in the class of stabilizing controls. Equation (6.1.27) has the quadratic solution $V(x) = x^T Sx$ if and only if S satisfies the algebraic Riccati equation (6.1.29) below, and in this case the corresponding control is

$$u^1(x) = -Mx$$

where

$$M = (B^T SB + F^T F)^{-1}(B^T SA + F^T D) \tag{6.1.28}$$

PROOF Let $\{x_k, u_k\}$ be any control/trajectory pair such that $\| x_k \| \to 0$ as $k \to \infty$ and write

$$\begin{aligned} V(x_N) - V(x_0) &= \sum_{k=0}^{N-1} V(x_{k+1}) - V(x_k) \\ &\geq \sum_{0}^{N-1} \|Dx_k + Fu_k\|^2 \qquad \text{(from (6.1.27))}. \end{aligned}$$

Thus

$$V(x_0) \leq \sum_{k=0}^{N-1} \|Dx_k + Fu_k\|^2 + V(x_N).$$

Now by the assumptions on V and x_k, $V(x_N) \to 0$ as $N \to \infty$ and hence

$$V(x_0) \leq \sum_{k=0}^{\infty} \|Dx_k + Fu_k\|^2 = J_\infty(u).$$

The same calculations hold with = replacing $\geq$ when $u = u^1$, and this

[†] A natural requirement since if $x = 0$ the control $u_k = 0$ is plainly optimal.

shows that

$$V(x_0) = J_\infty(u^1) = \min_u J_\infty(u).$$

Thus u^1 is optimal in the class of stabilizing controls (those for which $\|x_k\| \to 0$ as $k \to \infty$, x_k being the corresponding trajectory.)

Since the value function for the finite-horizon problem is a quadratic form, let us try a solution to (6.1.27) of the form $x^T Sx$ where S is a symmetric non-negative definite matrix. From (6.1.19), the minimum value on the right of (6.1.27) is then

$$x^T[A^T SA + D^T D - (A^T SB + D^T F)(B^T SB + F^T F)^{-1}(B^T SA + F^T D)]x$$

and $V(x) = x^T Sx$ is therefore a solution of (6.1.27) if and only if S satisfies the so-called *algebraic Riccati equation* (ARE):

$$S = A^T SA + D^T D - (A^T SB + D^T F)(B^T SB + F^T F)^{-1}(B^T SA + F^T D). \tag{6.1.29}$$

If S satisfies this then certainly $V(x) = x^T Sx$ is continuous and $V(0) = 0$. The corresponding minimizing u^1 is given as before by

$$u^1(x) = -Mx$$

where

$$M = (B^T SB + F^T F)^{-1}(B^T SA + F^T D).$$ □

If the matrix $A - BM$ is stable then $\|x_k\| \to 0$ as $k \to \infty$ where

$$x_{k+1} = Ax_k + Bu^1(x_k) = (A - BM)x_k.$$

The above proof thus shows that if S satisfies (6.1.29) and $A - BM$ is stable then the control $u^1(x_k) = -Mx_k$ is optimal in the class of all stabilizing controls. An important feature of this result is that the optimal control is *time-invariant* (does not depend explicitly on k), although time varying controls are not in principle excluded.

It is evident from Proposition 6.1.3 that the infinite time problem hinges on properties of the algebraic Riccati equation. These are somewhat technical and a full account will be found in Appendix B. Let us summarize the main results. The conditions required on the coefficient matrices A, B, D, F are as follows:

(a) The pair (A, B) is stabilizable.
(b) The pair $(\hat{D}, \hat{A})$ is detectable, where (6.1.30)

$$\hat{A} = A - B(F^T F)^{-1} F^T D$$
$$\hat{D} = [I - F(F^T F)^{-1} F^T]D.$$

The first of these conditions is a natural one since, as remarked before, it ensures the existence of at least one control giving finite cost. The motivation for condition (b) is less obvious, though it does seem clear that *some* condition involving D and F, in particular concerning the relation between states x_k and 'output' Dx_k, is required to justify limiting attention to stabilizing controls. Condition (b) takes the simpler form

(b′) (D, A) is detectable,

when $F^{\mathrm{T}}D = 0$; this is the case alluded to at the beginning of this section, in which the cost takes the form

$$\|Dx_k + Fu_k\|^2 = x_k^{\mathrm{T}}D^{\mathrm{T}}Dx_k + u_k^{\mathrm{T}}F^{\mathrm{T}}Fu_k.$$

Under conditions (6.1.30), the argument given in Appendix B shows that there is a unique non-negative definite matrix S satisfying the algebraic Riccati equation, that $A - BM$ is stable, where M is given by (6.1.28), and that the control $u^1(x_k) = -Mx_k$ is optimal in the sense of minimizing $J_\infty(u)$ over *all* control–trajectory pairs (x_k, u_k) satisfying the dynamic equation (6.1.1). (The less precise argument summarized in Proposition 6.1.3 only shows that $u^1(x)$ minimizes $J_\infty(u)$ over all such pairs satisfying $\|x_k\| \to 0$ as $k \to \infty$.)

The relation between the finite and infinite-time problems is also elucidated in Appendix B. In fact it is shown that under conditions (a) and (b),

$$S = \lim_{k \to \infty} S(-k) \tag{6.1.31}$$

where $S(-1), S(-2), \ldots$ is the sequence of matrices produced by the Riccati equation (6.1.19) with $S(0) = Q$ where Q is an arbitrary non-negative definite matrix. Now $x^{\mathrm{T}}S(-k)x$ is the minimal cost for the k-stage control problem (6.1.1)–(6.1.2) with terminal cost $x_k^{\mathrm{T}}Qx_k$. In view of (6.1.31) we see that as the time horizon recedes to infinity, the cost of the finite-horizon problem approaches that of the infinite horizon problem, whatever the terminal cost matrix Q. Q is unimportant because $\|x_k\|$ will be very small for large k when the optimal control is applied.

Generally, in the finite-horizon case, the optimal control $u_k = -M(k)x_k$ is time-varying. If, however, one selects $Q = S$ as the terminal cost, where S satisfies the algebraic Riccati equation, then $S(k) = S$ for all k, so that the time-invariant control $u_k = -Mx_k$ is optimal, and this is the same control that is optimal for the infinite-horizon problem. The situation is somewhat analogous to that of a transmission line terminated by a matched impedance. With this termination the line is indistinguishable from one of infinite length. In the control case, if the terminal cost is $x_k^{\mathrm{T}}Sx_k$ the controller is indifferent between paying it and stopping, or continuing optimally *ad infinitum.* In either case the total cost is the same, so it is reasonable to describe S as the 'matched' terminal cost matrix.

Finally, let us consider the infinite-time discounted cost problem, where the cost function is

$$J_\infty^\rho(u) = \sum_{k=0}^{\infty} \rho^k \|Dx_k + Fu_k\|^2.$$

Proceeding exactly as in the finite-horizon discounted case, we conclude that the optimal control is

$$u_k^\rho(x_k) = -M^\rho x_k.$$

Here

$$M^\rho = (B^{\rho\mathrm{T}}S^\rho B^\rho + F^{\mathrm{T}}F)^{-1}(B^{\rho\mathrm{T}}S^\rho A^\rho + F^{\mathrm{T}}D)$$

and S^ρ is the solution of the algebraic Riccati equation with A and B replaced by A^ρ and B^ρ respectively, where

$$A^\rho = \rho^{1/2}A, \qquad B^\rho = \rho^{1/2}B.$$

The conditions for existence of a solution S^ρ to the modified equation are the appropriately modified version of (6.1.30) above, namely

$$\begin{aligned}&\text{(c) } (A^\rho, B^\rho) \text{ is stabilizable.}\\ &\text{(d) } (\hat{D}, \hat{A}^\rho) \text{ is detectable } (\hat{A}^\rho = \rho^{1/2}\hat{A}).\end{aligned} \tag{6.1.32}$$

Note that if U is any $n \times n$ matrix with eigenvalues $\lambda_1, \ldots, \lambda_n$ then the eigenvalues of $\rho^{1/2}U$ are $\rho^{1/2}\lambda_1, \ldots, \rho^{1/2}\lambda_n$ since if x_i is an eigenvector corresponding to λ_i then

$$\rho^{1/2}Ux_i = \rho^{1/2}\lambda_i x_i. \tag{6.1.33}$$

Thus $A^\rho - B^\rho M = \rho^{1/2}(A - BM)$ is stable if $A - BM$ is stable. Similarly $\hat{A}^\rho - (\rho^{1/2}N)\hat{D} = \rho^{1/2}(\hat{A} - N\hat{D})$ is stable if $\hat{A} - N\hat{D}$ is stable. Thus conditions (6.1.30) imply conditions (6.1.32), so that S^ρ exists for any $\rho \leq 1$ if conditions (6.1.30) are satisfied. However, taking $U = A$ and $U = \hat{A}$ in (6.1.33) we see that, for sufficiently small ρ, A^ρ and $\hat{A}^\rho$ are both stable and, *a fortiori*, (A^ρ, B^ρ) and $(\hat{D}, \hat{A}^\rho)$ are stabilizable and detectable respectively. Thus an optimal solution to the discounted cost infinite-time problem *always* exists if the discount factor ρ sufficiently small. An optimal control with finite cost can, however, be obtained without discounting if the rather mild conditions (6.1.30) are met. This contrasts with the situation in the stochastic case considered in the next section, where discounting is always necessary to obtain finite costs in infinite-time problems.

This concludes our discussion of the deterministic optimal regulator problem. We need it as a stepping-stone to the stochastic case and also to isolate the duality relationships which connect the Riccati equations which arise here and in the Kalman filter. In Appendix B, the asymptotic behaviour of the Riccati equation is investigated by methods which rely heavily on its control-theoretic interpretation. But, thanks to the duality properties, these results apply equally to tell us something about asymptotic behaviour of the estimation error in the Kalman filter.

In recent years, techniques based on the linear/quadratic optimal regulator have become an important component of multivariable control system design methodology. It is outside the scope of this book to discuss such questions, but some references will be found in the Notes and References at the end of this chapter. The essential advantage of the linear/quadratic framework in this connection is that arbitrary dimensions m and p of input u_k and output Dx_k are allowed, whereas techniques which attempt to generalize the classical single-input, single-output methods are seriously complicated by the combinatorial fact that there are rp transfer functions to consider, one from each input to each output. A subsidiary advantage of the linear/quadratic framework is that time-varying systems are handled with relative ease.

6.2 The stochastic linear regulator

In this section we consider problems of optimal regulation when the state equation includes additive noise, as in the state-space stochastic model discussed in Section 2.4. Thus x_k satisfies

$$x_{k+1} = A(k)x_k + B(k)u_k + C(k)w_k \tag{6.2.1}$$

where $\{w_k\}$ is a sequence of l-vector random variables with mean 0 and covariance I. We will assume in this section that w_k and w_j are *independent* (rather than merely uncorrelated) for $k \neq j$. The initial state x_0 is a random vector independent of w_k with mean and covariance m_0, P_0 respectively. We suppose that the state x_k can be measured directly by the controller, so that controls will be feedback functions of the form $u_k = u_k(x_k)$. The objective is to minimize the cost criterion

$$C_N(u) = E\left[\sum_{k=0}^{N-1} \|D(k)x_k + F(k)u_k\|^2 + x_N^T Q x_N\right].$$

The value function $W_j(x)$ at time j for this problem is the minimum value of

$$E_{j,x}\left[\sum_{k=j}^{N-1} \|D(k)x_k + F(k)u_k\|^2 + x_N^T Q x_N\right]$$

where $E_{j,x}$ denotes the expectation given that the process starts off at $x_j = x$ (a fixed vector in $\mathbb{R}^n$). If $x_j = x$ and the control value $u_j = v$ is applied then the next state is

$$x_{j+1} = A(j)x + B(j)v + C(j)w_j$$

and, by definition, the minimal remaining cost for the rest of the problem from time $j+1$ to N is $W_{j+1}(x_{j+1})$. This, however, is now a random variable since x_{j+1} is determined partly by w_j. The *expected* minimal remaining cost is obtained by averaging this over the distribution of w_j, giving a value of

$$EW_{j+1}(A(j)x + B(j)v + C(j)w_j).$$

Thus the minimum expected cost starting at $x_j = x$, if control $u_j = v$ is used, is the sum of this and the cost $\|D(j)x + F(j)v\|^2$ paid at time j. This suggests that $W_j(x)$ should satisfy the stochastic Bellman equation

$$W_j(x) = \min_v [\|D(j)x + F(j)v\|^2 + EW_{j+1}(A(j)x + B(j)v + C(j)w_j)] \tag{6.2.2}$$

where again E means averaging over the distribution of w_j with x, v fixed. At the final time N no further control or noise enters the system, so that

$$W_N(x) = x^T Q x. \tag{6.2.3}$$

As before, (6.2.2)–(6.2.3) determine a sequence of functions $W_N, W_{N-1}, \ldots, W_0$ by backwards recursion. And, also as before, we do not rely on the above heuristic argument to conclude that these functions are indeed the value functions for the control problem, but provide independent direct verification.

Proposition 6.2.1

Suppose that $W_N, \ldots, W_0$ are given by (6.2.2), (6.2.3) and that $u_j^0(x)$ is the value of v that achieves the minimum in (6.2.2). Then the feedback control $u_k^* = u_k^1(x_k)$ minimizes the cost $C_N(u)$ over the class of all feedback control policies.

PROOF Let $u_k(x_k)$ be an arbitrary feedback control and let x_k be the process given by (6.2.1) with $u_k = u_k(x_k)$. Then

$$W_N(x_N) - W_0(x_0) = \sum_{k=0}^{N-1} (W_{k+1}(x_{k+1}) - W_k(x_k))$$

so that

$$E[W_N(x_N) - W_0(x_0)] = \sum_{k=0}^{N-1} E[W_{k+1}(x_{k+1}) - W_k(x_k)] \quad (6.2.4)$$

In calculating the expectations on the right we are entitled to introduce any intermediate conditional expectation. We therefore write

$$E[W_{k+1}(x_{k+1}) - W_k(x_k)] = E\{E[W_{k+1}(x_{k+1}) - W_k(x_k)|x_k]\}. \quad (6.2.5)$$

Now, given x_k, $W_k(x_k)$ is known and x_{k+1} is given by

$$x_{k+1} = A(k)x_k + B(k)u_k(x_k) + C(k)w_k.$$

The first two terms on the right are known and the third is a random vector independent of x_k. The conditional expectation of $W_{k+1}(x_{k+1})$ is therefore given by

$$E[W_{k+1}(x_{k+1})|x_k] = EW_{k+1}(A(k)x_k + B(k)u_k(x_k) + C(k)w_k)$$

where the expectation on the right is taken over the distribution of w_k for fixed x_k. Now, using (6.2.2) we obtain

$$\begin{aligned} E[W_{k+1}(x_{k+1}) - W_k(x_k)|x_k] &= EW_{k+1}(A(k)x_k + B(k)u_k(x_k) \\ &\quad + C(k)w_k) - W_k(x_k) \\ &\geq - \| D(k)x_k + F(k)u_k(x_k) \|^2. \quad (6.2.6) \end{aligned}$$

Combining (6.2.4)–(6.2.6) shows that

$$E[W_N(x_N) - W_0(x_0)] \geq - E \sum_{k=0}^{N-1} \|D(k)x_k + F(k)u_k(x_k)\|^2$$

and hence, since $W_N(x_N) = x_N^T Q x_N$, that

$$EW_0(x_0) \leq C_N(u). \tag{6.2.7}$$

On the other hand, the same argument holds with equality instead of inequality in (6.2.6) when $u_k(x) = u_k^1(x)$, so that

$$EW_0(x_0) = C_N(u^1). \tag{6.2.8}$$

Now (6.2.7) and (6.2.8) say that u^1 is optimal. □

The proof actually shows a little more than is claimed in the proposition statement. Indeed, since $W_0(x_0)$ is only a function of x_0, the expectation in (6.2.8) only involves the (arbitrary) distribution of the initial state x_0. In particular, if x_0 takes a fixed value, say $\bar{x}_0$, with probability one, then the corresponding optimal cost is just $W_0(\bar{x}_0)$. Thus $W_0(x_0)$ should be interpreted as the *conditional* optimal cost given the initial state x_0. The *overall* optimal cost is then obtained by averaging over x_0, as in (6.2.8). A similar interpretation applies to W_k, namely $W_k(x)$ is the optimal cost over stages $k, k+1, \ldots, N$ conditional on an initial state $x_k = x$.

The solution of (6.2.2) is related in a simple way to that of the 'deterministic' Bellman equation (6.1.4). In fact,

$$W_k(x) = x^T S(k) x + \alpha_k$$

where $S(N) = Q$, $S(N-1), \ldots, S(0)$ are given by the Riccati equation (6.1.20) as before, and α_k is a constant, to be determined below. Note that if $W_{k+1}(x) = x^T S(k+1) x + \alpha_{k+1}$ then for fixed x, v,

$$\begin{aligned}
&EW_{k+1}(A(k)x + B(k)v + C(k)w_k) \\
&= (A(k)x + B(k)v)^T S(k+1)(A(k)x + B(k)v) \\
&\quad + 2E(A(k)x + B(k)v)^T S(k+1) C(k) w_k \\
&\quad + E w_k^T C^T(k) S(k+1) C(k) w_k + \alpha_{k+1} \\
&= (A(k)x + B(k)v)^T S(k+1)(A(k)x + B(k)v) \\
&\quad + \operatorname{tr}[C^T(k) S(k+1) C(k)] + \alpha_{k+1}
\end{aligned}$$

where the last line follows from the facts that $Ew_k = 0$, $\operatorname{cov}(w_k) = I$. Notice that the final expression is identical to that obtained in the deterministic case except for the term $\operatorname{tr}[C^T(k)S(k+1)C(k)] + \alpha_{k+1}$, which does not depend on x or v and hence does not affect the minimization on the right-hand side of (6.2.2). Thus if $W_{k+1}(x) = x^T S(k+1) x + \alpha_{k+1}$ then the induction argument as used in the deterministic case shows that

$$W_k(x) = x^{\mathrm{T}}S(k)x + \alpha_{k+1} + \mathrm{tr}[C^{\mathrm{T}}(k)S(k+1)C(k)].$$

But $W_N(x) = x^{\mathrm{T}}Qx$, i.e. $\alpha_N = 0$, so working backwards from $k = n$ we see that

$$\alpha_k = \sum_{j=k}^{N-1} \mathrm{tr}[C^{\mathrm{T}}(j)S(j+1)C(j)].$$

Summarizing, we have the following result.

Theorem 6.2.2

For the stochastic linear regulator with complete observations, the optimal control is

$$u_k^1(x_k) = -M(k)x_k$$

where $M(k)$ is given by (6.1.16), i.e. is the same as in the deterministic case. The minimal cost is

$$C_N(u^0) = m_0^{\mathrm{T}}S(0)m_0 + \mathrm{tr}[S(0)P_0] + \sum_{k=0}^{N-1} \mathrm{tr}[C^{\mathrm{T}}(k)S(k+1)C(k)]. \tag{6.2.9}$$

PROOF The optimality of u^1 follows from Proposition 6.2.1. As to the cost, we note that

$$W_0(x) = x^{\mathrm{T}}S(0)x + \alpha_0$$

is the conditional minimal cost given that the process starts at $x_0 = x$. Taking the expectation over the distribution of x_0, and using Proposition 1.1.3(b), we obtain (6.2.9). □

Note that only the mean m_0 and covariance P_0 of the initial state are needed to compute the optimal cost, so it is not necessary to suppose that x_0 is normally distributed. The important feature of the above result is that the matrices $S(k)$ and $M(k)$ do not depend on the noise coefficients $C(k)$, so that in particular *the optimal control is the same as in the deterministic case.* Thus adding noise to the state equation as in (6.2.1) makes no difference to the optimal policy, but simply makes that policy more expensive. Indeed, if the system starts at a fixed state x_0 (so that $m_0 = x_0$ and $P_0 = 0$) then the additional cost is precisely

$$\sum_{k=0}^{N-1} \operatorname{tr}[C^{\mathrm{T}}(k)S(k+1)C(k)].$$

Let us now consider the *discounted cost case*. We will assume for simplicity of notation that the coefficient matrices A, B, D, F are time invariant but, with later applications in mind, time variation will be retained for $C(k)$. Thus the problem is to minimize

$$E\left(\sum_{k=0}^{N-1} \rho^k \|Dx_k + Fu_k\|^2 + \rho^N x_N^{\mathrm{T}} Q x_N\right).$$

We use the same device as before, namely rewriting the cost as

$$E\left(\sum_{k=0}^{N-1} \|Dx_k^\rho + Fu_k^\rho\|^2 + x_N^{\rho\mathrm{T}} Q x_N^\rho\right) \tag{6.2.10}$$

where $x_k^\rho = \rho^{k/2}x_k$, $u_k^\rho = \rho^{k/2}u_k$. Multiplying (6.2.1) by $\rho^{(k+1)/2}$ shows that x_k^ρ, u_k^ρ satisfy

$$x_{k+1}^\rho = A^\rho x_k^\rho + B^\rho u_k^\rho + C^\rho(k)w_k \tag{6.2.11}$$

where $A^\rho = \rho^{1/2}A$, $B^\rho = \rho^{1/2}B$, $C^\rho(k) = \rho^{(k+1)/2}C(k)$. Now (6.2.10) and (6.2.11) give the problem in non-discounted form. As noted above, the optimal control does not depend on $C^\rho(k)$; applying our previous results it is given by

$$u_k^\rho(x) = -M^\rho(k)x$$

where $M^\rho(k)$ is defined as in Section 6.1 above. The corresponding cost is, from (6.2.9)

$$\begin{aligned} C_N^\rho(u^\rho) &= m_0^{\mathrm{T}} S^\rho(0)m_0 + \operatorname{tr}[S^\rho(0)P_0] + \sum_{k=0}^{N-1} \operatorname{tr}[C^{\rho\mathrm{T}}(k)S^\rho(k+1)C^\rho(k)] \\ &= m_0^{\mathrm{T}} S^\rho(0)m_0 + \operatorname{tr}[S^\rho(0)P_0] + \sum_{k=0}^{N-1} \rho^{k+1}\operatorname{tr}[C^{\mathrm{T}}(k)S^\rho(k+1)C(k)]. \end{aligned}$$

The importance of the discount factor becomes apparent when we consider infinite-horizon problems. Suppose that conditions (6.1.30) are met and that S^ρ is the solution to the algebraic Riccati equation with coefficient matrices A^ρ, B^ρ. Such a solution exists for any $\rho \leq 1$. Now consider the N-stage problem as above, with terminal cost matrix $Q = S^\rho$. This is the 'matched impedance' case, discussed at the end of Section 6.1, for which $S^\rho(k) = S^\rho$ for all k. Thus the optimal control is the time-invariant feedback

$$u^\rho(x_k) = -M^\rho x_k \tag{6.2.12}$$

and the cost over N stages is

$$C_N^\rho(u^\rho) = m_0^{\mathrm{T}} S^\rho m_0 + \operatorname{tr}[S^\rho P_0] + \sum_{k=0}^{N-1} \rho^{k+1} \operatorname{tr}[C^{\mathrm{T}}(k) S^\rho C(k)]. \tag{6.2.13}$$

Note that if $\rho = 1$ (no discounting) and $C(k) \equiv C$ is constant, then $C_N^\rho \to \infty$ as $N \to \infty$ and hence the infinite-time problem has no solution (all controls give cost $+\infty$). This is not surprising. The reason that finite costs could be obtained in the deterministic case was that $\|x_k\|$ converged to zero sufficiently fast that

$$\sum_{k=0}^{\infty} \|x_k\|^2$$

was finite. However, in the present case $\|x_k\|$ does *not* converge to zero because at each stage it is being perturbed by the independent noise term Cw_k, and the controller has continually to battle against this disturbance to keep $\|x_k\|$ as small as possible. If, however, $\rho < 1$, then

$$\lim_{N \to \infty} C_N^\rho = m_0^{\mathrm{T}} S^\rho m_0 + \operatorname{tr}[S^\rho P_0] + \frac{\rho}{1-\rho} \operatorname{tr}[C^{\mathrm{T}} S^\rho C]. \tag{6.2.14}$$

Thus any amount of discounting, however little, leads to a finite limiting cost. One can show, by methods exactly analogous to those used in the previous section, that the time-invariant control u^ρ given by (6.2.12) does in fact minimize the cost

$$C_\infty^\rho(u) = E\left(\sum_{k=0}^{\infty} \rho^k \|Dx_k + Fu_k\|^2\right) \tag{6.2.15}$$

and that the minimal cost is precisely the expression given in (6.2.14). As to the conditions required, recall that if (A, B) is stabilizable then (A^ρ, B^ρ) is stabilizable for any $\rho \leq 1$; thus

(a) If conditions (6.1.30) are satisfied then the infinite time discounted problem is well-posed, and has the above solution, for any $\rho < 1$.
(b) If either of conditions (6.1.30) fails then we must take $\rho < \rho_0$ where ρ_0 is such that (A^ρ, B^ρ), $(\hat{D}, \hat{A}^\rho)$ are stabilizable and detectable respectively for any $\rho < \rho_0$. Generally, $\rho_0 < 1$.

If $C(k)$ is not constant then exactly similar results apply as long as

$$\sum_{k=0}^{\infty} \rho^{k+1} \operatorname{tr}[C^{\mathrm{T}}(k)S^{\rho}C(k)] < \infty$$

and this will certainly be the case for any $\rho < 1$ as long as the elements of $C^{\mathrm{T}}(k)$ are uniformly bounded, i.e. there is some constant c_1 such that for all i, j, k,

$$|C(k)_{ij}| \leq c_1.$$

This, in turn, is always true if the $C(k)$ sequence is convergent, i.e. there is a matrix C such that $C(k) \to C$ as $k \to \infty$. The same control is optimal but there is in general no closed-form expression, as in (6.2.14), for the minimal cost, which is now

$$m_0^{\mathrm{T}} S^{\rho} m_0 + \operatorname{tr}[S^{\rho} P_0] + \sum_{k=0}^{\infty} \rho^{k+1} \operatorname{tr}[C^{\mathrm{T}}(k) S^{\rho} C(k)]. \qquad (6.2.16)$$

Let us now consider minimizing the *average cost per unit time*,

$$C_{\mathrm{av}}(u) = \lim_{N \to \infty} \frac{1}{N} E\left[\sum_{k=0}^{N-1} \|Dx_k + Fu_k\|^2\right]. \qquad (6.2.17)$$

As before we assume that all coefficients are constant except for the noise matrices $C(k)$ which are supposed to be convergent: $C(k) \to C$ as $k \to \infty$. This is needed in the next section.

The limit in (6.2.17) may or may not exist for any particular control u, but it certainly does exist for all *constant, stabilizing* controls, i.e. controls of the form $u_k^K = -Kx_k$ where $\bar{A} := A - BK$ is stable. For then the closed-loop system is

$$x_{k+1} = \bar{A}x_k + C(k)w_k$$

and we know by a slight extension of results in Section 2.4 that $Q(k) := \operatorname{cov}(x_k) \to Q$ where Q satisfies

$$Q = \bar{A}Q\bar{A}^{\mathrm{T}} + CC^{\mathrm{T}}.$$

Thus

$$\begin{aligned} C_{\mathrm{av}}(u^K) &= \lim_{N \to \infty} \frac{1}{N} \sum_{k=0}^{N} \operatorname{tr}[(D - FK)Q(k)(D - FK)^{\mathrm{T}}] \\ &= \operatorname{tr}[(D - FK)Q(D - FK)^{\mathrm{T}}]. \end{aligned}$$

If the pair (A, B) is stabilizable then a stabilizing K exists and the problem of minimizing $C_{av}(u)$ is meaningful. We now show that $C_{av}(u)$ is minimized by the control $u_k = -Mx_k$ where M is given by (6.1.28). This is the same control policy that is optimal for the deterministic infinite-time problem.

Theorem 6.2.3

Suppose conditions (6.1.30) hold. Then, among all controls u for which $C_{av}(u)$ exists and $E\|x_k\|^2$ remains bounded, the minimal cost is achieved by the control $u_k^1(x) = -Mx$ where M is given by (6.1.28). The minimal value of the cost is

$$C_{av}(u^1) = \operatorname{tr}[C^{\mathrm{T}}SC]$$

where S is the unique solution of the algebraic Riccati equation (6.1.29).

PROOF It is shown in Appendix B that $A - BM$ is stable, so that $J_{av}(u^1)$ exists. Let S be the solution of the ARE (6.1.29) and consider the N-stage problem of minimizing

$$C_N(u) = E\left[\sum_{k=0}^{N-1} \|Dx_k + Fu_k\|^2 + x_N^{\mathrm{T}}Sx_N\right]$$

This is the 'matched terminal cost' problem for which, from Theorem 6.2.2, control u^1 is optimal. Thus for any control u,

$$C_N(u) \geq C_N(u^1) = m_0^{\mathrm{T}}Sm_0 + \operatorname{tr}[SP_0] + \sum_{k=0}^{N-1} \operatorname{tr}[C^{\mathrm{T}}(k)SC(k)]. \tag{6.2.18}$$

Thus

$$\lim_{N\to\infty} \frac{1}{N} C_N(u) \geq \lim_{N\to\infty} \frac{1}{N} C_N(u^1) = C_{av}(u^1)$$

as long as the left-hand limit exists. But if $C_{av}(u)$ exists and $E\|x_k\|^2$ is bounded, then

$$\lim_{N\to\infty} \frac{1}{N} C_N(u) = C_{av}(u) + \lim_{N\to\infty} \frac{1}{N} E[x_N^{\mathrm{T}}Sx_N] = C_{av}(u).$$

This shows that u^1 is optimal. From (6.2.18) its cost is

$$C_{av}(u^1) = \lim_{N\to\infty} \frac{1}{N} \sum_{k=0}^{N-1} \operatorname{tr}[C^{\mathrm{T}}(k)SC(k)] = \operatorname{tr}[C^{\mathrm{T}}SC]. \qquad \square$$

The control $u_k^1 = -Mx_k$ is not the only optimal control for the average cost per unit time problem. Indeed, for any integer j we can write

$$C_{av}(u) = \lim_{N\to\infty} \frac{1}{N} E\left[\sum_{k=0}^{j-1} \|Dx_k + Fu_k\|^2\right]$$
$$+ \lim_{N\to\infty} \frac{1}{N} E\left[\sum_{k=j}^{N-1} \|Dx_k + Fu_k\|^2\right].$$

Now for any given control u_k,

$$E\left[\sum_{0}^{j-1} \|Dx_k + Fu_k\|^2\right]$$

is a fixed number not depending on N. Thus the first limit is zero, and since $(N-j)/N \to 1$ as $N \to \infty$,

$$C_{av}(u) = \lim_{N\to\infty} \frac{1}{N-j} E\left[\sum_{k=j}^{N-1} \|Dx_k + Fu_k\|^2\right].$$

The expression on the right is the average cost from time j onwards starting in state x_j, and its minimal value does not depend at all on what controls u_k were used for $k < j$. Thus any control of the form

$$u_k = \begin{cases} \text{arbitrary}, & k < j \\ -Mx_k, & k \geq j \end{cases}$$

is optimal. Thus the average cost criterion is only relevant when one is mainly concerned with 'long-run performance'; the idea is that the system settles down to a statistically stationary state in which an average of precisely $\text{tr}[C^T S C]$ is added to the cost at each stage, and this is minimal. There is, however, nothing in the cost criterion which specifies just how long this settling-down period is supposed to last. The discounted cost formulation has the opposite effect: it emphasizes performance during some initial interval the length of which is effectively specified by the discount factor. In this case the optimal control is unique. Another advantage of discounted costs is that the stabilizability/detectability conditions can always be met by sufficiently rapid discounting, whereas with average costs little can be said if the original system matrices (A, B, D) do not satisfy these conditions.

6.3. Partial observations and the separation principle

We now consider control problems associated with the full state-space model

$$x_{k+1} = A(k)x_k + B(k)u_k + C(k)w_k \tag{6.3.1}$$

$$y_k = H(k)x_k + G(k)w_k. \tag{6.3.2}$$

As before, the initial state x_0 has mean and covariance m_0, P_0 and is uncorrelated with w_k. In this case the state x_k cannot be measured directly, but 'noisy observations' $y^k = (y_0, y_1, \ldots, y_k)$ are available at time k. Thus the control u_k will be a feedback function of the form

$$u_k = u_k(y^k). \tag{6.3.3}$$

This is the 'full LQG problem'. The difficulty here is, of course, that knowledge of y^k does not (except in special cases) determine x_k exactly, and the current state x_k is just what is needed for controlling the system at time k. We deal with this by replacing the state-space model (6.3.1), (6.3.2) by the corresponding *innovations representation*. As discussed in Section 3.4, this provides an equivalent model in the form

$$\hat{x}_{k+1|k} = A(k)\hat{x}_{k|k-1} + B(k)u_k + K(k)v_k \tag{6.3.4}$$

where the innovations process v_k is given by

$$v_k = y_k - H(k)\hat{x}_{k|k-1} \tag{6.3.5}$$

so that y_k satisfies

$$y_k = H(k)\hat{x}_{k|k-1} + v_k. \tag{6.3.6}$$

The Kalman gain $K(k)$ is given by (3.3.5). The new 'state' of the system is $\hat{x}_{k|k-1}$ and this *is* determined exactly by y^{k-1}. We thus reduce the situation to one in which the state is known, and can then apply the results of the previous section to determine optimal control policies. First, however, the status of the innovations representation (6.3.4), (6.3.6) must be clarified. We do this before continuing with our discussion of optimal control problems in Section 6.3.2 below.

6.3.1 The Kalman filter for systems with feedback control

In the derivation of the Kalman filtering formulae in Section 3.3 it was assumed that $\{w_k\}$ was a weak-sense white noise (w_k and w_l uncorrelated for $k \neq l$) and that $\{u_k\}$ was a *deterministic sequence*. Under these conditions $\hat{x}_{k|k-1}$ given by (6.3.4) is the best linear (more precisely, affine) estimator of x_k given y^{k-1}, and the input/output properties of the model (6.3.4), (6.3.6) are identical to those of the original model (6.3.1), (6.3.2). Now, however, we wish to consider controls u_k which are not deterministic but which are feedback functions as in (6.3.3). Further, there is no reason why $u_k(y^k)$ should be a linear function of y^k. Suppose in fact that this function is nonlinear. Combining (6.3.3)–(6.3.5), we see that $\hat{x}_{k|k-1}$ satisfies

$$\hat{x}_{k+1|k} = A(k)\hat{x}_{k|k-1} + B(k)u_k(y^k) + K(k)(y_k - H(x)\hat{x}_{k|k-1}). \quad (6.3.7)$$

Given the sequence $y^j = (y_0, y_1, \ldots, y_j)$, one can use this equation for $k = 0, 1, \ldots, j$ to compute $\hat{x}_{j+1|j}$. Thus $\hat{x}_{j+1|j}$ is a function of y^j, say

$$\hat{x}_{j+1|j} = g_j(y^j).$$

Now g_j is a nonlinear function, due to the nonlinearity of u_k in (6.3.7). So $\hat{x}_{j+1|j}$ cannot possibly be the best *linear* estimator of x_{j+1} given y^j, as it would be were u_k deterministic. To get round this apparently awkward fact, we use the alternative interpretation of the Kalman filter, namely that if the w_k are independent *normal* random vectors and x_0 is normal, then $\hat{x}_{j+1|j}$ is the *conditional expectation* of x_{j+1} given y^j. The advantage of this formulation is that there is no requirement that a conditional expectation should be a linear function of the conditioning random variables.

Theorem 6.3.1

Suppose that, in the model (6.3.1), (6.3.2), $x_0, w_0, w_1, \ldots$ are normally distributed and that u_k is a feedback control as in (6.3.3). Let $\hat{x}_{k|k-1}$ be generated by the Kalman filter equation of Theorem (3.3.1). Then

$$\hat{x}_{k|k-1} = E[x_k | y^{k-1}]. \quad (6.3.8)$$

The innovations process (6.3.5) is a *normal* white-noise sequence.

PROOF The proof relies on Proposition 1.1.6 which shows that

$$E[x_{j+1}|y^j] = E[x_{j+1}|\bar{y}^j]$$

if $y^j, \bar{y}^j$ are random vectors which are related to each other in a one-to-one way, i.e. there are functions h_j, h_j^{-1} such that

$$\bar{y}^j = h_j(y^j), \quad y^j = h_j^{-1}(\bar{y}^j).$$

As in Section 3.4, let us write the state x_k in (6.3.1) as $x_k = \bar{x}_k + x_k^*$, and correspondingly $y_k = \bar{y}_k + y_k^*$, where $\bar{x}_k$, x_k^*, $\bar{y}_k$, y_k^* satisfy:

$$\left.\begin{aligned} \bar{x}_{k+1} &= A(k)\bar{x}_k + C(k)w_k, \quad \bar{x}_0 = x_0 - m_0 \\ \bar{y}_k &= H(k)\bar{x}_k + G(k)w_k \end{aligned}\right\} \tag{6.3.9}$$

$$\left.\begin{aligned} x_{k+1}^* &= A(k)x_k^* + B(k)u_k(y^k), \quad x_0^* = m_0 \\ y_k^* &= H(k)x_k^*. \end{aligned}\right\} \tag{6.3.10}$$

Equations (6.3.9) are linear, so that $\bar{x}_{k+1}$, $\bar{y}_k$ are zero-mean normal random vectors for all k. x_{k+1}^* and y_k^* are random vectors which depend on $\bar{y}^k$ since $u_k(y^k) = u_k(\bar{y}^k + y^{*k})$. Applying the standard Kalman filter results from Section 3.3 we see that $\hat{\bar{x}}_{k+1|k} := E[\bar{x}_{k+1}|\bar{y}^k]$ satisfies

$$\hat{\bar{x}}_{k+1|k} = A(k)\hat{\bar{x}}_{k|k-1} + K(k)(\bar{y}_k - H(k)\hat{\bar{x}}_{k|k-1}) \tag{6.3.11}$$

where $K(k)$ is given by (3.3.5). We cannot obtain (6.3.4) immediately by adding (6.3.11) to (6.3.10) because the conditioning random variable is $\bar{y}^k$ and not y^k as required. However, $\bar{y}^k$ and y^k are equivalent in the sense mentioned earlier. Indeed, plainly from (6.3.10), x_k^*, and hence y_k^*, is determined by $y^{k-1} = (y_0, y_1, \ldots, y_{k-1})$. Thus

$$\bar{y}_k = y_k - y_k^* =: h_k(y^k).$$

Conversely, suppose $\bar{y}^k = (\bar{y}_0, \bar{y}_1, \ldots, \bar{y}_k)$ is given; then y_k is determined. We show this by induction. Suppose that for $j = 0, 1, \ldots, k$ there are functions f_j such that

$$y_j = f_j(\bar{y}^j). \tag{6.3.12}$$

Then given $\bar{y}^k$ we can calculate y_j, $0 \leq j \leq k$, and hence y_{k+1}^*, using (6.3.10). But now

$$y_{k+1} = \bar{y}_{k+1} + y_{k+1}^* =: f_{k+1}(\bar{y}^{k+1})$$

Thus (6.3.12) holds for $j = k + 1$. At time zero,

$$y_0^* = H(0)x_0^* = H(0)m_0$$

and m_0 is known, so that

$$y_0 = H(0)m_0 + \bar{y}_0 =: f_0(\bar{y}_0).$$

Thus (6.3.12) holds for all j, and $f_j = h_j^{-1}$.

This argument shows that $\bar{y}^k$ and y^k are obtained from each other in a one-to-one fashion, and hence that

$$\hat{\bar{x}}_{k+1|k} = E[\bar{x}_{k+1}|\bar{y}^k] = E[\bar{x}_{k+1}|y^k].$$

Now x^*_{k+1} is a function of y^k, so that

$$E[x^*_{k+1}|y^k] = x^*_{k+1}.$$

Combining these relations, we obtain

$$\begin{aligned}E[x_{k+1}|y^k] &= E[x^*_{k+1} + \bar{x}_{k+1}|y^k]\\ &= x^*_{k+1} + \hat{\bar{x}}_{k+1|k}.\end{aligned}$$

Adding the equations (6.3.10) and (6.3.11) shows that $\hat{x}_{k+1|k} := E[x_{k+1}|y^k]$ satisfies (6.3.4). Thus (6.3.4) is indeed the Kalman filter when u_k is a feedback control, as long as the disturbance process w_k is a normal white-noise process. As regards the innovations process v_k, note that

$$\begin{aligned}v_k &= y_k - H(k)\hat{x}_{k|k-1}\\ &= \bar{y}_k + y^*_k - H(k)(x^*_k + \hat{\bar{x}}_{k|k-1})\\ &= \bar{y}_k - H(k)\hat{\bar{x}}_{k|k-1}.\end{aligned}$$

Thus v_k coincides with the innovations process corresponding to the control-free system (6.3.9). It is therefore a normal white-noise process with covariance

$$E[v_k v_k^{\mathrm{T}}] = H(k)P(k)H^{\mathrm{T}}(k) + G(k)G^{\mathrm{T}}(k) \tag{6.3.13}$$

as in Section 3.4. □

It is perhaps worth pointing out that, even if w_k is a normal white-noise process, the state process x_k is not necessarily normal, since (6.3.1), (6.3.2) determine x_k as a possibly nonlinear function of w^{k-1}. However, the *conditional distribution* of x_k given y^{k-1} is normal, since x_k has the representation

$$\begin{aligned}x_k &= x^*_k + \hat{\bar{x}}_{k|k-1} + \tilde{\bar{x}}_{k|k-1}\\ &= \hat{x}_{k|k-1} + \tilde{\bar{x}}_{k|k-1}\end{aligned}$$

where $\tilde{\bar{x}}_{k|k-1} = \bar{x}_k - \hat{\bar{x}}_{k|k-1}$ is a normal random vector with mean 0 and covariance $P(k)$ given by (3.3.6). Thus the conditional distribution of x_k given y^{k-1} is $N(\hat{x}_{k|k-1}, P(k))$.

6.3.2 The linear regulator problem

Let us now return to the control problem of choosing u_k to minimize the cost

$$C_N(u) = E\left(\sum_{k=0}^{N-1} \|D(k)x_k + F(k)u_k\|^2 + x_N^{\mathrm{T}} Q x_N\right). \tag{6.3.14}$$

This is the same form of cost as in Section 6.2 but a different class of controls is involved. In this section we shall consider feedback controls of the form

$$u_k = u_k(y^{k-1}) \tag{6.3.15}$$

rather than $u_k(y^k)$ as discussed above. Controls (6.3.15) are of course a sub-class of those previously considered – we are now insisting that the control u_k should depend on the observations y_j for times j up to, but not including k, whereas previously dependence on y_k also was allowed. This restriction is introduced for two reasons. Practically, it means that 'instant' data processing is then not required: at time k we record the new observation y_k, and apply the control $u_k(y^{k-1})$ which can be computed somewhat in advance since it does not depend on y_k. Mathematically, controls (6.3.15) are related, as will be seen below, to our formulation of the Kalman filter as a predictor, giving the best estimate $\hat{x}_{k|k-1}$ of x_k given y^{k-1}. Analogous results can be obtained for controls $u_k(y^k)$, but these involve the Kalman filter in the form which computes the *current* state estimate $\hat{x}_{k|k}$, and this is somewhat more complicated.

The cost $C_N(u)$ in (6.3.14) is expressed in terms involving the state variables x_k; we wish, however, to use the innovations representation (6.3.4) in which the state variable is $\hat{x}_{k|k-1}$. The first task is therefore to re-express $C_N(u)$ in a way which involves $\hat{x}_{k|k-1}$ rather than x_k, and this is done by introducing conditional expectations as follows:

$$\begin{aligned} C_N(u) = E\Bigg(& \sum_{k=0}^{N-1} E[\,\|D(k)x_k + F(k)u_k\|^2 \,|\, y^{k-1}] \\ & + E[x_N^{\mathrm{T}} Q x_N | y^{N-1}]\Bigg). \end{aligned} \tag{6.3.16}$$

Now x_k can be expressed in the form

$$x_k = \hat{x}_{k|k-1} + \tilde{x}_{k|k-1}$$

where $\hat{x}_{k|k-1}$ is a function of y^{k-1} and the estimation error $\tilde{x}_{k|k-1}$ is independent of y^{k-1} with distribution $N(0, P(k))$. We can simplify the terms in (6.3.16) using this fact and properties of conditional expectations. The last term is:

$$\begin{aligned} E[x_N^{\mathrm{T}} Q x_N | y^{N-1}] &= E[(\hat{x}_{N|N-1} + \tilde{x}_{N|N-1})^{\mathrm{T}} Q (\hat{x}_{N|N-1} + \tilde{x}_{N|N-1}) | y^{N-1}] \\ &= \hat{x}_{N|N-1}^{\mathrm{T}} Q \hat{x}_{N|N-1} + E[\tilde{x}_{N|N-1}^{\mathrm{T}} Q \tilde{x}_{N|N-1} | y^{N-1}] \\ &= \hat{x}_{N|N-1}^{\mathrm{T}} Q \hat{x}_{N|N-1} + \mathrm{tr}[P(N)Q]. \end{aligned}$$

Similarly the kth term in the sum becomes

$$E[(D(k)\hat{x}_{k|k-1} + F(k)u_k + D(k)\tilde{x}_{k|k-1})^{\mathrm{T}} \cdot (D(k)\hat{x}_{k|k-1} + F(k)u_k + D(k)\tilde{x}_{k|k-1})|y^{k-1}] = \|D(k)\hat{x}_{k|k-1} + F(k)u_k\|^2 + \mathrm{tr}[P(k)D^{\mathrm{T}}(k)D(k)]$$

where we have used the fact that u_k is a function of y^{k-1}. Thus

$$C_N(u) = E\left(\sum_{k=0}^{N-1} \|D(k)\hat{x}_{k|k-1} + F(k)u_k\|^2 + \hat{x}^{\mathrm{T}}_{N|N-1}Q\hat{x}_{N|N-1}\right) + \sum_{k=0}^{N-1} \mathrm{tr}[D(k)P(k)D^{\mathrm{T}}(k)] + \mathrm{tr}[P(N)Q]. \tag{6.3.17}$$

This expresses $C_N(u)$ in a way which involves the state $\hat{x}_{k|k-1}$ of the innovations representation. The important thing to notice about this expression is that the first term is identical to the original expression (6.3.14) with x_k replaced by $\hat{x}_{k|k-1}$, and that the remaining two terms are constants which do not depend in any way on the choice of u_k. Thus minimizing $C_N(u)$ is equivalent to minimizing

$$E\left(\sum_{k=0}^{N-1} \|D(k)\hat{x}_{k|k-1} + F(k)u_k\|^2 + \hat{x}^{\mathrm{T}}_{N|N-1}Q\hat{x}_{N|N-1}\right) \tag{6.3.18}$$

where the dynamics of $\hat{x}_{k|k-1}$ are given by (6.3.4), namely

$$\hat{x}_{k|k-1} = A(k)\hat{x}_{k|k-1} + B(k)u_k + K(k)v_k. \tag{6.3.19}$$

Since the innovations process v_k is a sequence of independent normal random variables, the problem (6.3.18)–(6.3.19) is the standard 'completely observable' regulator problem considered in the previous section. All coefficients are as before except for the 'noise' term $K(k)v_k$ in (6.3.19). However, it was noted in Section 6.2 that the optimal control for the linear regulator does not depend on the noise covariance. Therefore the optimal control coefficients are the same as in the completely observable case. We have obtained the following result:

Theorem 6.3.2

The optimal control for the noisy observations problem (6.3.1), (6.3.2), (6.3.14) is

$$\hat{u}^1_k = -M(k)\hat{x}_{k|k-1} \tag{6.3.20}$$

where $M(k)$ is given as before by (6.1.16). The cost of this policy is

$$C_N(\hat{u}^1) = m_0^{\mathrm{T}} S(0) m_0 + \mathrm{tr}[P(N)Q] + \sum_{k=0}^{N-1} \mathrm{tr}[D(k)P(k)D^{\mathrm{T}}(k) + G(k)G^{\mathrm{T}}(k))K^{\mathrm{T}}(k)S(k+1)]. + K(k)(H(k)P(k)H^{\mathrm{T}}(k) \tag{6.3.21}$$

PROOF Only the expression (6.3.21) for the optimal cost remains to be verified. We use the expression (6.2.9) for the completely observable case. First, note that the initial condition for (6.3.19) is deterministic: $\hat{x}_{0|-1} = 0$. Next, consider the contribution of the 'noise' term $K(k)v_k$. Define

$$\tilde{v}_k = [H(k)P(k)H^{\mathrm{T}}(k) + G(k)G^{\mathrm{T}}(k)]^{-1/2} v_k$$

(the inverse exists since by our standing assumptions $G(k)G^{\mathrm{T}}(k) > 0$). From (6.3.13) we see that $E[\tilde{v}_k \tilde{v}_k^{\mathrm{T}}] = I$, so that $\tilde{v}_k$ is a normalized white-noise process, and (6.3.19) can be written

$$\hat{x}_{k+1|k} = A(k)\hat{x}_{k|k-1} + B(k)u_k + K(k)[H(k)P(k)H^{\mathrm{T}}(k) + G(k)G^{\mathrm{T}}(k)]^{1/2}\tilde{v}_k.$$

This is now in the standard form of (6.2.1) with a new 'C-matrix' $K[HPH^{\mathrm{T}} + GG^{\mathrm{T}}]^{1/2}$ and we can read off the optimal cost from (6.2.9). Remembering that the two constant terms from (6.3.17) must also be included, we obtain (6.3.21). □

Let us summarize the computations needed in order to implement the control policy described in Theorem 6.3.2. They are as follows:

(a) Solve the matrix Riccati equation of dynamic programming backwards from the terminal time to give matrices $S(N), \ldots, S(0)$:

$$\begin{aligned} S(k) = {} & A^{\mathrm{T}}(k)S(k+1)A(k) + D^{\mathrm{T}}(k)D(k) \\ & - [A^{\mathrm{T}}(k)S(k+1)B(k) + D^{\mathrm{T}}(k)F(k)] \\ & [B^{\mathrm{T}}(k)S(k+1)B(k) + F^{\mathrm{T}}(k)F(k)]^{-1} \\ & [B^{\mathrm{T}}(k)S(k+1)A(k) + F^{\mathrm{T}}(k)D(k)] \end{aligned} \tag{6.3.22}$$
$$S(N) = Q.$$

This determines the feedback matrices

$$M(k) = [B^{\mathrm{T}}(k)S(k+1)B(k) + F^{\mathrm{T}}(k)F(k)]^{-1} [B^{\mathrm{T}}(k)S(k+1)A(k) + F^{\mathrm{T}}(k)D(k)].$$

(b) Solve the matrix Riccati equation of Kalman filtering forwards from the initial time to give matrices $P(0), \ldots, P(N)$:

$$\begin{aligned} P(k+1) = {} & A(k)P(k)A^{\mathrm{T}}(k) + C(k)C^{\mathrm{T}}(k) - [A(k)P(k)H^{\mathrm{T}}(k) + C(k)G^{\mathrm{T}}(k)] \\ & H(k)P(k)H^{\mathrm{T}}(k) + G(k)G^{\mathrm{T}}(k)]^{-1} \\ & [H(k)P(k)A^{\mathrm{T}}(k) + G(k)C^{\mathrm{T}}(k)] \\ P(0) = {} & P_0. \end{aligned} \tag{6.3.23}$$

This determines the Kalman gain matrices

$$K(k) = [A(k)P(k)H^{\mathrm{T}}(k) + C(k)G^{\mathrm{T}}(k)][H(k)P(k)H^{\mathrm{T}}(k) + G(k)G^{\mathrm{T}}(k)]^{-1}.$$

It is important to notice that these computations refer *independently* to the control and filtering problems respectively, in that (a) involves the 'cost' parameters Q, $D(k)$, $F(k)$ but not the 'noise' parameters P_0, $C(k)$, $G(k)$, whereas the converse is true in the case of (b).

The property that the optimal control takes the form $\hat{u}^1(k) = -M(k)\hat{x}_{k|k-1}$ where $M(k)$ is the same as in the deterministic or complete observation cases, expresses the so-called 'certainty-equivalence principle' which, put in another way, states that, optimally, the controller *acts as if* the state estimate $\hat{x}_{k|k-1}$ were equal to the true state x_k *with certainty*. Of course, the controller knows that this is not the case, but no other admissible strategy will give better performance.

That $M(k)$ is unchanged in the presence of observation noise is entirely due to the quadratic cost criterion which ensures that the cost function for the problem in innovations form is, apart from a fixed constant, the same as that in the original form. On the other hand, the fact that the intermediate statistic to be computed is $\hat{x}_{k|k-1}$, regardless of cost parameters, is a property which extends to more general forms of cost function. To see this, recall that whatever admissible control is applied, the conditional distribution of x_k given y^{k-1} is $N(\hat{x}_{k|k-1}, P(k))$. Now suppose that the cost to be minimized takes a general form similar to (6.1.14), i.e.

$$C_N(u) = E\left(\sum_{k=0}^{N-1} l(k, x_k, u_k) + g(x_N)\right)$$

where l and g are, say, bounded functions. Introducing intermediate conditional expectations, we can express $C_N(u)$ as

$$C_N(u) = E\left(\sum_{k=0}^{N-1} E[l(k, x_k, u_k)|y^{k-1}] + E[g(x_N)|y^{N-1}]\right).$$

The conditional expectation can now be evaluated by integrating with respect to the conditional distribution. This gives

$$E[l(k, x_k, u_k)|y^{k-1}] = \hat{l}(k, \hat{x}_{k|k-1}, u_k)$$

and

$$E[g(x_N)|y^{N-1}] = \hat{g}(\hat{x}_{N|N-1})$$

where

$$\hat{l}(k, \hat{x}, u) = \int_{\mathbb{R}^n} l(k, z, u) \frac{1}{(2\pi)^{n/2}(\det(P(k)))^{1/2}} \cdot \exp((z-\hat{x})^{\mathrm{T}} P^{-1}(k)(z-\hat{x}))\, dz$$

$$\hat{g}(\hat{x}) = \int_{\mathbb{R}^n} g(z) \frac{1}{(2\pi)^{n/2}(\det(P(N)))^{1/2}} \cdot \exp((z-\hat{x})^{\mathrm{T}} P^{-1}(N)(z-\hat{x}))\, dz.$$

Thus

$$C_N(u) = E\left(\sum_{k=0}^{N-1} \hat{l}(k, \hat{x}_{k|k-1}, u_k) + \hat{g}(\hat{x}_{N|N-1}) \right). \qquad (6.3.24)$$

The problem (6.3.19), (6.3.24) is now in innovations form and can be solved by dynamic programming. Define functions $W_0, \ldots, W_N$ by

$$W_N(\hat{x}) = \hat{g}(\hat{x})$$
$$W_k(\hat{x}) = \min_v \{\hat{l}(k, \hat{x}, v) + E^{(v)} W_k(A\hat{x} + Bv + K(k)v_k)$$
$$k = N-1, \ldots, 0 \qquad (6.3.25)$$

where $E^{(v)}$ denotes expectation taken over the distribution of v_k, which is $N(0, HP(k)H^{\mathrm{T}} + GG^{\mathrm{T}})$. Let $\hat{u}^1(k, \hat{x})$ be a value of v which achieves the minimum in (6.3.25). Then the optimal control is

$$\hat{u}_k^1 = \hat{u}^1(k, \hat{x}_{k|k-1})$$

with minimal cost

$$C_N(\hat{u}^1) = W_0(m_0).$$

This can be checked by the same sort of 'verification theorem' proved earlier. Thus is this general problem the 'data processing' still consists of calculating $\hat{x}_{k|k-1}$ via the Kalman filter, but the control function $\hat{u}^1(k, \hat{x})$ is not related in any simple way to the control function $u^1(k, x)$ which is optimal in the case of complete observations.

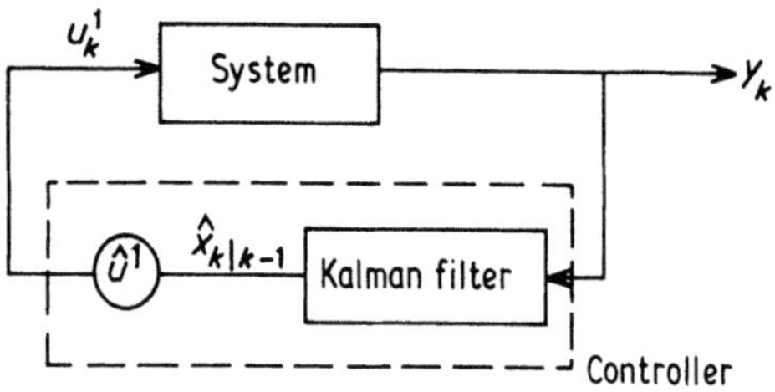

Fig. 6.2

In summary, we see that the optimal controller *separates* into two parts, a *filtering stage* and a *control stage* as shown in Fig. 6.2. The filtering stage is *always the same* regardless of the control objective. This is the *separation principle.* The certainty-equivalence principle applies when $\hat{u}^1(k, x_k)$ is the optimal completely observable control, but this is a much more special property which holds only in the quadratic cost case.

These results point to a general cybernetic principle, namely that when systems are to be controlled on the basis of noisy measurements the true 'state' of the system which is relevant for control is the *conditional distribution of the original state given the observations.* Note that in the LQG problem this is completely determined by $\hat{x}_{k|k-1}$ since the conditional distribution is $N(\hat{x}_{k|k-1}, P(k))$ and $P(k)$ does not depend on the observations. Thus the Kalman filter in effect updates the conditional distribution of x_k given y^{k-1}. The problem can be solved in an effective way because of the simple parametrization of the conditional density and the fact that there is an efficient algorithm – the Kalman filter – for updating the parameter $\hat{x}_{k|k-1}$. More general problems typically involve extensive computation due to the lack of any low-dimensional statistic characterizing the conditional distributions.

6.3.3 Discounted costs and the infinite-time problem

In this section we will assume that the system matrices A, B, H, C, G are time-invariant, that $D(k) = \rho^{k/2}D$, $F(k) = \rho^{k/2}F$, and that Q is replaced by $\rho^N Q$ for some $\rho < 1$, so that the cost function becomes

$$C_N^\rho(u) = E\left[\sum_{k=0}^{N-1} \rho^k \| Dx_k + Fu_k \|^2 + \rho^N x_N^T Q x_N\right].$$

In view of the 'separation property', the Kalman filter matrices $P(k)$, $K(k)$ are unaffected by the discount factor ρ. By specializing the preceding results, or by using an argument involving x_k^ρ, u_k^ρ as in Section 6.2, one can verify that the control which minimizes $C_N^\rho(u)$ is

$$\hat{u}_k^\rho = -M^\rho(k)\hat{x}_{k|k-1}$$

with $M^\rho(k)$ as before. The cost corresponding to $\hat{u}^\rho$ is

$$\begin{aligned} C_N^\rho(\hat{u}^\rho) &= m_0^{\mathrm{T}} S^\rho(0) m_0 + \rho^N \operatorname{tr}[P(N)Q] \\ &\quad + \sum_{k=0}^{N-1} \rho^k \operatorname{tr}[DP(k)D^{\mathrm{T}} \\ &\quad + \rho K(k)[HP(k)H^{\mathrm{T}} + GG^{\mathrm{T}}]K^{\mathrm{T}}(k)S^\rho(k+1)]. \end{aligned}$$

Thus if a discount factor is introduced, the filtering computation (b) is unchanged while, in the control computation (a), A and B are replaced by $\rho^{1/2}A$, $\rho^{1/2}B$ respectively.

Turning now to the minimization of the infinite-time cost,

$$C_\infty^\rho(u) = E\left[\sum_{k=0}^{\infty} \rho^k \|Dx_k + Fu_k\|^2\right],$$

we have to consider the asymptotic properties of *both* Riccati equations (6.3.32) and (6.3.23). The conditions required are as follows

$$\left.\begin{matrix}(A, B)\\ (\breve{A}, \breve{C})\end{matrix}\right\} \text{ stabilizable} \tag{6.3.26}$$

$$\left.\begin{matrix}(\hat{D}, \hat{A})\\ (H, A)\end{matrix}\right\} \text{ detectable}$$

where

$$\begin{aligned} \breve{A} &= A - CG^{\mathrm{T}}(GG^{\mathrm{T}})^{-1}H & \hat{A} &= A - B(F^{\mathrm{T}}F)^{-1}F^{\mathrm{T}}D \\ \breve{C} &= C[I - G^{\mathrm{T}}(GG^{\mathrm{T}})^{-1}G] & \hat{D} &= [I - F(F^{\mathrm{T}}F)^{-1}F^{\mathrm{T}}]D. \end{aligned}$$

These conditions simplify under the additional conditions, assumed at the outset in most treatments of LQG control, that $CG^{\mathrm{T}} = 0$ (no correlation between state and observation noise) and $F^{\mathrm{T}}D = 0$ (no 'cross-term' in the cost criterion). Under these conditions, $\breve{A} = \hat{A} = A$, $\breve{C} = C$ and $\hat{D} = D$; thus conditions (6.3.26) stipulate that the system be stabilizable from either the control or the noise input, and that it be detectable either via the output Hx_k or via the 'output' Dx_k appearing in the cost function.

According to the results in Appendix B, conditions (6.3.26) guarantee that the algebraic Riccati equations corresponding to (6.3.22), (6.3.23) have unique non-negative definite solutions S, P respectively and that the solutions of (6.3.22), (6.3.23) converge to S, P for arbitrary non-negative definite terminal condition Q and initial condition P_0 respectively. The optimal control for the infinite-time problem can now be obtained by applying the results of Section 6.2 concerning the completely observable case. Indeed, the innovations representation is, as above,

$$\hat{x}_{k+1|k} = A\hat{x}_{k|k-1} + Bu_k + \tilde{C}(k)\tilde{v}_k \tag{6.3.27}$$

where $\tilde{v}_k$ is the normalized innovations process and

$$\tilde{C}(k) = K(k)[HP(k)H^{\mathrm{T}} + GG^{\mathrm{T}}]^{1/2}.$$

Note that, as $k \to \infty$,

$$\tilde{C}(k) \to \tilde{C} = K[HPH^{\mathrm{T}} + GG^{\mathrm{T}}]^{1/2}.$$

where P is the solution of the algebraic Riccati equation and K the corresponding Kalman gain. As in (6.3.17) the cost expressed in terms of $\hat{x}_{k|k-1}$ is

$$C^{\rho}_{\infty}(u) = E\left[\sum_{k=0}^{\infty} \rho^k \| D\hat{x}_{k|k-1} + Fu_k \|^2\right] + \sum_{k=0}^{\infty} \rho^k \operatorname{tr}[DP(k)D^{\mathrm{T}}] \tag{6.3.28}$$

and the final sum is finite since $\operatorname{tr}[DP(k)D^{\mathrm{T}}] \to \operatorname{tr}[DPD^{\mathrm{T}}]$ as $k \to \infty$. We now apply the results of Section 6.2 to the infinite-time completely observable problem constituted by (6.3.27), (6.3.28), and conclude that the optimal control is

$$\hat{u}^{\rho}_k = -M^{\rho}\hat{x}_{k|k-1} \tag{6.3.29}$$

with cost, as in (6.2.16),

$$m_0^{\mathrm{T}} S^{\rho} m_0 + \sum_{k=0}^{\infty} \rho^{k+1} \operatorname{tr}[\tilde{C}^{\mathrm{T}}(k) S^{\rho} \tilde{C}(k)] + \sum_{k=0}^{\infty} \rho^k \operatorname{tr}[DP(k)D^{\mathrm{T}}].$$

Substituting for $\tilde{C}(k)$ gives the final cost expression

$$\begin{aligned} C^{\rho}_{\infty}(\hat{u}^{\rho}) = m_0^{\mathrm{T}} S^{\rho} m_0 + \sum_{k=0}^{\infty} \rho^k [DP(k)D^{\mathrm{T}} \\ + \rho K(k)[HP(k)H^{\mathrm{T}} + GG^{\mathrm{T}}]K^{\mathrm{T}}(k)S^{\rho}]]. \end{aligned}$$

Appearances to the contrary, $\hat{u}_k^\rho$ given by (6.3.29) is not a constant-coefficient controller since the gain $K(k)$ in the Kalman filter depends on $P(k)$ which is not constant unless P_0 happens to be equal to the stationary value P. A simpler control algorithm is obtained if $K(k)$ is replaced by its stationary value $K = [APH^T + CG^T]\ [HPH^T + GG^T]^{-1}$, that is we apply the control value

$$\hat{v}_k^\rho := -M^\rho z_k \tag{6.3.30}$$

where z_k is generated by

$$\begin{aligned} z_{k+1} &= Az_k - BM^\rho z_k + K(y_k - Hz_k) \\ z_0 &= m_0 \end{aligned} \tag{6.3.31}$$

(this is the Kalman filter algorithm with $P(k)$ replaced by P). Of course, z_k is in general not equal to $\hat{x}_{k|k-1}$. Control $\hat{v}^\rho$ is not optimal for the discounted cost problem, but $\hat{v}^1$ *is* optimal in the sense of minimizing the average cost per unit time,

$$C_{av}(u) = \lim_{N\to\infty} \frac{1}{N} E\left[\sum_{k=0}^{N} \|Dx_k + Fu_k\|^2\right]. \tag{6.3.32}$$

As remarked earlier, this criterion is insensitive to the behaviour of the process for small k; and, for large K, z_k and $\hat{x}_{k|k-1}$ are practically indistinguishable.

Theorem 6.3.3

Suppose conditions (6.3.26) hold. Then the control $\hat{v}^1$ given by (6.3.30), (6.3.31) with $\rho = 1$ minimizes $C_{av}(u)$ in the class of all output feedback controls such that $C_{av}(u)$ exists and $E\|x_k\|^2$ is bounded. The minimal cost is

$$C_{av}(\hat{v}^1) = \mathrm{tr}[DPD^T + K(HPH^T + GG^T)K^TS]. \tag{6.3.33}$$

PROOF It follows from the arguments above and Theorem 6.3.2 that the control $\hat{u}_k^1$ of (6.3.20) is optimal for C_{av} and that its cost is given by the expression in (6.3.33). Thus it remains to show that $\hat{v}^1$ is admissible and that its cost coincides with that of $\hat{u}^1$.

Define $\xi_k := x_k - z_k$. Recalling that $y_k = Hx_k + Gw_k$ and hence that $y_k - Hz_k = H(x_k - z_k) + Gw_k$, we see that the joint process (z_k, ξ_k) satisfies:

$$\begin{bmatrix} z_{k+1} \\ \xi_{k+1} \end{bmatrix} = \begin{bmatrix} A - BM & KH \\ 0 & A - KH \end{bmatrix} \begin{bmatrix} z_k \\ \xi_k \end{bmatrix} + \begin{bmatrix} KG \\ C + KG \end{bmatrix} w_k$$

$$=: \bar{A} \begin{bmatrix} z_k \\ \xi_k \end{bmatrix} + \bar{C} w_k. \tag{6.3.34}$$

Under conditions (6.3.26) both $A - BM$ and $A - KH$ are stable. This implies that $\bar{A}$ is stable since the eigenvalues of $\bar{A}$ are those of $(A - BM)$ together with those of $(A - KH)$. Thus the covariance matrix Ξ_k of (x_k, ξ_k) is convergent to Ξ satisfying $\Xi = \bar{A}\Xi\bar{A}^{\mathrm{T}} + \bar{C}\bar{C}^{\mathrm{T}}$. Since $x_k = \xi_k + z_k$, this shows that $E\|x_k\|^2$ is bounded and $C_{\mathrm{av}}(\hat{v}^1)$ exists. Note that

$$Dx_k + F\hat{v}_k^1 = \bar{D} \begin{bmatrix} z_k \\ \xi_k \end{bmatrix}$$

where $\bar{D} = (D - FM, D)$, so that

$$C_{\mathrm{av}}(\hat{v}^1) = \mathrm{tr}[\bar{D}\Xi\bar{D}^{\mathrm{T}}].$$

The process $\eta_k := \mathrm{col}\{\hat{x}_{k|k-1}, \tilde{x}_{k|k-1}\}$ satisfies (6.3.34) with $\bar{A}$ and $\bar{C}$ replaced by $\bar{A}(k)$ and $\bar{C}(k)$ obtained by substituting $K(k)$ for K in $\bar{A}$ and $\bar{C}$. Denote $\Gamma(k) := \mathrm{cov}(\eta_k)$. Then $\Gamma(k)$ satisfies

$$\Gamma(k+1) = \bar{A}(k)\Gamma(k)\bar{A}^{\mathrm{T}}(k) + \bar{C}(k)\bar{C}^{\mathrm{T}}(k) \tag{6.3.35}$$

We know that $\Gamma := \lim_{k \to \infty} \Gamma(k)$ exists and that

$$C_{\mathrm{av}}(\hat{u}^1) = \mathrm{tr}[\bar{D}\Gamma\bar{D}^{\mathrm{T}}].$$

Taking the limit as $k \to \infty$ in (6.3.35) we see that Γ satisfies $\Gamma = \bar{A}\Gamma\bar{A}^{\mathrm{T}} + \bar{C}\bar{C}^{\mathrm{T}}$, i.e. $\Gamma = \Xi$. This completes the proof. □

Finally, a remark on the stabilizability and detectability conditions (6.3.26). The conditions on (A, B), $(\hat{D}, A)$ ensure that S^ρ, the solution to the 'discounted' algebraic Riccati equation, exists for any $\rho < 1$, but if these conditions are not met then S^ρ may only exist for $\rho < \rho_0$ for some $\rho_0 < 1$. According to the separation principle, however, discounting has no effect on the Riccati equation (6.3.23) generating $P(k)$ so that no weakening of the conditions on $(\check{A}, \check{C})$ and (H, A) is possible. The reason for this minor asymmetry in the problem is of course that, while we are free to select the cost function coefficients D, F in any manner we choose, their counterparts C and G in the filtering problem are part of the system specification.

As in the complete observations case, little can be said about the average cost problem if conditions (6.3.26) are not met.

Notes

Dynamic programming was introduced in its modern form by Bellman (1957). Recent texts describing various aspects of it include Bertsekas (1976) and Whittle (1981). The linear regulator problem was solved by Kalman (1960) who also noted the filtering/control duality. For references on properties of the Riccati equation and the algebraic Riccati equation, see Chapter 3. The use of linear/quadratic control as a design methodology for multivariable systems has been pioneered by Harvey and Stein (1978); see also Kwakernaak (1976).

The 'certainty-equivalence principle' was first enunciated in the economics literature, by Simon (1956). The 'separation principle' is clearly presented (for continuous-time systems) in Wonham (1968) and is also discussed in Fleming and Rishel (1975). The stochastic linear regulator is discussed in one form or another in most texts on stochastic control, including Bertsekas (1976) and Whittle (1981).

References

Bellman, R. (1957) *Dynamic Programming*, Princeton University Press, New Jersey.

Bertsekas, D. P. (1976) *Dynamic Programming and Stochastic Control*, Academic Press, New York.

Fleming, W. H. and Rishel, R. W. (1975) *Deterministic and Stochastic Optimal Control*, Springer-Verlag, Berlin.

Harvey, C. A. and Stein, G. (1978) Quadratic weights for asymptotic regulator properties. *IEEE Trans. Automatic Control*, **AC-23**, 378–387.

Kalman, R. E. (1960) Contributions to the theory of optimal control. *Bol. Soc. Math. Mexicana*, **5**, 102–119.

Kwakernaak, H. (1976) Asymptotic root loci for multivariable linear optimal regulators. *IEEE Trans. Automatic Control*, **AC-21**, 378–381.

Simon, H. A. (1956) Dynamic programming under uncertainty with a quadratic criterion function. *Econometrica*, **24**, 74–81.

Whittle, P. (1981) *Optimization over Time*, vols 1 and 2, John Wiley, Chichester.

Wonham, W. M. (1968) On the separation theorem of stochastic control. *SIAM J. Control*, **6**, 312–326.

CHAPTER 7

Minimum-variance and self-tuning control

In Chapter 6 we have studied LQG control system design for state-space models. Since ARMAX models can be realized in state-space form, the results apply equally to ARMAX models. In either case, it is supposed that the parameters of the model are precisely known. On the other hand, we have presented in Chapter 4 techniques for identifying unknown systems from input/output data. Is it possible to combine these techniques and design controllers for 'unknown' systems involving some kind of on-line combination of identification and control? The general area to control system design for imperfectly known (and possible time-varying) systems is known as *adaptive control* and has been the subject of extensive study over many years. In this chapter we do not attempt any overall coverage of this area (which would require at least a whole book in itself) but restrict ourselves to discussing two key ideas – minimum-variance control and self-tuning regulators – which are closely related to the material of the preceding chapters. Both of these ideas are in their present form due to K.J. Åström and co-workers (1970, 1973) and have since burgeoned into a minor industry (quite literally, in that computer controllers incorporating these concepts are now commercially available). We also discuss the related ideas of pole-shifting regulators, which retain more links to classical control system design, and were introduced by Wellstead and co-workers (1979). Within the confines of a short chapter it is only possible to present the main theoretical results and we must refer the reader elsewhere for their ramifications in the context of practical control system design.

This chapter is concerned with regulator (minimizing output variance) and control (minimizing tracking error) for single-input single-output systems described by ARMAX models. The minimum-variance (m.v.) regulator can be viewed as the limiting case of LQG control when the cost for control energy is reduced to zero. The result is an extremely simple algorithm which, when it works, can provide effective regulation. There are cases, however, in which the m.v. regulator involves excessive use of control energy or even a loss of stability. In these cases one must resort to the 'full' LQG control (or some sub-optimal approximation to it) which gives better control at the expense of a vastly increased computational load. These topics are discussed in Section 7.1, and in Section 7.2 pole-shifting regulators are introduced; here the concept of 'optimality' is abandoned in favour of a qualitative specification of desired response expressed in terms of pole locations. In Section 7.3, adaptive versions of these algorithms are discussed. It is a remarkable fact that a combination of simple least-squares estimation and m.v. control can give a system whose long-run performance is as good as that which could be obtained if the system parameters were known exactly. The same is true for some classes of pole-shifting regulators. These algorithms must, however, be modified somewhat if one wishes to prove that the parameter estimates will actually converge under reasonably general conditions. We present one such algorithm, due to Goodwin, Ramadge and Caines (1981) in Section 7.4; a proof of convergence is given in Appendix C. This is a landmark result in stochastic adaptive control; much current research is based on similar ideas.

7.1 Regulation for systems with known parameters

7.1.1 Minimum-variance control

The minimum-variance controller is a simple scheme for regulating the output of an ARMAX model by a predictive cancellation procedure. The system model is given in standard ARMAX single-input, single-output form as

$$A(z^{-1})y_k = z^{-r}B(z^{-1})u_k + C(z^{-1})w_k. \tag{7.1.1}$$

Here $\{w_k\}$ is a white noise sequence with variance σ^2 and[†]

$$\begin{aligned} A(z^{-1}) &= 1 + a_1 z^{-1} + \cdots + a_n z^{-n} \\ B(z^{-1}) &= b_0 + b_1 z^{-1} + \cdots + b_{n-r} z^{-(n-r)} \\ C(z^{-1}) &= 1 + c_1 z^{-1} + \cdots + c_n z^{-n}. \end{aligned}$$

[†] For convenience we suppose in this section that $A, z^{-r}B$ and C have the same degree n. Some coefficients may vanish.

The number of steps of delay between input and output is r, so that $b_0 \neq 0$ by definition. A and C are assumed to be stable and it is supposed that $r \geq 1$; we shall comment below on the conditions required for B. The control objective here is regulation, i.e. we want to make the output y_k as small as possible. 'Minimum variance' means that we seek to minimize Ey_k^2 at each k. The key to minimum-variance control is the so-called predictor form of the ARMAX model. Consider first the control-free case $u_k = 0$ with time set $\mathbb{Z}$. Then according to the discussion in Chapter 2, the output y_k of the ARMAX system (7.1.1) is a stationary process which can be written as an infinite-order moving average:

$$y_k = Q(z^{-1})w_k = q_0 w_k + q_1 w_{k-1} + \cdots$$

where $Q(z^{-1}) = C(z^{-1})/A(z^{-1})$. The sum converges in quadratic mean. The model is invertible, in that w_k can be recovered from past outputs $y_k, y_{k-1}, \ldots$ by the formula

$$w_k = [Q(z^{-1})]^{-1} y_k.$$

Let us consider the r-step-ahead prediction problem of forming the best linear approximation at time k to y_{k+r}. We can write

$$y_{k+r} = \sum_{j=0}^{r-1} q_j w_{k+r-j} + \sum_{j=0}^{\infty} q_{j+r} w_{k-j}$$

where the two terms on the right are uncorrelated and the second is 'known' at time k. Take a general predictor in the form

$$X = \sum_{j=0}^{\infty} \alpha_j w_{k-j}.$$

Then the mean square error is

$$E[y_{k+r} - X]^2 = \sigma^2 \left\{ \sum_{j=0}^{r-1} q_j^2 + \sum_{j=0}^{\infty} (q_{j+r} - \alpha_j)^2 \right\}$$

and this is minimized by taking $\alpha_j = q_{j+r}$, so that $\sum_0^\infty q_{j+r} w_{k-j}$ is the best predictor, usually denoted $\hat{y}_{k+r|k}$. What remains is to develop an effective way of computing $\hat{y}_{k+r|k}$ given that what we observe is $\{y_k\}$, not $\{w_k\}$. The following proposition provides the solution.

Proposition 7.1.1

There are unique polynomials $F(z^{-1})$, $D(z^{-1})$ of degrees $r-1$, $n-1$ respectively, such that

$$C(z^{-1}) = A(z^{-1})F(z^{-1}) + z^{-r}D(z^{-1}). \tag{7.1.2}$$

PROOF F and D are obtained by equating coefficients of z^{-j} for $j = 1, \ldots, (n + r - 1)$ in (7.1.2). Write

$$F(z^{-1}) = 1 + f_1 z^{-1} + \cdots + f_{r-1} z^{-(r-1)}$$
$$D(z^{-1}) = d_0 + d_1 z^{-1} + \cdots + d_{n-1} z^{-(n-1)}.$$

Then we obtain

$$\begin{aligned}
c_1 &= a_1 + f_1 \\
c_2 &= a_2 + a_1 f_1 + f_2 \\
&\vdots \\
c_{r-1} &= a_{r-1} + a_{r-2} f_1 + \cdots + a_1 f_{r-2} + f_{r-1} \\
c_r &= a_r + a_{r-1} f_1 + \cdots + a_1 f_{r-1} + d_0 \\
&\vdots \\
0 &= a_n f_{r-1} + d_{n-1}.
\end{aligned}$$

The first $(r - 1)$ of these equations determine $f_1, \ldots, f_{r-1}$ recursively and the last n determine $d_0, \ldots, d_{n-1}$. □

Using the expression (7.1.2) for $C(z^{-1})$ in (7.1.1) (still with $u_k = 0$) we obtain

$$\begin{aligned}
y_{k+r} &= F(z^{-1}) w_{k+r} + \frac{D(z^{-1})}{A(z^{-1})} w_k \\
&= F(z^{-1}) w_{k+r} + \frac{D(z^{-1})}{C(z^{-1})} y_k.
\end{aligned}$$

The best predictor is thus

$$\hat{y}_{k+r|k} = \frac{D(z^{-1})}{C(z^{-1})} y_k$$

with prediction error

$$\sigma^2 \sum_{j=0}^{r-1} f_j^2$$

(the coefficients $f_1, \ldots, f_{r-1}$ coincide with $q_1, \ldots, q_{r-1}$ and $f_0 = q_0 = 1$). Note that this provides a very simple way of calculating $\hat{y}_{k+r|k}$: it is the output of the ARMA system

$$C(z^{-1}) \hat{y}_{k+r|k} = D(z^{-1}) y_k$$

driven by y_k. (Here, $z^{-1} \hat{y}_{k+r|k} = \hat{y}_{k-1+r|k-1}$.)

Let us now return to the controlled case with $u_k \neq 0$. A little algebra using the identity (7.1.2) shows that the system equation (7.1.1) can be written in the form

$$y_{k+r} = \frac{B(z^{-1})F(z^{-1})}{C(z^{-1})} u_k + \frac{D(z^{-1})}{C(z^{-1})} y_k + F(z^{-1})w_{k+r}, \qquad (7.1.3)$$

and, since F has degree $r-1$,

$$F(z^{-1})w_{k+r} = w_{k+r} + f_1 w_{k+r-1} + \cdots + f_{r-1} w_{k+1}. \qquad (7.1.4)$$

Now suppose that the control u_k is a linear function of present and past outputs $y_k, y_{k-1} \ldots$. In view of (7.1.4), the first two terms on the right of (7.1.3) are uncorrelated with $F(z^{-1})w_{k+r}$, and hence

$$Ey_{k+r}^2 = E\left(\frac{B(z^{-1})F(z^{-1})}{C(z^{-1})} u_k + \frac{D(z^{-1})}{C(z^{-1})} y_k\right)^2 + E(F(z^{-1})w_{k+r})^2. \qquad (7.1.5)$$

This expression is minimized by taking

$$B(z^{-1})F(z^{-1})u_k + D(z^{-1})y_k = 0, \qquad (7.1.6)$$

i.e.,

$$u_k = -\frac{D(z^{-1})}{B(z^{-1})F(z^{-1})} y_k. \qquad (7.1.7)$$

This is the *minimum-variance (m.v.) control law.* One of its advantages is that it is extremely easy to compute on-line, since the recursion (7.1.6) expresses the current control value u_k as a linear combination of a finite number of past u_j's and y_j's. If the minimum-variance controller is applied then the controlled process y_k satisfies the equation

$$y_k = F(z^{-1})w_k,$$

so that y_k is a moving-average process of order $r-1$. Its variance is

$$Ey_k^2 = (f_0^2 + f_1^2 + \cdots + f_{r-1}^2)\sigma^2.$$

Example 7.1.2

Let us consider the system

$$(1 + az^{-1})y_k = z^{-2}u_k + (1 + cz^{-1})w_k \qquad (7.1.8)$$

with $\sigma^2 = 1$. In this case $F = 1 + (c-a)z^{-1}$, $D = -a(c-a)$, and the minimum-variance controller is

$$u_k = \frac{a(c-a)}{1+(c-a)z^{-1}} y_k, \tag{7.1.9}$$

the output variance then being

$$Ey_k^2 = f_0^2 + f_1^2 = 1 + (c-a)^2. \tag{7.1.10}$$

An obvious drawback of the minimum variance controller is that it cannot be used if B has unstable zeros[†] (such systems are often called *non-minimum phase* systems). Indeed, since with m.v. control $\{y_k\}$ satisfies $y_k = F(z^{-1})w_k$ we see from (7.1.7) that $\{u_k\}$ is given in terms of the disturbance sequence $\{w_k\}$ by

$$u_k = -\frac{D(z^{-1})}{B(z^{-1})} w_k$$

If B has unstable zeros which are not cancelled by those of D then $\operatorname{var}(u_k) \to \infty$. One can easily see the mechanism of this instability by considering the deterministic system $y_k + y_{k-1} = u_k - 2u_{k-1}$. If $y_0 = 1$ and $u_0 = 0$ then the control sequence $u_k = 2^{k-1}$ gives $y_k = 0$ for $k \geq 1$; progressively larger control values are required to cancel out the effect of previous controls. Of course such a control policy is totally unacceptable but is not excluded by our formulation because the control values are uncosted.

For minimum-phase systems (B has no unstable zeros), the m.v. controller gives a control process $\{u_k\}$ which is asymptotically stationary; but this process may still have very large variance. An example is given in Section 7.1.3 below.

Minimum-variance controllers can also be designed with the objective of tracking a given (non-random) reference signal y_k^*. Indeed, subtracting y_{k+r}^* from both sides of (7.1.3) we can write, as in (7.1.5),

$$E(y_{k+r} - y_{k+r}^*)^2 = E\left(\frac{B(z^{-1})F(z^{-1})}{C(z^{-1})}u_k + \frac{D(z^{-1})}{C(z^{-1})}y_k - y_{k+r}^*\right)^2 + E\left(F(z^{-1})w_{k+r}\right)^2,$$

giving the modified m.v. controller recursively as follows:

$$B(z^{-1})F(z^{-1})u_k = C(z^{-1})y_{k+r}^* - D(z^{-1})y_k.$$

[†]We say zeros of a function $P(\sigma)$ are stable (unstable) if they lie outside (inside) the closed unit disc. Likewise we define stable and unstable poles.

As before, the error variance is

$$\sigma^2 \sum_0^{r-1} f_k^2.$$

The most frequently encountered practical case of this is $y_k^* \equiv y^*$ (a constant 'set point').

In many circumstances the m.v. controller is a satisfactory practical device. For a specific system with known parameters it is very easy to compute the coefficient of the m.v. controller; if these turn out to give an adequate stability margin then this design will provide effective control. If not, then some form of control costing must be incorporated but this involves a substantial increase in the amount of computation required. The main use of the m.v. controller in its simplest form is in fact in connection with self-tuning control as discussed in Section 7.3.

7.1.2 The minimum-variance regulator with control costs

The m.v. controller for the system (7.1.1) minimizes Ey_k^2 simultaneously for all k and hence minimizes the long-run average cost

$$\lim_{N\to\infty} \frac{1}{N} \sum_{k=1}^{N} Ey_k^2$$

or the discounted cost

$$E \sum_{k=1}^{\infty} \rho^k y_k^2$$

with $0 < \rho < 1$; in fact these quantities take the minimal values f^* and $\rho f^*/(1-\rho)$ respectively, where

$$f^* = \sigma^2 \sum_{k=0}^{r-1} f_k^2.$$

With control costing such simultaneous minimization is no longer possible since each u_k contributes to the output at several different times. Consider the cost function

$$J(u) = \lim_{N\to\infty} E \frac{1}{N} \sum_{k=1}^{N} (y_k^2 + \lambda u_k^2). \tag{7.1.11}$$

When $\lambda = 0$ control is uncosted and the minimizing u will be the m.v. controller. When $\lambda > 0$ we have a quadratic cost functional, and since the system equation (7.1.1) is linear the minimization of $J(u)$ is an LQG problem of the sort considered in the previous chapter. The parameter λ can be adjusted so as to penalize more or less severely the use of control energy.

In order to apply the LQG theory we have to realize the ARMAX model in state-space form. The standard realization as given in Chapter 2 is

$$\begin{aligned} x_{k+1} &= Ax_k + Bu_k + Cw_k \\ y_k &= [0,\ldots,0,1]x_k + w_k \end{aligned} \tag{7.1.12}$$

where

$$A = \begin{bmatrix} 0 & & & -a_n \\ 1 & & & -a_{n-1} \\ \cdot & \cdot & & \\ \cdot & & \cdot & \\ \cdot & & & \cdot \\ 0 & & 1 & -a_1 \end{bmatrix} \qquad B = \begin{bmatrix} b_{n-r+1} \\ \vdots \\ b_1 \\ 0 \\ \vdots \\ 0 \end{bmatrix}$$

$$C = \begin{bmatrix} c_n - a_n \\ \\ \\ \\ c_1 - a_1 \end{bmatrix}.$$

However this realization is not quite the appropriate one here since a state feedback control $u_k = Kx_k$ gives a control u_k depending only on $w_{k-1}, w_{k-2}, \ldots, w_0$, whereas the m.v. controller (for example) depends on $w_k, w_{k-1}, \ldots, w_0$. We should therefore include w_k as a state variable at time k. Define $v_k := w_{k+1}$ and $x_k^0 := w_k$. Then the state equations can be written in the form

$$\begin{aligned} \begin{bmatrix} x_{k+1}^0 \\ x_{k+1} \end{bmatrix} &= \begin{bmatrix} 0 & 0 \\ C & A \end{bmatrix} \begin{bmatrix} x_k^0 \\ x_k \end{bmatrix} + \begin{bmatrix} 0 \\ B \end{bmatrix} u_k + \begin{bmatrix} 1 \\ 0 \end{bmatrix} v_k \\ y_k &= [1,0,\ldots,0,1] \begin{bmatrix} x_k^0 \\ x_k \end{bmatrix}. \end{aligned} \tag{7.1.13}$$

With this realization, controls in state feedback form $u_k = K\bar{x}_k$ (where $\bar{x}_k^{\mathrm{T}} = (x_k^0, x_k^{\mathrm{T}})$) are sufficiently general to cover the minimum-variance controller as a special case.

The reader will object at this point that the correct class of controls for this problem is not state but output feedback and therefore that the solution is that obtained in Section 6.3 for the partially observable case, involving a Kalman filter to estimate the unobserved states.

Recall, however, that the realization (7.1.12) is in innovations form, which means that the only uncertainty is in the initial state x_0 (if this is known then the remaining states can be calculated exactly from the output) and consequently that the asymptotic estimation error covariance is zero. If we wished to minimize an infinite-time *discounted* cost then it would indeed be necessary to include a Kalman filter to provide accurate estimates of the initial state. But to minimize the average cost criterion (7.1.11) it is optimal, as shown in Section 6.3, to apply the optimal LQG control with the Kalman filter covariance set to its steady-state value, and this is equivalent to state feedback in the present context.

Denote by $\bar{A}, \bar{B}, \bar{C}$ the matrices in the state equation (7.1.13) and let $H := [1, 0, \ldots, 0, 1]$. The cost $J(u)$ can be written in standard form as

$$J(u) = \lim_{N\to\infty} \frac{1}{N} \sum_{k=1}^{N} |Dx_k + Fu_k|^2 \tag{7.1.14}$$

where

$$D = \begin{bmatrix} H \\ 0 \end{bmatrix}, \qquad F = \begin{bmatrix} 0 \\ \sqrt{\lambda} \end{bmatrix}.$$

The algebraic Riccati equation for the problem (7.1.13)–(7.1.14) is now given by (6.1.29) as

$$S = \bar{A}^{\mathrm{T}} S \bar{A} + H^{\mathrm{T}} H - \frac{1}{(\bar{B}^{\mathrm{T}} S \bar{B} + \lambda)} \bar{A}^{\mathrm{T}} S \bar{B} \bar{B}^{\mathrm{T}} S \bar{A} \tag{7.1.15}$$

and the optimal control is

$$u_k^0 = -Mx_k \tag{7.1.16}$$

where

$$M = (\bar{B}^{\mathrm{T}} S \bar{B} + \lambda)^{-1} \bar{B}^{\mathrm{T}} S \bar{A}.$$

These conclusions are valid under conditions (6.1.30), namely that $(\bar{A}, \bar{B})$ be stabilizable and $(D, \bar{A})$ detectable. It is easily checked that these conditions hold, in view of the fact that A is stable.

Use of control (7.1.16) is guaranteed to give a stable closed-loop system and to minimize $J(u)$. Computation of u_k^0 is, however, not elementary, since the algebraic Riccati equation (7.1.15) must be solved, and the form of the solution gives, it must be admitted, very little insight into the optimization process. In fact, u_k^0 given by (7.1.16) reduces to the minimum-variance controller when $\lambda = 0$. (Recall that the conditions ensuring closed-loop stability are not met when $\lambda = 0$, so the m.v. control may be unstable). Rather than showing this in general it is perhaps more illuminating to examine Example 7.1.2 again.

For this example the state space realization (7.1.13) is

$$\begin{bmatrix} x_{k+1}^0 \\ x_{k+1}^1 \\ x_{k+1}^2 \end{bmatrix} = \begin{bmatrix} 0 & 0 & 0 \\ 0 & 0 & 0 \\ c-a & 1 & -a \end{bmatrix} \begin{bmatrix} x_k^0 \\ x_k^1 \\ x_k^2 \end{bmatrix} + \begin{bmatrix} 0 \\ 1 \\ 0 \end{bmatrix} u_k + \begin{bmatrix} 1 \\ 0 \\ 0 \end{bmatrix} v_k$$

$$y_k = [1, 0, 1]x_k. \tag{7.1.17}$$

Let us denote the elements of the symmetric matrix S as

$$S = \begin{bmatrix} s_1 & s_2 & s_3 \\ s_2 & s_4 & s_5 \\ s_3 & s_5 & s_6 \end{bmatrix}.$$

Then the algebraic Riccati equation (7.1.15) becomes

$$\begin{bmatrix} s_1 & s_2 & s_3 \\ s_2 & s_4 & s_5 \\ s_3 & s_5 & s_6 \end{bmatrix} = \left[s_6 - \frac{s_5^2}{s_4 + \lambda} \right] \begin{bmatrix} (c-a)^2 & (c-a) & -a(c-a) \\ (c-a) & 1 & -a \\ -a(c-a) & -a & a^2 \end{bmatrix} + \begin{bmatrix} 1 & 0 & 1 \\ 0 & 0 & 0 \\ 1 & 0 & 1 \end{bmatrix}.$$

These six simultaneous equations are very easily solved when $\lambda = 0$, the solution then being $s_4 = 1$ and

$$\begin{aligned} s_1 &= (c-a)^2 s_4 + 1 & s_3 &= 1 - a(c-a)s_4 \\ s_2 &= (c-a)s_4 & s_5 &= -as_4 \\ & & s_6 &= 1 + a^2 s_4 \end{aligned} \tag{7.1.18}$$

(we retain the s_4-dependence here for later use). The control given by (7.1.16) is then

$$u_k^0 = a(c-a, 1, -a)\bar{x}_k.$$

Referring to (7.1.17) we see that in this realization $x_k^0 = w_k$, $x_k^1 = u_{k-1}^0$, $x_k^2 = y_k - w_k$, so that

$$\begin{aligned} u_k^0 &= a((c-a)w_k + u_{k-1}^0 - a(y_k - w_k)) \\ &= a(cw_k + u_{k-1}^0 - ay_k). \end{aligned}$$

Multiplying both sides by $(1 + cz^{-1})$ and using the basic ARMAX relation (7.1.8), we obtain

$$\begin{aligned}(1+cz^{-1})u_k^0 &= ac(1+cz^{-1})w_k + a(1+cz^{-1})z^{-1}u_k^0 - a^2(1+cz^{-1})y_k \\ &= ac[(1+az^{-1})y_k - z^{-2}u_k^0] \\ &\quad + a(1+cz^{-1})z^{-1}u_k^0 - a^2(1+cz^{-1})y_k \\ &= az^{-1}u_k^0 + a(c-a)y_k.\end{aligned}$$

But this shows that

$$u_k^0 = \frac{a(c-a)}{(1+(c-a)z^{-1})}y_k$$

which is the same as the m.v. controller (7.1.9). The cost of control u^0 is $J(u^0) = \bar{C}^{\mathrm{T}}S\bar{C} = s_1 = 1 + (c-a)^2$ which coincides with the expression given at (7.1.10).

The algebraic Riccati equation is also easily solved when $\lambda > 0$. We find that

$$s_4 = \tfrac{1}{2}\{1 - \lambda(1-a^2) + \sqrt{[(\lambda(1-a^2)-1)^2 + 4\lambda]}\}, \tag{7.1.19}$$

and that the s_i for $i \neq 4$ are given by (7.1.18) in terms of s_4. As $\lambda \to \infty$, s_4 converges to $1/(1-a^2)$ so that the optimal cost converges to $(c-a)^2/(1-a^2)+1$. Not surprisingly, this is precisely the steady-state variance of the system (7.1.8) with $u_k \equiv 0$: as $\lambda \to \infty$, control energy becomes so expensive that the optimal policy is not to use any control at all. Calculations similar to the above show that the optimal control u_k^λ corresponding to a given $\lambda \geq 0$ is given by

$$u_k^\lambda = \frac{a_\lambda(c-a)}{1+(c-a_\lambda)z^{-1}}y_k \tag{7.1.20}$$

where $a_\lambda := as_4/(s_4+\lambda)$, resulting in the following closed loop system description:

$$y_k = \frac{1+(c-a_\lambda)z^{-1}}{1+(a-a_\lambda)z^{-1}}w_k,\quad u_k^\lambda = \frac{a_\lambda(c-a)}{1+(a-a_\lambda)z^{-1}}w_k.$$

There are stable transfer functions since $|a - a_\lambda| < 1$ for all $\lambda \geq 0$. Note that it is not a requirement of the theory that the transfer function relating $\{u_k^\lambda\}$ to $\{y_k\}$ in (7.1.20) be stable; it may not be (take $a = -0.75$, $c = 0.75$ and λ sufficiently small, for example).

7.1.3 Minimum-variance control by frequency-domain methods

Consider a control given in transfer function form by

$$u_k = -R(z^{-1})y_k \tag{7.1.21}$$

where $R(z^{-1})$ is such that $\tilde{A}(z^{-1}) = A(z^{-1}) + z^{-r}B(z^{-1})R(z^{-1})$ has stable zeros. The closed-loop system (7.1.1) is then

$$\tilde{A}(z^{-1})y_k = C(z^{-1})w_k.$$

Thus $\{y_k\}$ is (asymptotically) a stationary process with spectral density function

$$\Psi(e^{i\omega}) = \frac{C(e^{-i\omega})C(e^{i\omega})}{\tilde{A}(e^{-i\omega})\tilde{A}(e^{i\omega})}. \tag{7.1.22}$$

The performance index (7.1.11) is equal to $E(y_k^2 + \lambda u_k^2)$ where $\{y_k\}$ is this stationary process and this can be calculated directly in terms of Ψ: the spectral density of $\{u_k\}$ is $|R(e^{i\omega})|^2\Psi(e^{i\omega})$ and hence using expression (2.3.10) we see that

$$J(u) = \frac{1}{2\pi i}\int_\Gamma \Psi(\zeta)(1 + \lambda R(\zeta)R(\zeta^{-1}))\zeta^{-1}\,d\zeta \tag{7.1.23}$$

(here Γ is the unit circle in the complex plane). Thus an equivalent formulation of the optimization problem is: choose R so as to minimize $J(u)$ given by (7.1.22), (7.1.23). This has been studied in a recent paper by Burt and Rigby (1982). They consider a slightly more general formulation than the above. The system model is given in transfer function form as

$$y_k = \frac{z^{-r}B(z^{-1})}{A(z^{-1})}u_k + \chi_k \tag{7.1.24}$$

where χ_k is a stationary noise process with known spectral density $\Phi(e^{i\omega})$. (7.1.1) is a special case of this with $\Phi(e^{i\omega}) = C(e^{i\omega})C(e^{-i\omega})/A(e^{i\omega})A(e^{-i\omega})$, but it is not necessary for χ_k to be generated by a finite-dimensional system in this way. The control u_k is given by (7.1.21) and one wishes to minimize

$$E(y_k^2 + \lambda v_k^2)$$

where

$$v_k = K(z^{-1})u_k$$

and K is a rational function in z^{-1}. This form of cost provides some additional flexibility: for example, if $K(z^{-1}) = 1 - z^{-1}$, then $v_k = u_k - u_{k-1}$ and we penalize *changes* in control value rather than the absolute value.

With control $u_k = -R(z^{-1})y_k$ the system model (7.1.24) becomes (we will often suppress the z^{-1} dependence of $A(z^{-1})$, etc., in the following)

$$y_k = \frac{A}{A + z^{-r}BR}\chi_k.$$

Now define

$$G = \frac{BR}{A + z^{-r}BR}.$$

Then

$$R = \frac{AG}{B(1 - z^{-r}G)}. \tag{7.1.25}$$

and

$$y_k = (1 - z^{-r}G)\chi_k.$$

Since G and R are in one-to-one correspondence, one can equally well regard G as the transfer function to be selected. The constraints on G are that it should be stable and that *its zeros should include all the unstable zeros of B*, so that G takes the form

$$G(z^{-1}) = H(z^{-1}) \prod_{|\beta_i|>1} (1 - \beta_i z^{-1})$$

where H is stable and β_i are the zeros of B. Since $u_k = -Ry_k = -(AG/B)\chi_k$, this ensures that $\{u_k\}$ is an asymptotically stationary process, as required to evaluate the cost (7.1.23). Burt and Rigby show that the G which satisfies these conditions and minimizes $E(y_k^2 + \lambda v_k^2)$ is given by

$$G(z^{-1}) = \frac{B(z^{-1})}{\Psi(z^{-1})N(z^{-1})} \sum_{v=0}^{\infty} z^{-v} \int_{\Gamma} \frac{\Psi(\zeta^{-1})B(\zeta)}{N(\zeta)} \zeta^{r+v-1} \, d\zeta. \tag{7.1.26}$$

In this expression $\Psi(\zeta^{-1})$ and $N(\zeta^{-1})$ are the stable spectral factors of Φ and $B\bar{B} + \lambda KA\bar{K}\bar{A}$ respectively, i.e. $\Psi(\zeta^{-1})$ and $N(\zeta^{-1})$ have all poles and zeros within the unit disc and

$$\Psi(z^{-1})\Psi(z) = \Phi(z^{-1})$$
$$N(z^{-1})N(z) = B(z^{-1})B(z) + \lambda K(z^{-1})A(z^{-1})K(z)A(z). \tag{7.1.27}$$

It is clear that when $\lambda > 0$, $K = 1$ and $\Psi(z^{-1}) = C(z^{-1})/A(z^{-1})$ the control given by (7.1.25), (7.1.26) must coincide with the optimal LQG controller (7.1.16), since these are the unique solutions to equivalent problems. It is a matter of computational convenience which solution is adopted: essentially the choice is between solving the algebraic

Riccati equation (7.1.15) or performing the spectral factorization (7.1.27). If $K \neq 1$ the solution can still be obtained by LQG theory but further augmentation of the state vector is necessary in order to produce v_k as an output.

When $\lambda = 0$ *and B has stable zeros* it is possible to show directly that (7.1.25)–(7.1.26) coincide with the minimum-variance controller, for then $N = B$ and

$$G(z^{-1}) = \frac{1}{\Psi(z^{-1})} \sum_{\nu=0}^{\infty} z^{-\nu} \frac{1}{2\pi i} \int_{\Gamma} \Psi(\zeta^{-1}) \zeta^{r+\nu-1} \, d\zeta. \quad (7.1.28)$$

We know that this choice of G minimizes Ey_k^2 where

$$\begin{aligned} y_{k+r} &= (1 - z^{-r} G(z^{-1})) \chi_{k+r} \\ &= \chi_{k+r} - G(z^{-1}) \chi_k. \end{aligned}$$

But this means that $G(z^{-1})\chi_k$ is the minimum mean square error predictor of χ_{k+r} given $\chi_k, \chi_{k-1}, \ldots$ and this fact precisely characterizes the m.v. controller.

To illustrate the above points, let us consider again Example 7.1.2 where the system is given by

$$(1 + az^{-1}) y_k = z^{-2} u_k + (1 + cz^{-1}) w_k.$$

Here $A = 1 + az^{-1}$, $B = 1$, $r = 2$ and $\chi_k = [(1 + cz^{-1})/(1 + az^{-1})] w_k$; thus $\Psi(z^{-1}) = (1 + cz^{-1})/(1 + az^{-1})$ as long as $|c| < 1$. From (7.1.28) the G for the minimum variance controller is

$$G(z^{-1}) = \frac{1 + az^{-1}}{1 + cz^{-1}} \sum_{\nu=0}^{\infty} z^{-\nu} \frac{1}{2\pi i} \int_{\Gamma} \frac{1 + c\zeta^{-1}}{1 + a\zeta^{-1}} \zeta^{\nu+1} \, d\zeta.$$

Using the method of residues, we find that the νth term in the sum is $z^{-\nu}(c - a)(-a)^{\nu+1}$ and hence that the sum is

$$-a(c - a) \sum_{\nu=0}^{\infty} (-az^{-1})^{\nu} = \frac{-a(c - a)}{(1 + az^{-1})}.$$

Thus $G(z^{-1}) = -a(c - a)/(1 + cz^{-1})$; from (7.1.24) the corresponding R is $R(z^{-1}) = -a(c - a)/(1 + (c - a)z^{-1})$, and this is the m.v. controller.

With $\lambda > 0$, one has to compute $N(z^{-1})$. This is a first-degree polynomial; denoting it $N(z^{-1}) = \sqrt{\gamma}(1 + \beta z^{-1})$ we see from (7.1.27) with $K = 1$ that

$$\gamma(1 + \beta z^{-1})(1 + \beta z) = 1 + \lambda(1 + az^{-1})(1 + az).$$

Solving for γ, β we find that

$$\gamma = s_4 + \gamma, \qquad \beta = \lambda a/(s_4 + \lambda)$$

where s_4 is given by (7.1.19). We can now compute G and R as before, and after a lot of laborious algebra we find that $R(z^{-1}) = -a_\lambda(c-a)/(1+(c-a_\lambda)z^{-1})$, in agreement with (7.1.20).

Finally, we should point out that there may be significant advantages in using the modified m.v. controller ($\lambda > 0$) even when simple m.v. control gives a stable closed loop system. Sometimes the modified controller expends vastly less energy to give an output variance only slightly greater than that of the strict m.v. controller. This is illustrated by Burt and Rigby for the system

$$y_k = \frac{z^{-1}(1+1.3z^{-1}+0.75z^{-2})}{(1-1.157z^{-1}+0.81z^{-2})}u_k + \frac{0.435}{1-0.9z^{-1}}w_k$$

which is stable and minimum phase. The variance of the uncontrolled system is $Ey_k^2 = 1$. Figure 7.1 shows Ey_k^2 plotted against Eu_k^2 for optimal controllers with values of λ increasing from 0 to ∞. Under m.v. control $Ey_k^2 = 0.19$ and $Eu_k^2 = 1.6$. (Point T in the figure.) When

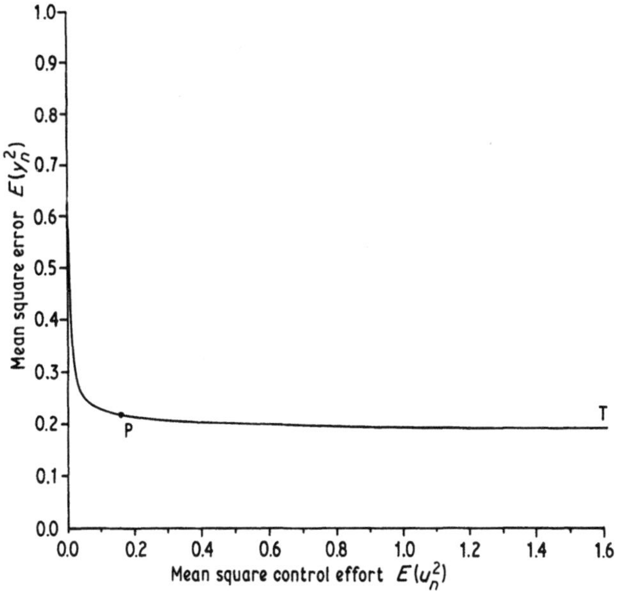

Fig. 7.1

$\lambda = \infty$ we have $Ey_k^2 = 1$ since the optimal control is then $u_k = 0$, as discussed earlier. But consider point P on the curve, where $Eu_k^2 = 0.16$, $Ey_k^2 = 0.23$: as compared to m.v. control, the control energy has been reduced by a factor of 10 for an increase in output variance of only 15%. Thus the m.v. controller is using enormous amounts of energy to squeeze the last 15% of performance out of the system.

7.2 Pole/zero shifting regulators

As we have seen above, minimum-variance regulation has certain disadvantages in respect of stability and excessive use of control energy, which can be overcome by the use of LQG-based regulators. The latter, however, have themselves two severe drawbacks: firstly, they are hard to compute, and secondly, they do not have the 'self-tuning property' discussed in Section 7.3 below. For these reasons it is worth investigating different sorts of regulator design, based on 'classical' control system design rather than optimal control theory. The objective of 'optimality' in a well-defined sense is abandoned in favour of obtaining qualitatively satisfactory closed-loop system behaviour. For a time-invariant linear system the response is entirely determined by the closed-loop transfer function, i.e. by the positions of the poles and zeros of the closed-loop system, and the objective of classical control system design is to locate these in positions corresponding to satisfactory dynamic response. This is a subject with many ramifications and we content ourselves here with presenting some algorithms for pole and zero shifting for the ARMAX model (7.1.1). These are algorithms on which self-tuning controllers can successfully be based, as will be shown in Section 7.3.

The basic ARMAX model is, as before

$$A(z^{-1})y_k = z^{-r}B(z^{-1})u_k + C(z^{-1})w_k. \tag{7.2.1}$$

In this section and subsequently we wish to have a little more flexibility about the degrees of the polynomials A, B, C. These will now be denoted n_A, n_B, n_C and may all be different (as opposed to the values n, $n-r$, n assumed previously). The corresponding predictor model is

$$y_{k+r} = \frac{B(z^{-1})F(z^{-1})}{C(z^{-1})}u_k + \frac{D(z^{-1})}{C(z^{-1})}y_k + F(z^{-1})w_{k+r}. \tag{7.2.2}$$

The polynomial F always has degree $(r-1)$ but for the degree n_D of D

there are two cases:

$$n_D = \begin{cases} n_A - 1 & \text{if } n_C \leqq n_A + r - 1 \\ n_C - r & \text{if } n_C \geqq n_A + r - 1 \end{cases}.$$

Controls are determined by dynamic feedback as follows:

$$u_k = \frac{H(z^{-1})}{J(z^{-1})} y_k \tag{7.2.3}$$

Here

$$H(z^{-1}) = h_0 + h_1 z^{-1} + \cdots + h_{n_H} z^{-n_H}$$
$$J(z^{-1}) = 1 + j_1 z^{-1} + \cdots + j_{n_J} z^{-n_J}.$$

Combined with (7.2.2) this gives the closed-loop system description

$$[J(C - z^{-r}D) - z^{-r}BFH]y_k = JFCw_k. \tag{7.2.4}$$

The objective now is to choose H, J so that this coincides with a specified w_k-to-y_k transfer function $Z(z^{-1})/T(z^{-1})$, or alternatively so that the output spectral density is the specified function

$$\frac{Z(e^{-i\omega})Z(e^{i\omega})}{T(e^{-i\omega})T(e^{i\omega})}.$$

For this it is necessary, from (7.2.4), that

$$Z[J(C - z^{-r}D) - z^{-r}BFH] = JFCT. \tag{7.2.5}$$

Equating the coefficients of $z^0, z^{-1}, z^{-2}, \ldots$ on either side of (7.2.5) gives us a set of linear equations for the controller coefficients h_i, j_i, and conditions must be such that there is at least one solution to this set of equations. We discuss below a few specific cases.

One could equally well compute the closed-loop system directly from the original system (7.2.1) together with control (7.2.3) and this would give a condition similar to (7.2.5) but expressed in terms of A, B, C, rather than B, C, D, F. The reason for preferring the predictor form is that it is the normal parametrization of the system used in self-tuning control.

7.2.1 Minimum-variance control

Under m.v. control the closed-loop system is $y_k = F(z^{-1})w_k$ i.e. $Z = F$, $T = 1$. Thus (7.2.5) is satisfied if

$$DJ + BFH = 0$$

so that

$$H = -\frac{1}{b_0}D$$

$$J = \frac{1}{b_0}BF$$

(we normalize so that $j_0 = 1$). Evidently the appropriate degrees are

$$n_H = n_D \qquad n_J = n_B + r - 1.$$

When $T = 1$ the transfer function Z/T can be thought of as having all its poles at infinity. An alternative to this is to use a so-called 'detuned' m.v. control, introduced by Wellstead *et al.* (1979) in which some of these poles are placed elsewhere than at infinity, giving a transfer function F/T for some non-degenerate T. Experimental evidence in Wellstead *et al.* shows that this can produce 'better' system response than strict m.v. control, though of course under steady-state conditions the output variance will be increased. Detuned m.v. control is particularly simple to apply if the polynomial T takes the form

$$T(z^{-1}) = 1 + z^{-r}T^*(z^{-1}).$$

Then (7.2.5) becomes

$$JC - z^{-r}[JD + BFH] = JC[1 + z^{-r}T^*].$$

The coefficients of z^{-i} for $i < r$ agree automatically and we merely require that

$$JD + BFH = JCT^*$$

which is satisfied by

$$H = -\frac{1}{b_0}(D - CT^*)$$

$$J = \frac{1}{b_0}BF.$$

This differs from m.v. control only in the replacement of D by $D - CT^*$. In the 'white-noise case' $C(z^{-1}) = 1$, important for self-tuning, we obtain simply

$$H = -\frac{1}{b_0}(D - T^*)$$

and the degree of H is $n_H = n_A - 1$ assuming, as is usually the case, that $n_{T*} \leq n_D$.

7.2.2 Pole-shifting regulators

As the name implies, the objective of a pole-shifting regulator is to place the closed-loop poles in positions determined by a specified polynomial T while leaving the zeros to be determined by the design algorithm. In particular, an effective algorithm is obtained if we set $Z = J$ (i.e. the closed-loop zeros are the poles of the control transfer function), in which case (7.2.5) reduces to

$$J(C - z^{-r}D) - z^{-r}BFH = FCT. \tag{7.2.6}$$

The degrees of the three polynomials in this equality must generically agree. Denoting the common degree by q, we have (assuming that $n_C \leq n_A + r - 1$)

$$q = n_J + n_A + r - 1 = n_B + 2r - 1 + n_H = r - 1 + n_C + n_T.$$

The number of unknown parameters h_i, j_i is $n_H + n_J + 1$, so that for solvability we must have

$$q \leq n_H + n_J + 1.$$

One choice which satisfies this with equality is

$$\begin{aligned} n_H &= n_A - 1 \\ n_J &= n_B + r - 1. \end{aligned} \tag{7.2.7}$$

Then $q = n_A + n_B + 2r - 2$ and the degree n_T of the specified polynomial T is limited by

$$n_T \leq n_A + n_B + r - 1 - n_C.$$

The values given by (7.2.7) are in fact the only values such that (7.2.6) has a unique solution. Obtaining this solution represents a considerable computational burden since the q equations involved are not in triangular form. In Wellstead *et al.* (1979) some special structures are introduced for which the computational problem is somewhat reduced.

7.3 Self-tuning regulators

The most important property of the minimum-variance controller and some pole-shifting regulators is that it is possible to apply them in a simple and effective way to systems described by the ARMAX model (7.1.1) with *unknown parameters.*

In general, controlling systems with unknown parameters is a formidable task. In some situations it may be possible to identify the system off-line, using the techniques described in Chapter 4. Then a controller can be designed using the parameters of the fitted model (assuming that these are time-invariant). Often, however, it is not feasible to isolate a system for off-line experimentation, and some form of joint estimation and control must be employed. The most thorough-going approach to this would be to regard both estimation and control as part of an overall optimization problem. Suppose the parameter vector $\theta = (a_1, \ldots, a_{n_A}, b_0, \ldots b_{n_B}, c_0, \ldots, c_{n_C})$ of the model (7.1.1) is regarded as a random vector with a known prior density function, independent of the system noise w_k. We can then adjoin to the state-space realization (7.1.13) an additional constant state θ_k satisfying

$$\theta_{k+1} = \theta_k, \quad \theta_0 = \theta \tag{7.3.1}$$

and consider the control problem for the joint system (7.1.13), (7.3.1) of choosing a control u_k depending only on the observations y_k, $y_{k-1}, \ldots$ so as to minimize, say, the average cost criterion (7.1.14) where the expectation is taken over the joint distribution of (θ, w_0, $w_1, \ldots$). Generally, such a problem is impossibly complicated: the system is no longer linear since products of the state variable appear in (7.1.13), so the LQG theory of Chapter 6 does not apply. In particular, it cannot be expected that any straightforward form of separation principle will hold. For these reasons, attempts to solve the overall optimization problem generally have to be abandoned.

In these circumstances, a very natural idea is to adopt a 'certainty-equivalence' approach, consisting of the following steps: supposing an estimate $\hat{\theta}_k$ of θ is available at time k, we apply the control u_k which would be optimal if $\hat{\theta}_k$ were the true parameter value. The output y_{k+1} is observed and the estimate $\hat{\theta}_k$ updated to $\hat{\theta}_{k+1}$. Now the procedure is repeated. Thus the parameters are estimated recursively and at each stage a controller is designed assuming that the current parameter estimate is actually the true value.

Such a procedure will never be optimal in the sense of, say, discounted cost, but may be optimal in the sense of long-run average cost per unit time or some other asymptotic sense. We say that a procedure has the *self-tuning property* if its performance coincides with the performance that would be obtained if the system parameters were known exactly. This statement will be made more precise in the context of specific cases examined below. Note that it is *not* part of the self-tuning property that the parameter estimates should converge to their true values; this may not be necessary.

In designing a proposed self tuning algorithm one has to choose:

(a) A class of models to represent the system,
(b) An estimation procedure,
(c) A control algorithm.

One's first thought is that the class of models should be one including the 'true' system and that the estimation procedure should be one that gives consistent parameter estimates for this class of models. In the case of the ARMAX model (7.1.1), this would mean using, say, recursive maximum likelihood identification, which involves a very substantial amount of real-time computation. This first thought, however, ignores the effect that the control algorithm might have on the estimation. It is a striking fact, uncovered by Åström and Wittenmark (1973), that in some circumstances a combination of simple least-squares estimation and minimum-variance control has the self-tuning property. This is true even if the system is represented by (7.1.1) where $C(z^{-1})$ has degree $n_C > 0$, when least-squares estimation would be expected to give biased estimates. The effect of the control algorithm is somehow to 'unbias' the estimates. This is a very attractive result since, least squares being by far the simplest form of recursive identification, it opens up the possibility of designing self-tuning controllers of very modest computational complexity.

In this section we shall discuss self-tuning control based on least-squares estimation for a system represented accurately by the ARMAX model (7.1.1) where the system order and time delay, but not the parameter values, are known. The control algorithms will be minimum-variance regulators or pole-shifting controllers of some sort. Of course it is generally unrealistic to suppose that the system order is known *a priori*; we will comment on this further below.

In this section we are not concerned with establishing convergence of self-tuning algorithms, but rather with examining what happens *if* convergence takes place, i.e. investigating whether the limiting system then has the self-tuning property. Below we study some properties of least-squares estimation and give in Proposition 7.3.2 a general condition for self-tuning. This is then applied to various specific control algorithms.

A self-tuning algorithm with guaranteed convergence is presented in Section 7.4.

7.3.1 Some properties of least-squares estimation

In the special case when $C(z^{-1}) = 1$ the predictor form (7.1.3) of the ARMAX model (7.1.1) can be written as

$$y_{k+r} = \mathscr{A}(z^{-1})y_k + \mathscr{B}(z^{-1})u_k + \varepsilon_{k+r} \tag{7.3.2}$$

where $\mathscr{A} = D, \mathscr{B} = BF$ and $\varepsilon_k = F(z^{-1})w_k$. The polynomials $\mathscr{A}$ and $\mathscr{B}$ have degrees m and l respectively, where

$$m = n_A - 1, \quad l = n_B + r - 1. \tag{7.3.3}$$

We write

$$\mathscr{A}(z^{-1}) = 1 + \alpha_1 z^{-1} + \cdots + \alpha_m z^{-m}$$
$$\mathscr{B}(z^{-1}) = \beta_0 + \beta_1 z^{-1} + \cdots + \beta_l z^{-l}.$$

Now suppose that u_k is generated by feedback from y_k through a transfer function H/J, i.e.

$$u_k = \frac{H(z^{-1})}{J(z^{-1})} y_k \tag{7.3.4}$$

with $j_0 = 1$. We suppose that H and J have no common factors. Then the closed-loop system becomes

$$L(z^{-1})y_k = J(z^{-1})\varepsilon_k \tag{7.3.5}$$

where

$$L = J(1 - z^{-r}\mathscr{A}) - z^{-r}\mathscr{B}H. \tag{7.3.6}$$

For compatibility we suppose that the degree of H and J are

$$n_H = m \qquad \text{and} \qquad n_J = l,$$

so that the degree of L is in general

$$n_L = l + m + r.$$

We now regard (7.3.2) as a *model set* and estimate the parameters α_i, β_i by ordinary least squares. (This is a convenient way of parametrizing the system since many control algorithms are more directly related to $\mathscr{A}$ and $\mathscr{B}$ than to the original parameters A, B, C: for example the minimum-variance controller is given simply by $\mathscr{B}u_k = -\mathscr{A}y_k$). Write the observations for $k = 0$ to N in the standard least-squares form as $y = X\theta + \varepsilon$, i.e.

$$\begin{bmatrix} y_r \\ y_{r+1} \\ \vdots \\ y_{r+N} \end{bmatrix} = \begin{bmatrix} y_0 & y_{-1} & \cdots & y_{-m} & u_0 & u_{-1} & \cdots & u_{-l} \\ y_1 & y_0 & & y_{1-m} & u_1 & u_0 & & u_{1-l} \\ \vdots & \vdots & & \vdots & \vdots & \vdots & & \vdots \\ y_N & y_{N-1} & & y_{N-m} & u_N & u_{N-1} & & u_{N-l} \end{bmatrix} \begin{bmatrix} \alpha_0 \\ \vdots \\ \alpha_m \\ \beta_0 \\ \vdots \\ \beta_l \end{bmatrix} + \begin{bmatrix} \varepsilon_r \\ \varepsilon_{r+1} \\ \vdots \\ \varepsilon_{r+N} \end{bmatrix}.$$

The starting values y_k, u_k for $k < 0$ will not be important. The normal equations characterizing the least-squares estimates $\hat{\alpha}_i$, $\hat{\beta}_i$ are $(1/N)X^{\mathrm{T}}y = (1/N)X^{\mathrm{T}}X\hat{\theta}$, or

$$\begin{bmatrix} \frac{1}{N}\Sigma y_k y_{k+r} \\ \frac{1}{N}\Sigma y_{k-1} y_{k+r} \\ \vdots \\ \frac{1}{N}\Sigma y_{k-m} y_{k+r} \\ \frac{1}{N}\Sigma u_k y_{k+r} \\ \vdots \\ \frac{1}{N}\Sigma u_{k-l} y_{k+r} \end{bmatrix} =$$

$$\left[\begin{array}{ccc|ccc} \frac{1}{N}\Sigma y_k^2 & \frac{1}{N}\Sigma y_k y_{k-1} & \frac{1}{N}\Sigma y_k y_{k-m} & \frac{1}{N}\Sigma y_k u_k & \cdots & \frac{1}{N}\Sigma y_k u_{k-l} \\ \vdots & \frac{1}{N}\Sigma y_{k-1}^2 & \vdots & & & \vdots \\ \frac{1}{N}\Sigma y_{k-m} y_k & \ddots & \frac{1}{N}\Sigma y_{k-m}^2 & \frac{1}{N}\Sigma y_{k-m} u_k & \cdots & \frac{1}{N}\Sigma y_{k-m} u_{k-l} \\ \hline \frac{1}{N}\Sigma u_k y_k & \cdots & \frac{1}{N}\Sigma u_k y_{k-m} & \frac{1}{N}\Sigma u_k^2 & & \frac{1}{N}\Sigma u_k u_{k-l} \\ \vdots & & & & & \vdots \\ \frac{1}{N}\Sigma u_{k-l} y_k & & \frac{1}{N}\Sigma u_{k-l} y_{k-m} & \frac{1}{N}\Sigma u_k u_{k-l} & & \frac{1}{N}\Sigma u_{k-l}^2 \end{array}\right] \begin{bmatrix} \hat{\alpha}_0 \\ \vdots \\ \hat{\alpha}_m \\ \hat{\beta}_0 \\ \vdots \\ \hat{\beta}_l \end{bmatrix}. \tag{7.3.7}$$

The factor $1/N$ is introduced so that all the coefficients are sample averages over $k = 1, 2, \ldots, N$. Taking, say, the first row of (7.3.7) we see that

$$\frac{1}{N}\sum_{k=0}^{N} y_k[y_{k+r} - \hat{\alpha}_0 y_k - \cdots - \hat{\alpha}_m y_{k-m} - \hat{\beta}_0 u_k - \cdots - \hat{\beta}_l u_{k-l}] = 0$$

i.e.

$$\frac{1}{N}\sum_{k=0}^{N} y_k \varepsilon_{k+r} = 0,$$

where ε_k is the residual sequence defined by

$$\varepsilon_{k+r} := y_{k+r} - \hat{\mathscr{A}}(z^{-1})y_k - \hat{\mathscr{B}}(z^{-1})u_k.$$

Similarly, taking the other rows of (7.3.7) we obtain

$$\begin{aligned}
&\frac{1}{N}\sum_{k=0}^{N} y_{k-j}\varepsilon_{k+r} = 0 \quad j = 0, 1, \ldots, m \\
&\frac{1}{N}\sum_{k=0}^{N} u_{k-j}\varepsilon_{k+r} = 0 \quad j = 0, 1, \ldots, l.
\end{aligned} \tag{7.3.8}$$

In general, the parameter estimates $\hat{\alpha}_i(N)$, $\hat{\beta}_i(N)$, and hence the residuals ε_k, depend on N and there is no guarantee that they will converge as $N \to \infty$. However, we wish to *examine what happens if they do converge*. Let us therefore make the *ad hoc* assumptions that the parameter estimates converge and that the resulting closed-loop system (7.3.5) is stable. Then y_k and ε_k are asymptotically stationary processes whose covariances are the limits of the corresponding sample averages. We describe this situation by saying the 'parameter estimates have converged'. From (7.3.8) we then obtain the following information.

Proposition 7.3.1

If the parameter estimates have converged then

$$\begin{aligned}
&E y_k \varepsilon_{k+j} = 0 \qquad j = r, r+1, \ldots, r+m \\
&E u_k \varepsilon_{k+j} = 0 \qquad j = r, r+1, \ldots, r+l.
\end{aligned} \tag{7.3.9}$$

Indeed, these expressions are obtained from (7.3.8) by replacing the sample averages there by the corresponding expected values. This result does not depend in any way on the control transfer function but is a simple consequence of the structure of least-squares estimation. The next result is really the heart of the self-tuning property.

Proposition 7.3.2

Suppose that the true system is described by the ARMAX model (7.1.1), that parameter estimates $\hat{\alpha}_i, \hat{\beta}_i$ in the predictor model (7.3.2) are estimated by least squares, where the degrees m, l of $\mathscr{A}$, $\mathscr{B}$ satisfy (7.3.3), and that these estimates have converged. The control u_k is given by (7.3.4) where $n_H = m$ and $n_J = l$ and H and J are assumed to have no common factors. Then the residual sequence ε_k is a moving average of order $r-1$ if

$$n_C \leq n_A + n_B + 2r - 2 - n_L \tag{7.3.10}$$

where n_L is the degree of the polynomial L given by (7.3.6).

REMARK Note from (7.3.6) that the 'generic' degree of L is $n_L = l + m + r = n_A + n_B - 2r - 2$ in which case condition (7.3.10) says $n_C = 0$, i.e. the noise in the true system model is white. This is precisely the situation in which least-squares estimates can be expected to 'behave'. What the result says is that it is possible to obtain the moving-average property of the residuals even with non-white system noise ($n_C > 0$) if the control parameters H, J are chosen, as functions of the model estimates $\mathscr{A}, \mathscr{B}$, in such a way that L has less than its generic degree, i.e. some cancellation of higher-order coefficients occurs. For instance, in the m.v. regulator we take $H = -(1/\beta_0)\mathscr{A}$ and then $L = J$ with degree $n_J = n_B + r - 1$, so that (7.3.10) specifies $n_C \leq n_A + r - 1$.

PROOF Suppose that convergence has taken place so that all processes are stationary and (7.3.9) holds. Define an auxiliary process ζ_k by

$$\zeta_k = \frac{1}{L(z^{-1})}\varepsilon_k.$$

Then

$$y_k = J(z^{-1})\zeta_k$$
$$u_k = H(z^{-1})\zeta_k.$$

Thus in view of (7.3.9) we have

$$0 = Ey_k\varepsilon_{k+j} = E[\zeta_k\varepsilon_{k+j} + j_1\zeta_{k-1}\varepsilon_{k+j} + \cdots + j_l\zeta_{k-l}\varepsilon_{k+j}],$$
$$j = r, \ldots, r+m$$

and

$$0 = Eu_k\varepsilon_{k+j} = E[h_0\zeta_k\varepsilon_{k+j} + \cdots + h_m\zeta_{k-m}\varepsilon_{k+j}], \quad j = r, \ldots, r+l.$$

Define $R_{\zeta\varepsilon}(j) = E\zeta_k\varepsilon_{k+j}$ and assemble these equalities in matrix form. This gives the following:

$$\left[\begin{array}{cccccccccc} 1 & j_1 & \cdot & \cdot & \cdot & j_l & & & & \\ 0 & 1 & j_1 & & & & j_l & & & \\ & & \cdot & & & & & \cdot & & \\ & & & \cdot & & & & & \cdot & \\ & & & & \cdot & & & & & \cdot \\ & & & & & 1 & j_1 & & & j_l \\ \hline h_0 & h_1 & \cdot & \cdot & \cdot & h_m & & & & \\ & \cdot & & & & & \cdot & & & \\ & & \cdot & & & & & \cdot & & \\ & & & \cdot & & & & & \cdot & \\ & & & & h_0 & & h_1 & & & h_m \end{array}\right] \begin{bmatrix} R_{\zeta\varepsilon}(r) \\ R_{\zeta\varepsilon}(r+1) \\ \vdots \\ \\ R_{\zeta\varepsilon}(r+m+l) \end{bmatrix} = 0.$$

The $(m+l+2) \times (m+l+1)$ matrix on the left has full rank as long as H, J have no common factors. Thus

$$R_{\zeta\varepsilon}(j) = 0, \qquad j = r, \ldots, r+m+l. \tag{7.3.11}$$

We want to show that in fact (7.3.11) holds for all $j \geq r$. Now the 'true' closed-loop system is given by (7.1.1) with $u_k = (H/J)y_k$, i.e.

$$[JA - z^{-r}BH]y_k = JCw_k$$

and hence

$$\zeta_k = \frac{1}{J(z^{-1})}y_k = \frac{C}{[JA - z^{-r}BH]}w_k. \tag{7.3.12}$$

Thus ζ_k is an ARMA (n_C, n^*) process, where $n^* = n_A + n_B + r - 1$. Referring to Section 2.3, this means that its covariance function $R_\zeta(p)$ satisfies

$$R_\zeta(p) = \phi_1 R_\zeta(p-1) + \cdots + \phi_{n^*}R_\zeta(p-n^*) \qquad \text{for } p > n_C \tag{7.3.13}$$

where the denominator polynomial in (7.3.12) is $(1 - \sum_1^{n^*}\phi_i z^{-i})$. Bearing in mind that $\varepsilon_k = L(z^{-1})\zeta_k$ we have

$$\zeta_k\varepsilon_{k+p+1} = \zeta_k(L\zeta_{k+p+1}) = \zeta_k\zeta_{k+p+1} + l_1\zeta_k\zeta_{k+p} + \cdots + l_{n_L}\zeta_k\zeta_{k+p-n_L+1}$$

and hence

$$R_{\zeta\varepsilon}(p+1) = R_\zeta(p+1) + l_1 R_\zeta(p) + \cdots + l_{n_L} R_\zeta(p - n_L + 1). \tag{7.3.14}$$

If $p - n_L + 1 > n_C$, then from (7.3.13), $R_\zeta(j) = \sum \phi_i R_\zeta(j-i)$ for $j = p - n_L + 1, \ldots, p+1$. Substituting in (7.3.14) we obtain

$$R_{\zeta\varepsilon}(p+1) = \sum_{i=1}^{n^*} \phi_i L R_\zeta(p - i + 1).$$

Now

$$\begin{aligned} E\zeta_k \varepsilon_{k+j} &= E\zeta_k[\zeta_{k+j} + l_1 \zeta_{k+j-1} + \cdots] \\ &= R_\zeta(j) + l_1 R_\zeta(j-1) + \cdots = LR_\zeta(j) \end{aligned}$$

so that

$$R_{\zeta\varepsilon}(p+1) = \sum_{i=1}^{n^*} \phi_i R_{\zeta\varepsilon}(p - i + 1).$$

Taking $p = r + m + l$ we see from (7.3.11) that the right-hand side is equal to zero, so that

$$R_{\zeta\varepsilon}(r + m + l + 1) = 0.$$

We can repeat the argument with $r + m + l + 1$ replacing $r + m + l$, and so on, to conclude that $R_{\zeta\varepsilon}(j) = 0$ for all $j \geq r$. The condition required is $p - n_L + 1 > n_C$ where $p = r + m + l$, and this coincides with (7.3.10).

Finally, $\varepsilon_k = L(z^{-1})\zeta_k$, so that

$$\begin{aligned} R_\varepsilon(j) = E\varepsilon_k \varepsilon_{k+j} &= E[\zeta_k \varepsilon_{k+j} + l_1 \zeta_{k-1} \varepsilon_{k+j} + \cdots + l_{n_L} \zeta_{k-n_L} \varepsilon_{n+j}] \\ &= 0 \qquad \text{for } j \geq r. \end{aligned}$$

Thus ε_k is a moving-average process of order $(r-1)$. □

Example 7.3.3

Let us consider the system

$$(1 - az^{-1}) y_k = b u_{k-1} + (1 + cz^{-1}) w_k$$

where $b \neq 0$ and $|a|$, $|c| < 1$. The delay is $r = 1$. Condition (7.3.10) becomes $n_L = 0$. The degrees of the polynomials involved are $n_J = n_H = m = l = 0$ so that

$$L(z^{-1}) = 1 - (1 + \beta_0 h_0) z^{-1}.$$

Thus the *only* controller that satisfies (7.3.10) is $H = -1/\beta_0, J = 1$ and this is the m.v. regulator for the system. Note in particular that (7.3.10) is not satisfied when $H = 0$, and this is just as well since we know that least-squares estimates for A in the ARMA model $A(z^{-1})y_k = C(z^{-1})w_k$ are biased unless $C(z^{-1}) = 1$.

7.3.2 Minimum-variance regulators

As already pointed out above, the minimum-variance regulator corresponding to the model (7.3.2) is

$$\mathscr{A}(z^{-1})y_k + \mathscr{B}(z^{-1})u_k = 0, \tag{7.3.15}$$

i.e.

$$u_k = -\frac{1}{\beta_0}[\beta_1 u_{k-1} + \cdots + \beta_l u_{k-l} + \mathscr{A}(z^{-1})y_k].$$

The controller (7.3.4) is thus $H/J = -\mathscr{A}/\mathscr{B}$, and $L(z^{-1}) = J$. The conclusions of Proposition 7.3.2 are thus valid if $n_C \leq n_A - 1$. In this case we have the following result.

Proposition 7.3.4

Suppose that the conditions of Proposition 7.3.2 hold, and that the control is given by (7.3.15), so that condition (7.3.10) becomes $n_C \leq n_A + r - 1$. Then the controller coincides with the m.v. regulator (7.1.7) for the system (7.1.1) with known parameters, i.e.

$$u_k = \frac{-D(z^{-1})}{B(z^{-1})F(z^{-1})} y_k$$

where D, F satisfy (7.1.2). In particular, the asymptotic variance of the output is $\sigma^2(f_0^2 + f_1^2 + \cdots + f_{r-1}^2)$.

PROOF With the m.v. regulator, $y_k = \varepsilon_k$, so that by Proposition 7.3.2 the output y_k is a moving average of order $r - 1$. Now the closed-loop system is

$$[JA - z^{-r}BH]y_k = JCw_k \tag{7.3.16}$$

so it must be the case that

$$\frac{JC}{JA - z^{-r}BH} = \tilde{F}$$

where $\tilde{F}$ is a polynomial of order $r-1$. Thus $y_k = \tilde{F}w_k$ and from (7.3.16)

$$A\tilde{F} - z^{-r}\frac{BH\tilde{F}}{J} = C.$$

Denoting $\tilde{D} = BH\tilde{F}/J$ this becomes

$$A\tilde{F} - z^{-r}\tilde{D} = C.$$

But this equality is satisfied by unique polynomials $\tilde{F} = F$, $\tilde{D} = D$. Thus $H/J = D/BF$ and this coincides with the m.v. regulator (7.1.7) designed for the known system (7.1.1). □

Note that this result does *not* say that the combination of least-squares estimation and the controller given by (7.3.15) *will* converge to the m.v. regulator. It says that *if* convergence takes place to some controller such that the closed-loop system is stable, then that controller must be the m.v. regulator. Some separate argument has to be employed to show that convergence actually does take place; this question is discussed in Section 7.4. Nonetheless, Proposition 7.3.4 is a striking result because it means that one can get away with using simple least-squares estimation in a context in which one expects that something more sophisticated would be required.

Wellstead *et al.* (1979) recommend use of the *detuned m.v. regulator*. This was introduced in Section 7.2.1 above and is a kind of compromise between m.v. control and regulation by pole shifting. Everything is the same as before except that the control algorithm (7.3.15) is replaced by

$$\begin{aligned} H &= \frac{1}{\beta_0}[-\mathscr{A} + T^*] \\ J &= \frac{1}{\beta_0}\mathscr{B} \end{aligned} \qquad (7.3.17)$$

where T^* is an arbitrary polynomial of degree n_{T^*}. In this case the self-tuning result is as follows.

Proposition 7.3.5

Suppose that the conditions of Proposition 7.3.2 hold and that the control is given by (7.3.17). Then condition (7.3.10) becomes

$$n_C \le n_A - 1 - n_{T^*} \qquad (7.3.18)$$

and under this condition the asymptotic closed-loop system is given by

$$y_k = \frac{F(z^{-1})}{1 + z^{-1}T^*(z^{-1})} w_k \tag{7.3.19}$$

where $F(z^{-1})$ is the same $(r-1)$th degree polynomial that appears in the m.v. regulator.

PROOF Here

$$L = \mathscr{B}(1 - z^{-r}T^*)$$

with degree $n_B + r + n_{T^*}$ so that the condition (7.3.10) becomes (7.3.18). The residuals ε_k are given by

$$\varepsilon_k = (1 - z^{-1}T^*)y_k$$

and ε_k is a moving average of order $(r-1)$ under the stated conditions. The closed-loop system is given by (7.3.16), so that

$$\varepsilon_k = (1 - z^{-r}T^*)y_k = \frac{(1 - z^{-r}T^*)JC}{JA - z^{-r}BH} w_k = \tilde{F} w_k$$

where $\tilde{F}$ is a polynomial of degree $(r-1)$. Rearranging, we see that

$$C = \tilde{F}A - z^{-r}\tilde{D}$$

where

$$\tilde{D} = \frac{\tilde{F}BH}{J} - T^*C.$$

As before, $\tilde{F}$, $\tilde{D}$ must coincide with F, D from the m.v. regulator, and

$$\frac{H}{J} = \frac{D + T^*C}{BF}.$$

From (7.2.1) and (7.3.16), we obtain (7.3.19) as the closed-loop system.

The closed-loop system for detuned m.v. control is an ARMA system with the same zeros as the m.v. regulated system but with poles given by the roots of $1 + \zeta^r T^*(\zeta) = 0$ instead of all poles being at infinity. Wellstead *et al.* (1979) adduce some evidence that the detuned regulator has better stability properties than the strict m.v. regulator. However, the scope of 'detuning' is severely limited unless the system has almost white noise, since the condition $n_C + n_{T^*} \leq n_A - 1$ must be satisfied to guarantee the self-tuning property.

Finally, a note on system order. If $m \geq n_A$ and $l \geq n_B + r$, i.e. we have overestimated the order of the true system, then $\mathscr{A}$ and $\mathscr{B}$ must contain common factors. For m.v. regulation $H = -(1/\beta_0)/\mathscr{A}$ and $J = (1/\beta_0)\mathscr{B}$ so that H and J have common factors, which violates a condition of Proposition 7.3.2. To obtain the self-tuning property, we must take $H/J = -\mathscr{A}|\mathscr{B}$ *cancelling all common factors*. In practice this is not a simple thing to do, since one needs a numerical procedure to decide on the presence of common factors when $\mathscr{A}$ and $\mathscr{B}$ are not given in factored form.

7.3.3 LQG regulators

We now turn to m.v. controllers with control costs, i.e. to LQG regulators of the sort discussed in Section 7.1. The main result is that these regulators do *not* have the self-tuning property except, possibly, when $C(z^{-1}) = 1$. Even if they did, they would not constitute a very practical form of control algorithm because of the necessity of performing a spectral factorization at every step in order to compute the required control value. In the known-parameter case this only has to be done once and the LQG regulator is a viable way of handling difficulties with m.v. control associated with non-minimum phase systems, etc. In the self-tuning case, some other way of handling these difficulties must be found; hence the interest in detuned m.v., pole-shifting control and other non-optimal algorithms.

The discussion here will be limited to the unit-delay case $r = 1$ since only for this case is the solution of the LQG problem given directly for the predictor model (7.2.2). Similar results may however be expected when $r > 1$.

As for m.v. regulation, the self-tuning LQG regulator is designed for the predictor model on the (possibly erroneous) assumption that the process noise is white. Now when $C(z^{-1}) = 1$ and $r = 1$ the system model (7.1.1) becomes

$$A(z^{-1})y_k = z^{-1}B(z^{-1})u_k + w_k. \tag{7.3.20}$$

The coefficients of this and of the predictor model (7.3.2) are therefore related by

$$\mathscr{A} = z(1 - A)$$
$$\mathscr{B} = B.$$

The putative self-tuning LQG controller is therefore obtained by the following procedure:

(a) Estimate the parameters $\mathscr{A}$ and $\mathscr{B}$ of the predictor model (7.3.2) by recursive least squares;
(b) Apply the LQG control designed for the system (7.1.1) with

$$\begin{aligned} A &= \tilde{\mathscr{A}} := 1 - z^{-1}\mathscr{A} \\ B &= \mathscr{B} \\ C &= 1. \end{aligned} \tag{7.3.21}$$

Since we want the control in transfer function form $u_k = (H/J)y_k$ it is convenient to use the frequency-domain solution of the LQG problem as presented in Section 7.1.3, which gives the control directly in this form. This solution can be expressed in a more explicit way for the problem at hand, in which the system is, with parameters given by (7.3.21) above,

$$y_k = z^{-1}\frac{\mathscr{B}(z^{-1})}{\tilde{\mathscr{A}}(z^{-1})}u_k + \frac{1}{\tilde{\mathscr{A}}(z^{-1})}w_k$$

Thus, in the notation of Section 7.1.3, $\Psi(z^{-1}) = 1/\tilde{\mathscr{A}}(z^{-1})$ and the LQG controller is given by

$$\frac{H}{J} = \frac{-\tilde{\mathscr{A}}\mathscr{G}}{\mathscr{B}(1 - z^{-1}\mathscr{G})}$$

where

$$\mathscr{G}(z^{-1}) = \frac{\tilde{\mathscr{A}}(z^{-1})\mathscr{B}(z^{-1})}{N(z^{-1})}\sum_{\nu=0}^{\infty} z^{-\nu}\frac{1}{2\pi i}\int_{\Gamma}\frac{1}{\tilde{\mathscr{A}}(\zeta^{-1})}\frac{\mathscr{B}(\zeta)}{N(\zeta)}\zeta^{\nu}\mathrm{d}\zeta \tag{7.3.22}$$

and

$$N(z^{-1})N(z) = \mathscr{B}(z^{-1})\mathscr{B}(z) + \lambda\tilde{\mathscr{A}}(z^{-1})\tilde{\mathscr{A}}(z).$$

(we take $K(z^{-1}) = 1$, which means that control energy is being costed directly).

We can express $\mathscr{G}$ in more explicit form if we suppose, as is generically the case, that $z \to \tilde{\mathscr{A}}(z^{-1})$ has distinct zeros $p_1, \ldots, p_{n_A}$. Then there are constants $\delta_1, \ldots, \delta_{n_A}$ such that

$$\frac{1}{\tilde{\mathscr{A}}(z^{-1})} = \sum_{j=1}^{n_A}\frac{\delta_j}{1 - p_j z^{-1}}$$

and the sum in (7.3.22) can be evaluated by the method of residues as follows:

$$\begin{aligned}
&\sum_{v=0}^{\infty} z^{-v}\frac{1}{2\pi i}\int_{\Gamma}\frac{1}{\tilde{\mathscr{A}}(\zeta^{-1})}\frac{\mathscr{B}(\zeta)}{N(\zeta)}\zeta^{v}\,\mathrm{d}v \\
&=\sum_{j=1}^{n_A}\sum_{v=0}^{\infty} z^{-v}\frac{1}{2\pi i}\int_{\Gamma}\frac{\delta_j}{(1-p_j\zeta^{-1})}\frac{\mathscr{B}(\zeta)}{N(\zeta)}\zeta^{v}\,\mathrm{d}\zeta \\
&=\sum_{j=1}^{n_A}\sum_{v=0}^{\infty} z^{-v}\frac{\delta_j p_j\mathscr{B}(p_j)}{N(p_j)}p_j^{v} \\
&=\sum_{j=1}^{n_A}\frac{\delta_j p_j\mathscr{B}(p_j)}{N(p_j)}\frac{1}{1-p_j z^{-1}} \\
&=\frac{\Delta(z^{-1})}{\tilde{\mathscr{A}}(z^{-1})}
\end{aligned} \tag{7.3.23}$$

where $\Delta(z^{-1})$ is a polynomial of order n_A-1. Thus (7.3.22) becomes

$$\mathscr{G}(z^{-1})=\frac{\Delta(z^{-1})\mathscr{B}(z^{-1})}{N(z^{-1})}$$

and the controller is

$$\frac{H}{J}=\frac{-\tilde{\mathscr{A}}\Delta}{N-z^{-1}\Delta\mathscr{B}}. \tag{7.3.24}$$

The closed-loop predictor model equation (7.3.2) with this control is

$$\tilde{\mathscr{A}}Ny_k=(N-z^{-1}\Delta\mathscr{B})\varepsilon_k. \tag{7.3.25}$$

Thus $L=\tilde{\mathscr{A}}N$ and $n_L=n_A+\max(n_A,n_B)$, so that condition (7.3.10) is satisfied only if $n_B\geq n_A$ and $n_C=0$.

Let us now see in what way the situation is different in the minimum-variance, minimum-phase case when $\lambda=0$, $N=\mathscr{B}$. We know that the m.v. regulator is given by $u_k=-(\mathscr{A}/\mathscr{B})y_k$, so it must be the case that

$$\Delta=\mathscr{G}=\mathscr{A}.$$

(This can, with some difficulty, be checked directly from (7.3.23).) Thus

$$(N-z^{-1}\Delta\mathscr{B})=\tilde{\mathscr{A}}\mathscr{B}$$

so that here $L=J=\mathscr{B}$. Thus the order of L has been reduced to n_B and this allows scope for self-tuning as described in Proposition 7.3.4.

If $n_B\geq n_A$, $n_C=0$ and convergence takes place, then we conclude from (7.3.20) and (7.3.25) that, asymptotically, the residual sequence ε_k

and w_k are related by

$$\varepsilon_k = \frac{\tilde{\mathscr{A}}N}{J} y_k = \frac{\tilde{\mathscr{A}}N}{(JA - z^{-1}BH)} w_k.$$

It follows from Proposition 7.3.2 that $\tilde{\mathscr{A}}N/[JA - z^{-1}BH] = f_0$ for some constant f_0. i.e.

$$JA - z^{-1}BH = (1/f_0)\tilde{\mathscr{A}}N. \tag{7.3.26}$$

On the other hand, from (7.3.24) we have

$$J\tilde{\mathscr{A}} - z^{-1}\mathscr{B}H = (1/n_0)\tilde{\mathscr{A}}N. \tag{7.3.27}$$

If we regard (7.3.26), (7.3.27) as equations for 'unknowns' $\tilde{\mathscr{A}}, \mathscr{B}$, then one solution is certainly $\tilde{\mathscr{A}} = A, \mathscr{B} = B$, implying convergence of $\tilde{\mathscr{A}}, \mathscr{B}$ to the true parameter values and hence of H/J to the true LQG controller. We have not, however, succeeded in showing that this is only possible limit point of H/J. The conclusion is therefore that the conditions for self-tuning are not met unless the system noise is white and that even in this case there is some possible ambiguity as to the convergence point of the algorithm.

7.3.4 Pole-shifting regulators

Following the discussion in Section 7.2, a self-tuning pole-shifting regulator is determined in the following way. A polynomial $T(z^{-1})$ of degree n_T is selected and the control parameters H, J are related to the coefficients $\mathscr{A}, \mathscr{B}$ of the predictor model (7.3.2) by

$$L = J(1 - z^{-1}\mathscr{A}) - z^{-r}\mathscr{B}H = TP \tag{7.3.28}$$

where P is a polynomial of degree $(r-1)$, to be determined:

$$P(z^{-1}) = 1 + p_1 z^{-1} + \cdots + p_{r-1} z^{-(r-1)}.$$

We thus set the denominator of the closed-loop transfer function of the predictor model equal to TP. If T has (maximum) degree $n_T = l + m + 1$, then both sides (7.3.28) have degree $l + m + r$, which is equal to the number of parameters $j_1, \ldots, j_l$, $h_0, \ldots, h_m$, $p_1, \ldots, p_{r-1}$. Therefore (7.3.28) is (in general) satisfied by *unique* H, J, P for given $\mathscr{A}, \mathscr{B}, T$. According to Proposition 7.3.2, since $n_L = n_T + r - 1$ we require

$$n_T \le n_A + n_B + r - 1 - n_C = l + m + 1 - n_C \tag{7.3.29}$$

for the self-tuning property, which is that $\varepsilon_k = (L/J)y_k = (TP/J)y_k$ is an

$(r-1)$th-order moving average. In this case the behaviour of the resulting closed-loop system is given by the following proposition.

Proposition 7.3.6

Suppose the conditions of Proposition 7.3.2 hold (in particular, (7.3.29) is satisfied) and that the control parameters H, J satisfy identity (7.3.28). Then the closed-loop system is the ARMA system

$$y_k = \frac{J(z^{-1})}{T(z^{-1})} w_k.$$

Thus the closed-loop zeros are the poles of the controller and the closed-loop poles are those of the specified polynomial $T(z^{-1})$.

PROOF From (7.3.16) we have

$$\varepsilon_k = \frac{L}{J} y_k = \frac{TP}{J} y_k = \frac{TPC}{[JA - z^{-r}BH]} w_k.$$

Therefore

$$\frac{TPC}{[JA - z^{-r}BH]} = \tilde{F} \tag{7.3.30}$$

where $\tilde{F}$ is a polynomial of degree $(r-1)$, i.e.

$$\tilde{F}JA - z^{-r}\tilde{F}BH = TPC. \tag{7.3.31}$$

Under condition (7.3.29) the left and right sides are polynomials of degrees $(n_A + n_B + 2r - 2)$, and this is the number of parameters in $\tilde{F}, H, J$. Therefore (7.3.31) is satisfied by unique $\tilde{F}, H, J$ for given A, B, T, C, P. Consider, on the other hand, the identity

$$JA - z^{-r}BH = TC. \tag{7.3.32}$$

Again, this is satisfied by unique H, J for given A, B, T, C. But if H, J satisfy (7.3.32) then P, H, J satisfy (7.3.31), so $\tilde{F} = P$. From (7.3.30),

$$JA - z^{-r}BH = TC$$

and hence the closed-loop system is $y_k = (J/T)w_k$, as required. □

7.4 A self-tuning controller with guaranteed convergence

The results in the previous section give conditions under which the self-tuning phenomenon can occur, and help us to identify possible candidates for self-tuning regulation. However, the results only refer to asymptotic properties of the closed-loop systems *assuming* that the parameter estimates converge; there is no guarantee that they actually will converge. Until fairly recently it had been a long- standing open problem in adaptive control to give an algorithm with guaranteed convergence properties: that is, a controller which when applied to an 'unknown' system would under reasonably general conditions give at least a stable closed-loop system. In the last few years, however, there has been considerable progress in this area and convergence proofs have been given for a number of adaptive control algorithms. Nevetheless, this is still an area of active research which has certainly not reached its final form. In this section we will discuss the simplest case of an algorithm due to Goodwin, Ramadge and Caines (1981). This is closely related, but not identical, to the minimum-variance self-tuning regulators discussed earlier. The convergence proof of Goodwin, Ramadge and Caines (1981), which we give in Appendix C, was the first general such result given for stochastic systems.

As before, the system to be controlled is described by the ARMAX model

$$A(z^{-1})y_k = z^{-r}B(z^{-1})u_k + C(z^{-1})w_k \tag{7.4.1}$$

with polynomial degrees n_A, n_B, n_C. A and C are monic, i.e. $A(0) = C(0) = 1$. w_k is a sequence of independent random variables with $Ew_k = 0$, $\text{var}(w_k) = \sigma^2$. A property of w_k which will be useful later is this: we know by the strong law of large numbers that with probability one,

$$\bar{w}_N^2 := \frac{1}{N}\sum_{k=1}^{N} w_k^2 \to \sigma^2 \qquad \text{as } n \to \infty.$$

It follows in particular that, with probability one, for any realization of the process the sequence $\bar{w}_k^2$ is bounded, i.e.

$$|\bar{w}_N^2| \le K \qquad \text{for all } N \tag{7.4.2}$$

where K may depend on the realization.

A non-random reference signal y_k^* is given and this is supposed to be bounded

$$|y_k^*| \le M \qquad \text{for all } k.$$

The objective of the controller is to 'track' y_k^*. The algorithm we will describe applies only to the unit delay case $r = 1$. (Other algorithms for $r > 1$ are described in Goodwin *et al.* 1981). As in self-turning m.v. regulation, the idea is to estimate the control parameters directly rather than identifying the system and then calculating the appropriate control. The algorithm is as follows.

Algorithm 7.4.1 (*Unit delay algorithm*)

Let $m^* = \max(n_A, n_B, n_C)$. For $k \geq m^* + 1$ define

$$\phi_{k-1}^{\mathrm{T}} = [y_{k-1}, \ldots, y_{k-n_A+1}, u_{k-1}, \ldots, u_{k-n_B}, -y_{k-1}^*, \ldots, -y_{k-n_C}^*].$$

$\hat{\theta}_k$ is an n^*-vector of control parameter estimates, where $n^* = n_A + n_B + n_C - 1$, which is generated recursively together with control values u_k by:

$$\hat{\theta}_k = \hat{\theta}_{k-1} + \frac{\bar{a}}{r_{k-1}} \phi_{k-1}[y_k - \phi_{k-1}^{\mathrm{T}} \hat{\theta}_{k-1}] \quad k \geq m^* + 1 \tag{7.4.3a}$$

$$r_{k-1} = r_{k-2} + \phi_{k-1}^{\mathrm{T}} \phi_{k-1}, \qquad r_{m^*-1} = 1 \tag{7.4.3b}$$

$$\phi_k^{\mathrm{T}} \hat{\theta}_k = y_{k+1}^*. \tag{7.4.3c}$$

The initial estimate $\hat{\theta}_{m^*}$ is an arbitrary constant, as are the first m^* control values $u_1, \ldots, u_{m^*}$. $\bar{a}$ is a constant whose value is fixed in Theorem 7.4.2 below. Note that (7.4.3c) specifies u_k recursively, since written out explicitly it states

$$\begin{aligned} u_k = -\frac{1}{\hat{\theta}_k^{n_A}} [\hat{\theta}_k^1 y_k + \cdots + \hat{\theta}_k^{n_A-1} y_{k-n_A+2} + \hat{\theta}_k^{n_A+1} u_{k-1} + \cdots \\ + \hat{\theta}_k^{n_A+n_B-1} u_{k-n_B+1} - y_{k+1}^* - \hat{\theta}_k^{n_A+n_B} y_k^* - \cdots \\ - \theta_k^{n_A+n_B+n_C-1} y_{k-n_C+1}^*]. \end{aligned}$$

The recursion formulae (7.4.3a) and (7.4.3b) belong to a class of algorithms known as *stochastic approximation algorithms*. They are clearly related to recursive least-squares estimators, but have no direct statistical interpretation since, for example, an arbitrary parameter $\bar{a}$ is involved. Note that (7.4.3) is computationally extremely simple to implement since r_k is a scalar sequence (in recursive least-squares (7.4.3b) is replaced by a matrix equation). The choice of control (7.4.3c) is analogous to the m.v. controller for output tracking given in Section 7.1.1 above for the known-parameter case. This was shown to be given by

$$BFu_k = Cy_{k+r}^* - Dy_k. \tag{7.4.4}$$

With unit delay $r=1$ we have $F(z^{-1})=1$ and $D(z^{-1})=z(C(z^{-1})-A(z^{-1}))$. We can write (7.4.4) in the form

$$\phi_k^{\mathrm{T}}\theta_0 = y_{k+1}^* \tag{7.4.5}$$

where ϕ_k is as before and θ_0 contains the parameters of B, C, D. With this control the output y_k is equal to $y_k^* + w_k$ and $\mathrm{var}(y_k)=\sigma^2$. Denoting by $\hat{y}_{k|k-1}$ the best predictor of y_k give the past up to $(k-1)$, we know that σ^2 is also equal to the prediction error $E[y_k - \hat{y}_{k|k-1}]^2$. The adaptive controller (7.4.3) simply uses the 'estimate' $\hat{\theta}_k$ in place of the true parameter θ_0. Under certain conditions, stated in Theorem 7.4.2 below, this algorithm has a performance which is asymptotically equivalent to that of the m.v. controller (7.4.4).

One of the conditions of Theorem 7.4.2 is a so-called *positive-real condition.* A polynomial (or transfer function) $P(z^{-1})$ is said to be (*strictly*) *positive real* if there is some number $\delta>0$ such that

$$\mathrm{Re}\{P(e^{i\omega})\} \geq \delta > 0 \qquad \text{for all } |\omega| \leq \pi. \tag{7.4.6}$$

In the first-order case this is equivalent to stability, since if $P(z^{-1}) = 1+pz^{-1}$ then $\mathrm{Re}\{P(e^{i\omega})\} = 1 + p\cos\omega$, so that P is positive real if and only if $|p|<1$. In general, positive realness is a stronger condition than stability: by the Nyquist criterion, a positive real polynomial has stable zeros, but on the other hand, a polynomial with stable zeros need not be positive real. Consider for example the second-order case $P(z^{-1}) = (1+p_1 z^{-1})(1+p_2 z^{-1})$; then

$$\mathrm{Re}\{P(e^{i\omega})\} = 1 + (p_1+p_2)\cos\omega + p_1 p_2 \cos 2\omega.$$

If we take $p_1 = -0.8$, $p_2 = -0.7$, then at $\omega = \pi/4$ we have $\mathrm{Re}\{P(e^{i\omega})\} = 1 - 1.5/\sqrt{2} < 0$ so that (7.4.6) is violated.

The property of positive real polynomials that we need is the so-called positive real lemma, Lemma C.3 of Appendix C. Some further comments will be found there.

We denote by $\mathscr{Y}_k$ the collection of random variables $\{y_1, y_2, \ldots, y_k, w_0, \ldots, w_{m^*}\}$; thus for any random variable R with finite expectation $E[R|\mathscr{Y}_k] = E[R|y_1, \ldots, y_k, w_0, \ldots, w_{m^*}]$. Note from (7.4.3) that u_k is a nonlinear function of $y_1, \ldots, y_k$ so that if the noise w_k is normal the output y_k will not in general be normal. However, it is clear from (7.4.1) that the *conditional* distribution of y_k given $\mathscr{Y}_{k-1}$ is normal, and in general, whatever the distribution of w_k the best predictor $\hat{y}_{k|k-1} = E[y_k|\mathscr{Y}_{k-1}]$ is a linear function of $\{y_j, u_j, j=k-1, k-2, \ldots\}$ and $\{w_0, \ldots, w_{m^*}\}$. Indeed, in the unit delay case the system model can be written

$$y_k - w_k = (1 - A)y_k + z^{-1}Bu_k + (C - 1)w_k. \tag{7.4.7}$$

Now if the 'initial conditions' $\{w_0, \ldots, w_{m^*}\}$ are known then w_j can be calculated recursively for $j > m^*$ given $\{y_k, u_k, k \leq j\}$. Thus (7.4.7) takes the form

$$y_k - w_k = \sum_{j=1}^{k-1} \alpha_j y_j + \sum_{j=1}^{k-1} \beta_j u_j + \sum_{j=0}^{m^*} \gamma_j w_j$$

for some constants α_j, β_j, γ_j. Since w_k and $\mathscr{Y}_{k-1}$ are independent, the right-hand side of (7.4.7) is equal to $E[y_k|\mathscr{Y}_{k-1}]$, and the prediction error is

$$w_k = y_k - E[y_k|\mathscr{Y}_{k-1}]$$

with

$$E(y_k - E[y_k|\mathscr{Y}_{k-1}])^2 = \sigma^2.$$

Since $w_0, \ldots, w_{m^*}$ are not outputs of the system it might seem more natural to define $\mathscr{Y}_k = \{y_1, \ldots, y_k\}$. The theorem below is true with this definition but the calculations become a little more complicated as we have to take account of the (asymptotically negligible) unknown initial conditions.

Here then is the main result.

Theorem 7.4.2

Suppose that the true system is given by (7.4.1) where $r = 1$ and $n_A \leq n_A^0$, $n_B \leq n_B^0$ and $n_C \leq n_C^0$ where n_A^0, n_B^0, n_C^0 are known constants. Suppose also that

$$C(z^{-1}) - \frac{\bar{a}^0}{2}$$

is strictly positive real for some $\bar{a}^0 > 0$. Let the control u_k be generated by the unit delay algorithm 7.4.1 with $\bar{a} = \bar{a}^0$, $n_A = n_A^0$, $n_B = n_B^0$, $n_C = n_C^0$. Then with probability one,

$$\limsup_{N \to \infty} \frac{1}{N} \sum_{k=1}^{N} y_k^2 < \infty \tag{7.4.8}$$

$$\limsup_{N \to \infty} \frac{1}{N} \sum_{k=1}^{N} u_k^2 < \infty \tag{7.4.9}$$

and

$$\lim_{N\to\infty} \frac{1}{N} \sum_{k=1}^{N} E[(y_k - y_k^*)^2 | \mathscr{Y}_{k-1}] = \sigma^2 \tag{7.4.10}$$

The proof of this result is given in Appendix C.

Properties (7.4.8), (7.4.9) constitute a form of stability for the closed-loop system. They are violated if $|y_k| \to \infty$ or $|u_k| \to \infty$ as $k \to \infty$, but they do not by themselves imply that $|y_k|$ and $|u_k|$ are bounded: for example, the sequence

$$y_k = \begin{cases} 0 & k \neq 2^n \text{ for some integer } n \\ n^{1/2} & k = 2^n \end{cases}$$

satisfies (7.4.8) but is not bounded. Thus occasional large deviations are allowed.

As regards property (7.4.10), we know that $E[(y_k - y_k^*)^2 | \mathscr{Y}_{k-1}] = \sigma^2$ when the system is controlled by the known-parameter m.v. controller. Thus (7.4.10) states that the unit delay algorithm asymptotically achieves the best performance that could be obtained if the system were identified exactly, in the sense that the conditional variance of $y_k - y_k^*$ given $\mathscr{Y}_{k-1}$ asymptotically coincides with that of the best one-step predictor. (Again, occasional large deviations are not theoretically excluded.) In Section 7.1.1 a somewhat stronger result was obtained for the known-parameter case, namely that $E[(y_k - y_k^*)^2] = \sigma^2$ (no conditioning). No similar claim is made here, but it is in fact possible to show that under stronger hypotheses on the noise process w_k (for example, $Ew_k^4 \leq M < \infty$), (7.4.10) can be replaced by

$$\lim_{N\to\infty} \frac{1}{N} \sum_{k=1}^{N} E(y_k - y_k^*)^2 = \sigma^2. \tag{7.4.11}$$

It will be seen in Appendix C that the convergence property is more easily established in the form (7.4.10) since we can essentially analyse the system along a single realization whereas to establish (7.4.11), 'ensemble' properties must be investigated.

As mentioned earlier, Theorem 7.4.2 was the first general result establishing convergence of a stochastic adaptive control algorithm. It has been followed by analysis of other algorithms including, for example, some designed to cover the important case of non-constant parameters. These results represent a major advance in adapative control. Nevertheless, they all suffer from the serious disadvantage that *an upper bound for the order of the system must be known in advance*. In other words, the complexity of the model must be at least as great as that of the system to be controlled. This is of course unrealistic since physical systems are complex objects which are only approximately represented by low-order ARMAX models. An important and satisfying feature of the prediction error formulation of (off-line) system identification, as presented in Chapters 4 and 5, is that the possibility that the true system may not be contained in the model set is explicitly allowed for. So far no formulation of adaptive control with analogous features has been given. What is involved is a careful analysis of the 'robustness' properties (stability margins, etc.) of algorithms such as the unit delay algorithm presented here. Such questions are the subject of current research.

That the conditions required by theory are not generally met in practice does not mean that adaptive control ideas cannot be successfully used in applications. The last few years have seen, in parallel with theoretical developments, a greatly enhanced understanding of the practical issues involved in the implementation of adaptive controllers. A state-of-the-art discussion of these issues will be found in the survey papers of Åström (1983) and Wittenmark and Åström (1984). Two key problems are the following:

(a) *Persistent excitation* This concept was introduced in Definition 5.3.5 and is an essential condition for consistency results such as Theorem 5.3.7, where the system input is supposed to contain an exogenous persistently exciting component. If the input is generated entirely by feedback, as in many adaptive control schemes, it is difficult to guarantee that it will be persistently exciting. The important thing is that updating of parameter estimates should be carried out only when there is sufficient excitation in the input signal. A variety of detection procedures have been devised which can be incorporated in the control loop to ensure this.

(b) *Unmodelled high-frequency dynamics* Typically, a low-order model is designed to capture only the dominant modes of a system, and it is well-known that standard 'classical' control system design techniques are rather insensitive to the elementary 'model reduction' procedure of simply ignoring system poles which are close to the origin. In adaptive control, one possible source of instability is excitation of the system at frequencies at which the unmodelled dynamics play some significant role. If one has some *a priori* idea as to what the dominant modes of a system are, such instabilities can be eliminated by introducing a low-pass filter into the control loop.

Adaptive control is in a state of rapid development, and has already advanced to the point where controllers based on the self-tuning principle (and incorporating the sort of safeguards mentioned above) are commercially available. On the theoretical side, guaranteed stabilization under progressively more realistic conditions is being demonstrated; on the practical side, the mechanisms of instability are much better understood. Perhaps a grand synthesis is not too far around the corner.

Notes

The literature on adaptive and self-tuning control is voluminous. The survey article by Åström (1983) gives an up-to-date overview and a lengthy list of references. For a general introduction see Harris and Billings (1981); this book contains several articles discussing both basic theory and practical issues. Landau (1981) describes Model Reference Adaptive Control, a somewhat different approach to that discussed here.

Section 7.1 Minimum-variance controllers appear in Åström's book (1970) and were also derived by Box and Jenkins (1962). The relation between m.v. regulators and LQG control has been discussed by several authors, for example Caines (1972). The frequency-domain approach to control of stationary processes is treated thoroughly by Whittle (1963). Another recent reference is Youla, Bongiorno and Jabre (1976). We follow Burt and Rigby (1982).

Section 7.2 Pole/zero shifting regulators were introduced in this context by Wellstead and co-workers (1979a, b); see also Åström and Wittenmark (1980) and Wellstead and Prager's article in Harris and Billings (1981).

Section 7.3 The self-tuning regulator is due to Åström and Wittenmark (1973). The self-tuning argument we give in Section 7.3.2 essentially follows Wellstead *et al.* (1979b), as does the discussion of pole-shifting regulators. Other non-optimal algorithms, designed to introduce control costs and to have the self-tuning property, have been given by Clarke and Gawthrop (1979).

Section 7.4 This section is taken from Goodwin, Ramadge and Caines (1981). As mentioned in the text, this is an area of active research and the results have already been extended in various ways; see for example Chen and Caines (1983) and Sin and Goodwin (1982). Stochastic approximation algorithms and the positive real condition are discussed at length in Kushner and Clarke (1978) and Ljung (1977). All of these matters are discussed in the recent book of Goodwin and Sin (1984).

References

Åström, K. J. (1970) *Stochastic Control Theory*, Academic Press, New York.

Åström, K. J. (1983) Theory and applications of adaptive control – a survey. *Automatica*, **19**, 471–486.

Åström, K. J. and Wittenmark, B. (1973) On self-tuning regulators. *Automatica*, **9**, 185–199.

Åström, K. J., Borisson, U., Ljung, L. and Wittenmark, B. W. (1977) Theory and applications of self-tuning regulators. *Automatica*, **13**, 457–476.

Box, G. E. P. and Jenkins, G. M. (1962) Some statistical aspects of adaptive optimization and control. *JRSS*, **B24**, 297–343.

Burt, E. G. C. and Rigby, L. (1982) Constrained minimum-variance control for minimum and non-minimum phase processes. Preprint, Department of Electrical Engineering, Imperial College, London.

Caines, P. E. (1972) Relationships between Box–Jenkins–Åström control and Kalman linear regulators. *Proc. IEE*, **119**, 615–620.

Chen, H. F. and Caines, P. E. (1983) On the adaptive control of a class of systems with random parameters and disturbances, Preprint, McGill University, Montreal.

Clarke, D. W. and Gawthrop, P. J. (1979) Self-tuning control. *Proc. IEE*, **126**, 633–640.

Goodwin, G. C., Ramadge, P. J. and Caines, P. E. (1981) Discrete-time stochastic adaptive control. *SIAM J. Control and Optimization*, **19**, 829–853.

Goodwin, G. C. and Sin, K. S. (1984) *Adaptive Filtering, Prediction and Control*, Prentice Hall, Englewood Cliffs, NJ.

Harris, C. J. and Billings, S. A. (eds) (1981) *Self-tuning and Adaptive Control*, Peter Peregrinus, Stevenage.

Kushner, H. J. and Clarke, D. S. (1978) *Stochastic Approximation Methods for Constrained and Unconstrained Systems*, Springer-Verlag, New York.

Landau, Y. D. (1981) Deterministic and stochastic model reference adaptive control. In *Stochastic Systems* (ed. M. Hazewinkel and J. C. Willems) D. Reidel, Dordrecht.

Ljung, L. (1977) On positive real transfer functions and the convergence of some recursive schemes. *IEEE Trans. Automatic Control*, **AC-22**, 539–551.

Neveu, J. (1975) *Discrete Parameter Martingales*, North Holland, Amsterdam.

Sin, K. S. and Goodwin, G. C. (1982) Stochastic adaptive control using a modified least squares algorithm. *Automatica*, **18**, 315–321.

Wellstead, P. E., Prager, D. and Zanker, P. (1979a) Pole assignment self-tuning regulator. *Proc. IEE*, **126**, 781–787.

Wellstead, P. E., Edmunds, J. M., Prager, D. and Zanker, P. (1979b) Self-tuning pole/zero assignment regulators. *Int. J. Control*, **30**, 1–26.

Whittle, P. (1963) *Prediction and Regulation*, English Universities Press (reprinted (1984) Basil Blackwell, Oxford).

Wittenmark, B. and Åström, K. J. (1984) Practical issues in the implementation of self-tuning control. *Automatica*, **20**, 595–605.

Youla, D. C., Bongiorno, J. J. and Jabr, H. A. (1976) Modern Wiener–Hopf design of optimal controllers. *IEEE Trans. Automatic Control*, **AC-21**, 3–13.

APPENDIX A

A uniform convergence theorem and proof of Theorem 5.2.1

Consider an identification experiment in the framework of Section 5.1 in which the data is generated by a stable system, the models supply uniformly stable predictors and the identification criterion is quadratically bounded.

The definitions of 'stability', 'uniform stability' and 'quadratic boundedness' (see Definitions 5.1.1, 5.1.3 and 5.1.5) each refer to an open neighbourhood of the parameter constraint set D. We assume that the conditions of the definitions are satisfied for some common neighbourhood $\mathscr{D}$ of D (this can always be arranged by choosing $\mathscr{D}$ to be intersection of the three neighbourhoods).

Let d_N, $N = 1, 2, \ldots$, be a collection of matrix-valued functions all defined on a subset $\mathscr{F}$ of $\mathbb{R}^q$ and let F be a subset of $\mathscr{F}$. The functions are said to be *equicontinuous* on F if, given any $\delta > 0$, there exists $\varepsilon > 0$ such that $\| d_N(s) - d_N(s') \| \leq \delta$ for $N = 1, 2, \ldots$ and for all $s, s' \in D$ such that $\| s - s' \| < \varepsilon$. (Any matrix norm may be used; see Appendix D.2.) The functions are said to be *uniformly bounded* on F if there exists a constant c such that $\| d_N(s) \| \leq c$, for $N = 1, 2, \ldots$ and for all $s \in D$.

We shall first build up through several steps a result we refer to as the 'uniform convergence theorem'. This concerns the relationship between the random variable

$$Q_N(\theta; y^N, u^{N-1}) = \frac{1}{N} \sum_{k=1}^{N} l_k(\theta, \varepsilon_k(\theta))$$

and its expected value in the limit as $N \to \infty$, and the equicontinuity and uniform boundedness of the functions $EQ_N(\theta; y^N, u^{N-1})$, $N = 1, 2, \ldots$. Theorem 5.2.1 will follow as a simple consequence.

It is convenient initially to prove the uniform convergence theorem in the special case when the functions $l_k(\theta, \varepsilon)$, $k = 1, 2, \ldots$, are scalar valued. In this case our proof hinges on application of the ergodic theorem (Theorem 1.1.15) to a sequence of random variables $\{\eta_k\}$ of the form:

$$\eta_k(\alpha, \bar{\theta}) = \sup_{\theta \in B_\alpha(\bar{\theta})} \{l_k(\theta, \varepsilon_k(\theta))\}, \qquad k = 1, 2, \ldots \tag{A.1}$$

Here $B_\alpha(\bar{\theta})$ denotes the ball $\{\theta: \| \theta - \bar{\theta} \| < \alpha\}$ in $\mathbb{R}^q$, $\bar{\theta}$ is an appropriate element in D and α is some positive number such that $B_\alpha(\bar{\theta}) \subset \mathscr{D}$.

The following lemma collects together a number of properties of $\eta_k(\alpha, \theta)$ and $l_k(\theta, \varepsilon_k(\theta))$ which will be required.

Lemma A.1

Suppose that $l_k(\theta, \varepsilon)$ is scalar valued for $k = 1, 2, \ldots$ and let $\eta_k(\alpha, \bar{\theta})$ be defined by (A.1). Then there exist $\lambda \in (0, 1)$ and $c > 0$ such that, for any $\bar{\theta} \in D$ and α which satisfies $B_\alpha(\bar{\theta}) \subset \mathscr{D}$ we have:

(a) $\operatorname{cov}\{\eta_k(\alpha, \bar{\theta}), \eta_j(\alpha, \bar{\theta})\} \leq c\lambda^{k-j}$, for all k and j such that $k \geq j$;
(b) $E\eta_k(\alpha, \bar{\theta}) - El_k(\theta, \varepsilon_k(\theta)) \leq c\alpha$, for all $\theta \in B_\alpha(\bar{\theta})$ and for all k;
(c) $E|l_k(\bar{\theta}, \varepsilon_k(\bar{\theta}))| \leq c$, for all k; and,
(d) $E\left\| \dfrac{\partial}{\partial\theta} l_k(\bar{\theta}, \varepsilon_k(\bar{\theta})) \right\| \leq c$, for all k.

PROOF Fix $\bar{\theta} \in D$ and let α be such that $B_\alpha(\bar{\theta}) \in \mathscr{D}$. For convenience we write η_k for $\eta_k(\alpha, \bar{\theta})$.

In what follows numbers λ, λ_1 in the interval (0, 1) and positive numbers $c_1, c_2, \ldots$ are introduced; it is understood that they do not depend on $\bar{\theta}$ or α.

Let k, j be positive numbers such that $k \geq j$ and define

$$y_j^k = (y_{k,j}, y_{k-1,j}, \ldots, y_{j+1,j}, 0, \ldots, 0)$$

and

$$u_j^k = (u_{k,j}, u_{k-1,j}, \ldots, u_{j+1,j}, 0, \ldots, 0).$$

In these expressions $y_{k,j}, \ldots, y_{j+1,j}, u_{k,j}, \ldots, u_{j+1,j}$ are the random variables associated with the stability of the system (see Definition 5.1.1).

Now define

$$\varepsilon_{k,j}(\theta) = y_{k,j} - f_k(\theta; y_j^{k-1}, u_j^{k-1}).$$

We begin by establishing the following bounds:

$$E \sup_\theta \| \varepsilon_k(\theta) \|^4 \leq c_1, \qquad \text{for all } k, \tag{A.2}$$

$$E \sup_{\theta} \|\varepsilon_{k,j}(\theta)\|^4 \le c_1, \qquad \text{for all } k,j \text{ with } k \ge j, \tag{A.3}$$

$$E \sup_{\theta} \|\varepsilon_k(\theta) - \varepsilon_{k,j}(\theta)\|^4 \le c_1 \lambda^{k-j}, \qquad \text{for all } k,j \text{ with } k \ge j \tag{A.4}$$

and

$$E \sup_{\theta} \left\| \frac{\partial}{\partial \theta} \varepsilon_k(\theta) \right\|^4 \le c_1, \qquad \text{for all } k. \tag{A.5}$$

In each case, the supremum is taken over θ's in $B_\alpha(\bar{\theta})$.

Consider (A.2). We have

$$\sup_{\theta} \|\varepsilon_k(\theta)\|^4 = \sup_{\theta} \|y_k - f_k(\theta; y^{k-1}, u^{k-1})\|^4$$

$$= \sup_{\theta} \|y_k - f_k(\theta; 0, 0) - f_k(\theta; y^{k-1}, u^{k-1}) + f_k(\theta; 0, 0)\|^4$$

$$\le \sup_{\theta} (\|y_k\| + \|f_k(\theta; 0, 0)\| + c_2 \sum_{i=1}^{k-1} \lambda_0^{k-i}(\|y_i\| + \|u_i\|))^4$$

since the predictors are uniformly stable (see Definition 5.1.3),

$$\le c_3 \left[1 + \left(\sum_{i=1}^{k} \lambda_1^{k-i}(\|y_i\| + \|u_i\|) \right)^4 \right]$$

since the system is stable and since $|a+b|^4 \le 8(|a|^4 + |b|^4)$,

$$\le c_3 \left[1 + \left(\sum_{i=0}^{k} \lambda_1^{k-1} \right)^3 \sum_{i=0}^{k} \lambda_1^{k-i}(\|y_i\| + \|u_i\|)^4 \right]$$

by the generalized Hölder inequality (see Appendix E),

$$\le c_4 \left[1 + \left(\sum_{i=0}^{k} \lambda_1^{k-i} \right)^3 \sum_{i=0}^{k} \lambda_1^{k-i}(\|y_i\|^4 + \|u_i\|^4) \right].$$

Taking expectations and noting that, since the system is stable, $E\|y_i\|^4, E\|u_i\|^4, i = 0, 1, \ldots$ are uniformly bounded, we obtain

$$E \sup_{\theta} \|\varepsilon_k(\theta)\|^4 \le c_4 \left[1 + \left(\sum_{i=0}^{k} \lambda_1^{k-i} \right)^4 \right] \le c_6.$$

The proofs of (A.3) and (A.5) are along similar lines to that of (A.2). In the case of (A.3), we note that $\varepsilon_{k,j}(\theta)$ is obtained from the formula for $\varepsilon_k(\theta) = y_k - f_k(\theta; y^{k-1}, u^{k-1})$, by substitution of $y_{k,j}$ in place of y_k, $u_{k,j}$ in place of u_k, etc. So the earlier arguments hold good again, provided that the substituted random variables are appropriately bounded;

specifically, we require

$$E\|y_{j,i}\|^4 + E\|u_{j,i}\|^4 \le c_7 \qquad \text{for all } j, i \text{ with } j \ge i.$$

However, we readily deduce this bound from stability of the system (see Definition 5.1.1).

As for (A.5), we use the fact that

$$\frac{\partial}{\partial \theta} \varepsilon_k(\theta)$$

is obtained from the formula for $\varepsilon_k(\theta) = y_k - f_k(\theta; y^{k-1}, u^{k-1})$, by substitution of 0 in place of y_k and $\partial f_k/\partial\theta$ in place of f_k. In view of the conditions placed upon $\partial f_k/\partial\theta$ in the definition of a uniformly stable predictor, the earlier arguments can be used a further time, to yield (A.5).

Finally we consider (A.4). We define $y_{i,j} = 0$, $u_{i,j} = 0$ for $i \le j$.

$$\begin{aligned}
&\sup_\theta \|\varepsilon_k(\theta) - \varepsilon_{k,j}(\theta)\|^4 \\
&\quad = \sup_\theta \|y_k - y_{k,j} - f_k(\theta; y^{k-1}, u^{k-1}) + f_k(\theta; y_j^{k-1}, u_j^{k-1})\|^4 \\
&\quad \le c_8 \left(\sum_{i=0}^{k} \lambda_1^{k-i} (\|y_i - y_{i,j}\| + \|u_i - u_{i,j}\|) \right)^4
\end{aligned}$$

since the predictors are uniformly stable

$$\le c_9 \left(\sum_{i=0}^{k} \lambda_1^{k-i} \right)^3 \left(\sum_{i=0}^{k} \lambda_1^{k-i} (E\|y_i - y_{i,j}\|^4 + E\|u_i - u_{i,j}\|^4) \right).$$

by the generalized Hölder inequality and since $E|a+b|^4 \le 8(E\|a\|^4 + E\|b\|^4)$.

Taking expectations and noting that $E\|y_i - y_{i,j}\|^4 + E\|u_i - u_{i,j}\|^4$ is bounded by $c_{10}\lambda_1^{i-j}$ for $i > j$ and by c_{10} for $i \le j$, we obtain

$$\begin{aligned}
E\sup_\theta \|\varepsilon_k(\theta) - \varepsilon_{k,j}(\theta)\|^4 &\le c_{11} \left[\sum_{i=0}^{j} \lambda_1^{k-i} + \sum_{i=j+1}^{k} \lambda_1^{k-j} \right] \\
&\le c_{11} \lambda^{k-j} \left(\sum_{i=0}^{j} \lambda_1^{j-i} + (k-j) \right) \le c_{12} \lambda^{k-j}
\end{aligned}$$

for any $\lambda \in (\lambda_1, 1)$ and for some appropriately chosen number c_{12}. We have proved (A.4).

We are now ready to prove claims (a), ..., (d) of the lemma.

(a) Define

$$e_{k,j} = \sup_\theta l_k(\theta; \varepsilon_{k,j}(\theta)).$$

In view of the properties of the random variables y_j^k and u_j^k used in the construction of $\varepsilon_{k,j}(\theta)$, it is not difficult to see that $\varepsilon_{k,j}(\theta)$ and $\varepsilon_j(\theta)$ are independent for arbitrary θ. It follows that $e_{k,j}$ and η_j are independent, so that by Schwarz's inequality

$$\begin{aligned} \operatorname{cov}\{\eta_k, \eta_j\} &= \operatorname{cov}\{\eta_k - e_{k,j}, \eta_j - l_j(\bar{\theta}; 0)\} \\ &\leq (E|\eta_k - e_{k,j}|^2)^{1/2}(E|\eta_j - l_j(\bar{\theta}; 0)|^2)^{1/2}. \end{aligned} \tag{A.6}$$

We now bound the terms in the product (A.6). Application of the mean value theorem yields

$$\begin{aligned} |\eta_j - l_j(\bar{\theta}; 0)|^2 &= |\sup_\theta \{l_j(\theta; \varepsilon_j(\theta)) - l_j(\bar{\theta}; 0)\}|^2 \\ &= \left|\sup_\theta \left\{\frac{\partial}{\partial\theta} l_j(\theta - \bar{\theta}) + \frac{\partial}{\partial\varepsilon} l_j \varepsilon_j(\theta)\right\}\right|^2 \end{aligned}$$

in which $\partial l/\partial\theta$ and $\partial l/\partial\varepsilon$ are evaluated at $((1-\sigma)\bar{\theta} + \sigma\theta, \sigma\varepsilon_j(\theta))$, for some $\sigma \in [0, 1]$,

$$\leq c_{13}\left(\sup_\theta \|\varepsilon_j(\theta)\|^2\right)^2$$

(since the identification criterion is quadratically bounded (see Definition 5.1.5))

$$= c_{13} \sup_\theta \|\varepsilon_j(\theta)\|^4.$$

Taking expectations, we conclude now from (A.2) that

$$E|\eta_j - l_j(\bar{\theta}; 0)|^2 \leq c_{14}. \tag{A.7}$$

Examine next $\eta_k - e_{k,j}$. We have

$$\begin{aligned} \eta_k - e_{k,j} &\leq \sup_\theta \{l_k(\theta; \varepsilon_k(\theta)) - l_k(\theta; \varepsilon_{k,j}(\theta))\} \\ &= \sup_\theta \left\{\frac{\partial}{\partial\varepsilon} l_k(\theta; \sigma\varepsilon_k(\theta) + (1-\sigma)\varepsilon_{k,j}(\theta))(\varepsilon_k(\theta) - \varepsilon_{k,j}(\theta))\right\} \end{aligned}$$

for some $\sigma \in [0, 1]$ (which is a function possibly of θ) by the mean value theorem,

$$\leq c_{15} \sup_\theta \{(\|\varepsilon_k(\theta)\| + \|\varepsilon_{k,j}(\theta)\|)\|\varepsilon_k(\theta) - \varepsilon_{k,j}(\theta)\|\}$$

(we have used again quadratic boundedness). Likewise, we use the mean value theorem to show that

$$\eta_k - e_{k,j} \geq \inf_\theta \{l_k(\theta; \varepsilon_k(\theta)) - l_k(\theta; \varepsilon_{k,j}(\theta))\}$$

$$\geq -c_{15} \sup_\theta \{(\|\varepsilon_k(\theta)\| + \|\varepsilon_{k,j}(\theta)\|)\|\varepsilon_k(\theta) - \varepsilon_{k,j}(\theta)\|\}.$$

It follows that

$$|\eta_k - e_{k,j}|^2 \leq c_{16} \sup_\theta \{(\|\varepsilon_{k,j}(\theta)\| + \|\varepsilon_k(\theta)\|)^2 \|\varepsilon_k(\theta) - \varepsilon_{k,j}(\theta)\|^2\}$$

$$\leq c_{16} \sup_\theta \{(\|\varepsilon_{k,j}(\theta)\| + \|\varepsilon_k(\theta)\|)^2\} \sup_\theta \{\|\varepsilon_k(\theta) - \varepsilon_{k,j}(\theta)\|^2\}.$$

Taking expectations and applying Schwarz's inequality yields

$$\begin{aligned} E|\eta_k - e_{k,j}|^2 &\leq c_{16}[E\sup_\theta(\|\varepsilon_{k,j}(\theta)\| + \|\varepsilon_k(\theta)\|^4]^{1/2} \\ &\quad \cdot [E\sup_\theta \|\varepsilon_k(\theta) - \varepsilon_{k,j}(\theta)\|^4]^{1/2} \\ &\leq c_{17}[E\sup_\theta \|\varepsilon_{k,j}(\theta)\|^4 + E\sup_\theta \|\varepsilon_k(\theta)\|^4]^{1/2} \\ &\quad \cdot [E\sup_\theta \|\varepsilon_k(\theta) - \varepsilon_{k,j}(\theta)\|^4]^{1/2} \\ &\leq c_{18}\lambda^{k-j} \end{aligned}$$

by (A.2), (A.3) and (A.4). This inequality together with inequalities (A.6) and (A.7) establishes (a).

(b) For arbitrary $\theta \in B_\alpha(\bar{\theta})$

$$\eta_k - l_k(\theta; \varepsilon_k(\theta)) = \sup_{\theta'} \{l_k(\theta'; \varepsilon_k(\theta') - l_k(\theta; \varepsilon_k(\theta))\}$$

(the sup, as usual, is taken over $B_\alpha(\bar{\theta})$),

$$= \sup_{\theta'} \left\{ \left(\frac{\partial}{\partial \theta} l_k(\theta''; \varepsilon_k(\theta'')) + \frac{\partial}{\partial \varepsilon} l_k(\theta''; \varepsilon_k(\theta'')) \frac{\partial}{\partial \theta} \varepsilon_k(\theta'') \right) (\theta' - \theta) \right\}$$

by the mean value theorem (θ'' is a point in $B_\alpha(\bar{\theta})$ which depends, possibly, on θ'),

$$\leq c_{19}\alpha \left[\sup_{\theta'} \|\varepsilon_k(\theta')\|^2 + \sup_{\theta'} \|\varepsilon_k(\theta')\| \sup_{\theta'} \left\| \frac{\partial}{\partial \theta} \varepsilon_k(\theta') \right\| \right]$$

by the quadratic boundedness of the identification criterion

$$\leq c_{20}\alpha\left[\sup_{\theta'}\|\varepsilon_k(\theta')\|^2+\sup_{\theta'}\left\|\frac{\partial}{\partial\theta}\varepsilon_k(\theta')\right\|^2\right].$$

Taking expectations and applying Schwarz's inequality, we obtain

$$\begin{aligned}E\eta_k-El_k(\theta;\varepsilon_k(\theta))&\leq c_{20}\alpha\left[(E\sup_{\theta'}\|\varepsilon_k(\theta')\|^4)^{1/2}\right.\\&\quad\left.+\left(E\sup_{\theta'}\left\|\frac{\partial}{\partial\theta}\varepsilon_k(\theta')\right\|^4\right)^{1/2}\right]\\&\leq c_{21}\alpha\end{aligned}$$

by (A.2) and (A.5).

(c) We have

$$|l_k(\bar{\theta};\varepsilon_k(\bar{\theta}))|=\left|l_k(\bar{\theta};0)+\frac{\partial}{\partial\varepsilon}l_k(\bar{\theta};\sigma\varepsilon_k(\bar{\theta}))\varepsilon_k(\bar{\theta})\right|$$

by the mean value theorem, for some $\sigma\in[0,1]$,

$$\leq c_{22}[1+\|\varepsilon_k(\bar{\theta})\|^2]$$

sincc thc identification criterion is quadratically bounded. Now take expectations. There results

$$\begin{aligned}E|l_k(\bar{\theta};\varepsilon_k(\bar{\theta}))|&\leq c_{22}[1+E\|\varepsilon_k(\bar{\theta})\|^2]\\&\leq c_{22}[1+(E\|\varepsilon_k(\bar{\theta})\|^4)^{1/2}]\leq c_{23}\end{aligned}$$

by (A.2). This is the required inequality.

(d) Note that

$$\begin{aligned}\left\|\frac{\partial}{\partial\theta}l_k(\bar{\theta};\varepsilon_k(\bar{\theta}))\right\|&=\left\|\frac{\partial}{\partial\theta}l_k(\bar{\theta};\varepsilon_k(\bar{\theta}))+\frac{\partial}{\partial\varepsilon}l_k(\bar{\theta};\varepsilon_k(\bar{\theta}))\frac{\partial}{\partial\theta}\varepsilon_k(\bar{\theta})\right\|\\&\leq c_{24}\left(\|\varepsilon_k(\bar{\theta})\|^2+\|\varepsilon_k(\bar{\theta})\|\left\|\frac{\partial}{\partial\theta}\varepsilon_k(\bar{\theta})\right\|\right)\\&\leq c_{25}\left(\|\varepsilon_k(\bar{\theta})\|^2+\left\|\frac{\partial}{\partial\theta}\varepsilon_k(\bar{\theta})\right\|^2\right).\end{aligned}$$

(we have used the property that the identification criterion is quadratically bounded). Taking expectations and applying Schwarz's inequality, we obtain:

$$E\left\|\frac{\partial}{\partial\theta}l_k(\bar{\theta};\varepsilon_k(\bar{\theta}))\right\| \leq c_{25}\left((E\|\varepsilon_k(\bar{\theta})\|^4)^{1/2} + \left(E\left\|\frac{\partial}{\partial\theta}\varepsilon_k(\bar{\theta})\right\|^4\right)^{1/2}\right)$$
$$\leq c_{26}$$

by (A.2) and (A.5). The proof is complete. □

Next we note a simple criterion for equicontinuity.

Lemma A.2

Let F be a compact subset of $\mathbb{R}^q$ and let $\mathscr{F}$ be an open subset of $\mathbb{R}^q$ which contains F. Let $d_N, N = 1, 2, \ldots$ be a family of continuously differentiable, scalar-valued functions on $\mathscr{F}$ and suppose that there exists a constant c such that

$$\left|\frac{\partial}{\partial\theta}d_N(\theta)\right| \leq c, \qquad \theta\in\mathscr{F}, N = 1, 2, \ldots \tag{A.8}$$

Then the functions $d_N, N = 1, 2, \ldots$ are equicontinuous on F.

PROOF Proof is by contradiction. Suppose that the functions $d_N, N = 1, 2, \ldots$ are not equicontinuous on F. This means that there exists an increasing sequence of positive integers $\{N_i\}$, sequences $\{\theta_i\}$ and $\{\theta'_i\}$ in F and $\varepsilon > 0$ such that

$$\theta_i - \theta'_i \to 0 \qquad \text{as} \quad i \to \infty$$

and

$$\|d_{N_i}(\theta_i) - d_{N_i}(\theta'_i)\| > \varepsilon, \qquad i = 1, 2, \ldots \tag{A.9}$$

Since F is compact, we can extract subsequences with the property that

$$\theta_{i_j} \to \bar{\theta}, \theta'_{i_j} \to \bar{\theta}. \qquad j \to \infty, \tag{A.10}$$

for some $\bar{\theta}\in F$. $\bar{\theta}$ lies in the open set $\mathscr{F}$. An open ball B can be chosen therefore, with centre $\bar{\theta}$ and which is contained in $\mathscr{F}$. We now choose an integer J such that $\theta_{i_j}, \theta'_{i_j}$ lie in B for $j \geq J$. The line segment s_j which joins θ_{i_j} and θ'_{i_j} lies in B for $j \geq J$ (this follows from the convexity of B). We deduce from the mean value theorem that

$$d_{N_{i_j}}(\theta_{i_j}) - d_{N_{i_j}}(\theta'_{i_j}) = \frac{\partial}{\partial\theta}d_{N_{i_j}}(\bar{\theta}_j)(\theta_{i_j} - \theta'_{i_j}), \quad j \geq J, \tag{A.11}$$

for some $\bar{\theta}_j\in s_j$.

Properties (A.8), (A.10) and (A.11) imply that

$$d_{N_{ij}}(\theta'_{i_j}) - d_{N_{ij}}(\theta_{i_j}) \to 0 \qquad \text{as} \quad j \to \infty.$$

This contradicts (A.9). The functions $d_N, N = 1, 2, \ldots$ must therefore be equicontinuous. □

We are now ready to prove the uniform convergence theorem.

Theorem A.3

(a) The random variables $Q_N(\theta; y^N, u^{N-1}), \theta \in D, N = 1, 2, \ldots$ have the property

$$Q_N(\theta; y^N, u^{N-1}) - EQ_N(\theta; y^N, u^{N-1}) \to 0 \qquad \text{as} \quad N \to \infty$$

uniformly in $\theta \in D$, almost surely; and,

(b) The functions $\theta \to EQ(\theta; y^N, u^{N-1})$ are uniformly bounded and equicontinuous on D.

PROOF We consider first of all the case when the $l_k(\cdot, \cdot)$ are scalar valued.

(a) Let ε be an arbitrary positive number. Part (a) of the theorem can be restated: there exists a number $N(\omega)$, which depends on the sample point ω and which is almost surely finite, such that

$$\sup_{\theta \in D} \left| \frac{1}{N} \sum_{k=1}^{N} (l_k(\theta, \varepsilon_k(\theta)) - El_k(\theta, \varepsilon_k(\theta)) \right| < \varepsilon \tag{A.12}$$

whenever $N > N(\omega)$. It is convenient to prove it in this form.

Let $B_{\bar{\alpha}}(\bar{\theta})$ be an open ball (with centre $\bar{\theta}$, radius $\bar{\alpha}$) in $\mathscr{D}$. We draw from the ergodic theorem, Theorem 1.1.15, and from Lemma A.1 which tells (among other things) us that the functions $\eta_k(\bar{\alpha}, \bar{\theta})$, $k = 1, 2, \ldots$ satisfy the hypotheses of the ergodic theorem, the following conclusions: given $\bar{\varepsilon} > 0$, there exists a positive integer $N(\alpha, \bar{\theta}, \bar{\varepsilon}, \omega)$ which depends on the ball $B_{\bar{\alpha}}(\bar{\theta})$, $\bar{\varepsilon}$ and the sample point ω and which is almost surely finite, such that

$$\left| \frac{1}{N} \sum_{k=1}^{N} (\eta_k(\bar{\alpha}, \bar{\theta}) - E\eta_k(\bar{\alpha}, \bar{\theta})) \right| < \bar{\varepsilon} \tag{A.13}$$

whenever $N > N(\bar{\alpha}, \bar{\theta}, \bar{\varepsilon}, \omega)$. Here, we recall,

$$\eta_k(\bar{\alpha}, \bar{\theta}) = \sup_{\theta \in B_{\bar{\alpha}}(\bar{\theta})} l_k(\theta; \varepsilon_k(\theta)).$$

Notice that

$$\sup_{\theta \in B_{\bar{\alpha}}(\bar{\theta})} \frac{1}{N} \sum_{k=1}^{N} (l_k(\theta; \varepsilon_k(\theta)) - El_k(\theta; \varepsilon_k(\theta)))$$

$$\leq \frac{1}{N} \sum_{k=1}^{N} (\eta_k(\bar{\alpha}, \bar{\theta}) - \inf_{\theta \in B_{\bar{\alpha}}(\bar{\theta})} El_k(\theta; \varepsilon_k(\theta)))$$

$$\leq \frac{1}{N} \sum_{k=1}^{N} (\eta_k(\bar{\alpha}, \bar{\theta}) - E\eta_k(\bar{\alpha}, \bar{\theta})) + c\bar{\alpha}$$

by Lemma A.1,

$$\leq \bar{\varepsilon} + c\bar{\alpha} \tag{A.14}$$

provided $N > N(\bar{\alpha}, \bar{\theta}, \bar{\varepsilon}, \bar{\omega})$ by (A.13). Here c is positive number which does not depend on $\bar{\alpha}$ or $\bar{\theta}$.

Now the set of open balls

$$\mathscr{S} = \left\{ B_\alpha(\theta) : 0 < \alpha < \frac{\varepsilon}{2c}, \theta \in D, B_\alpha(\theta) \subset \mathscr{D} \right\}$$

covers D. It follows from the compactness of D that there exists a finite collection of balls in $\mathscr{S}$ such that

$$D \subset \bigcup_{i=1}^{n} B_{\alpha_i}(\theta_i).$$

We have from (A.14) that

$$\sup_{\theta \in D} \left\{ \frac{1}{N} \sum_{k=1}^{N} ((l_k(\theta; \varepsilon_k(\theta)) - El_k(\theta; \varepsilon_k(\theta))) \right\}$$

$$\leq \sup_{\theta \in \cup_i B_{\alpha_i}(\theta_i)} \{\cdots\} = \max_{i=1,\ldots,n} \sup_{\theta \in B_{\alpha_i}(\theta_i)} \{\cdots\}$$

$$\leq \frac{\varepsilon}{2} + \frac{c\varepsilon}{2c} = \varepsilon \tag{A.15}$$

whenever $N > N_1(\omega)$. Here

$$N_1(\omega) = \max_{i=1,\ldots,n} \left\{ N\left(\alpha_i, \theta_i, \frac{\varepsilon}{2}, \omega \right) \right\}.$$

Exactly the same arguments apply when $-l$ replaces l. It follows that there exists a positive integer $N_2(\omega)$, which depends on the sample point and which is almost surely finite, such that

$$\inf_{\theta \in D} \left\{ \frac{1}{N} \sum_{k=1}^{N} (l_k(\theta; \varepsilon_k(\theta)) - El_k(\theta; \varepsilon_k(\theta))) \right\} > -\varepsilon \tag{A.16}$$

whenever $N > N_2(\omega)$. Inequalities (A.15) and (A.16) imply (A.12) when $N(\omega)$ is taken to be max $\{N_1(\omega), N_2(\omega)\}$.

(b) For $N = 1, 2, \ldots$, the function d_N, with domain $\mathscr{D}$, is defined to be

$$d_N(\theta) = \frac{1}{N} \sum_{k=1}^{N} El_k(\theta; \varepsilon_k(\theta)). \tag{A.17}$$

We must show that the d_N are uniformly bounded and equicontinuous on D.

Uniform boundedness follows from Lemma A.1. Bearing in mind that D is a compact subset of $\mathscr{D}$, we can deduce from Lemma A.2 that the functions d_N, $N = 1, 2, \ldots$ are equicontinuous on D provided we can show that they are continuously differentiable on $\mathscr{D}$ and that the derivatives

$$\frac{\partial}{\partial \theta} d_N, \qquad N = 1, 2, \ldots,$$

are uniformly bounded on $\mathscr{D}$.

However, in view of the smoothness of the f_k and l_k,

$$\left.\begin{array}{l}\text{the function } \theta \to l_k(\theta; \varepsilon_k(\theta)) \text{ is continuously differentiable for} \\ \text{given samples } y^k, u^{k-1} \text{ of the input and output, } k = 1, 2, \ldots\end{array}\right\} \tag{A.18}$$

By Lemma A1,

$$E\left|\frac{\partial}{\partial \theta} l_k(\theta; \varepsilon_k(\theta))\right| \leq c, \qquad \text{for all } \theta \in \mathscr{D}, \qquad k = 1, 2, \ldots \tag{A.19}$$

for some positive number c.

It is known that (A.18) and (A.19) imply that the following interchange of the expectation and differentiation operators is valid:

$$E \frac{\partial}{\partial \theta} l_k(\theta; \varepsilon_k(\theta)) = \frac{\partial}{\partial \theta} El_k(\theta; \varepsilon_k(\theta)), \theta \in \mathscr{D}, \qquad k = 1, 2, \ldots$$

and that these expressions depend continuously on θ. Recalling the definition (A.17) of d_N, we see that the functions $d_N, N = 1, 2, \ldots$ are continuously differentiable on $\mathscr{D}$, and the derivatives

$$\frac{\partial}{\partial \theta} d_N, \qquad N = 1, 2, \ldots,$$

are uniformly bounded on $\mathscr{D}$. It follows that the d_N are equicontinuous on D.

The theorem has been proved in the special case when the $l_k(\cdot,\cdot)$ are scalar valued. However, it is not difficult to see that, by applying the special case of the theorem when l_k is replaced by an arbitrary component of l_k, we can deduce that the assertions of the theorem are true in general. □

Proof of Theorem 5.2.1

We shall write $Q_N(\theta; y^N, u^{N-1})$ briefly as $Q_N(\theta)$.

Suppose that

$$Q_N(\theta) - EQ_N(\theta) \to 0 \qquad \text{as } N \to \infty \text{ uniformly over } \theta \in D. \tag{A.20}$$

By Theorem A.3, this event has probability 1.

The functions $\theta \to EQ_N(\theta)$, $N = 1, 2, \ldots$ are uniformly bounded and equicontinuous on D, by Theorem A.3. The function h is continuous and therefore uniformly continuous on some open ball which contains the point $EQ_N(\theta)$ for $N = 1, 2, \ldots$ and for all $\theta \in D$; in view of (A.20) this ball also contains $Q_N(\theta)$ for all $\theta \in D$ and for all N sufficiently large. We deduce from these properties, and (A.20), that

$$\left.\begin{array}{l}\text{the functions } \theta \to h(EQ_N(\theta)), \quad N = 1, 2, \ldots \text{ are} \\ \text{uniformly bounded and equicontinuous on } D\end{array}\right\} \tag{A.21}$$

and

$$h(Q_N(\theta)) - h(EQ_N(\theta)) \to 0 \qquad \text{as } N \to \infty \text{ uniformly over } \theta \in D. \tag{A.22}$$

Now let $\{\theta_{N_i}\}$ be an arbitrary convergent subsequence of $\{\theta_N\}$. Let $\bar{\theta} = \lim_i \theta_{N_i}$, and let ψ be an arbitrary element in D. The theorem will be proved if we can show that

$$\liminf_{N \to \infty} \{h(EQ_N(\bar{\theta})) - h(EQ_N(\psi))\} \leq 0 \tag{A.23}$$

since the maximum over ψ of the left-hand side of this inequality is obviously non-negative. But

$$\begin{aligned}\liminf_{N \to \infty} \{h(EQ_N(\bar{\theta})) - h(EQ_N(\psi))\} &\leq \liminf_{i \to \infty} \{h(EQ_{N_i}(\bar{\theta})) - h(EQ_{N_i}(\psi))\} \\ &= \liminf_{i \to \infty} \{h(EQ_{N_i}(\bar{\theta}_{N_i})) \\ &\quad - h(EQ_{N_i}(\psi))\}\end{aligned}$$

by (A.21) and since $\theta_{N_i} \to \bar{\theta}$,

$$= \liminf_{i\to\infty} \{h(Q_{N_i}(\theta_{N_i})) - h(Q_{N_i}(\psi))\}$$

by (A.22)

$$\leq 0$$

since θ_{N_i} minimizes the identification criterion $\theta \to h(Q_{N_i}(\theta))$. Inequality (A.23) is proved. □

APPENDIX B

The algebraic Riccati equation

The purpose of this appendix is to establish various properties of the algebraic Riccati equation (ARE), as required for application to the Kalman filter and the linear/quadratic control problem. We shall consider the ARE in its 'control' form since the proofs are based on control rather than filtering ideas. The corresponding results for the 'filtering' form of the ARE are obtained by using the duality relationships given in Section 6.1.

Let A, B, D, F be matrices of dimensions respectively $n \times n$, $n \times m$, $p \times n$, $p \times m$, where $p \geq m$. We suppose throughout that $F^{\mathrm{T}}F$ is strictly positive definite, i.e. there exists $\delta > 0$ such that

$$u^{\mathrm{T}}F^{\mathrm{T}}Fu \geq \delta \|u\|^2$$

for all $u \in \mathbb{R}^m$. Then $F^{\mathrm{T}}F$ is non-singular. We denote

$$\begin{aligned}\Theta &= (F^{\mathrm{T}}F)^{-1}\\ \hat{A} &= A - B\Theta F^{\mathrm{T}}D\\ \hat{D} &= [I - F\Theta F^{\mathrm{T}}]D.\end{aligned}$$

In addition, we sometimes write $\|x\|_Q^2$ for the quadratic form $x^{\mathrm{T}}Qx$ when x is a vector and Q is non-negative definite matrix.

The equations in question are:

The discrete-time Riccati equation

$$\begin{aligned}S(k) = A^{\mathrm{T}}S(k+1)A + D^{\mathrm{T}}D - (A^{\mathrm{T}}S(k+1)B + D^{\mathrm{T}}F)\\ \cdot (B^{\mathrm{T}}S(k+1)B + F^{\mathrm{T}}F)^{-1}(B^{\mathrm{T}}S(k+1)A + F^{\mathrm{T}}D). \qquad \text{(B.1)}\end{aligned}$$

This generates matrices $S(N-1)$, $S(N-2), \ldots$ from a given terminal condition $S(N) = S_0$. It follows from Theorems 6.1.1 and 6.1.2 that if S_0 is symmetric and non-negative definite then the same is true of $S(k)$ for all $k \leq N$.

The algebraic Riccati equation

$$\begin{cases} S = A^{\mathrm{T}}SA + D^{\mathrm{T}}D - (A^{\mathrm{T}}SB + D^{\mathrm{T}}F)(B^{\mathrm{T}}SB + F^{\mathrm{T}}F)^{-1}(B^{\mathrm{T}}SA + F^{\mathrm{T}}D) \\ S = S^{\mathrm{T}}, S \geq 0. \end{cases} \tag{B.2}$$

The results are as follows.

Theorem B.1

Suppose that (A, B) is stabilizable and that $S(-1)$, $S(-2), \ldots$ is the sequence of matrices defined by (B.1) with $S(0) = 0$. Then as $i \to -\infty, S(i) \to S$ where S is a non-negative definitive symmetric matrix satisfying (B.2). Now suppose also that $(\hat{D}, \hat{A})$ is detectable. Then $(A - BK)$ is stable, where

$$K = (B^{\mathrm{T}}SB + F^{\mathrm{T}}F)^{-1}(B^{\mathrm{T}}SA + F^{\mathrm{T}}D) \tag{B.3}$$

and S is the only non-negative definite solution of B.2. Further, $S(i) \to S$ as $i \to -\infty$, where $S(i)$ is the sequence defined by (B.1) with $S(0)$ an *arbitrary* non-negative definite symmetric matrix.

Theorem B.2

For each $x \in \mathbb{R}^n$, define

$$\eta(x) = \inf \left\{ \sum_{k=0}^{\infty} \|Dx_k + Fu_k\|^2 \right\}$$

where the infimum is taken over all sequences $\{x_k, u_k\}$ satisfying

$$\begin{aligned} x_0 &= x \\ x_{k+1} &= Ax_k + Bu_k \qquad k = 0, 1, \ldots \end{aligned} \tag{B.4}$$

Suppose that (A, B) is stabilizable and $(\hat{D}, \hat{A})$ is detectable. Let S be the solution to (B.2) and let K be given by (B.3).

Then

$$\eta(x) = x^{\mathrm{T}}Sx$$

and $u_k = -Kx_k$ is optimal in that

$$x^{\mathrm{T}}Sx = \sum_{k=0}^{\infty} \|D\bar{x}_k + F\bar{u}_k\|^2$$

where $\{\bar{x}_k, \bar{u}_k\}$ satisfy (B.4) with $\bar{u}_k = -K\bar{x}_k$.

These two theorems summarize the results of the sequence of lemmas stated and proved below.

Lemma B.3

Take integers M, N with $N > M$. Suppose $S(M), S(M+1), \ldots, S(N)$ satisfy (B.1) and that $\{x_k, u_k, M \leq k \leq N\}$ are sequences satisfying (B.4). Suppose also that $S(N) = S^{\mathrm{T}}(N)$, $S(N) \geq 0$. Then

$$\sum_{k=M}^{N-1} \|Dx_k + Fu_k\|^2$$

$$= x_M^{\mathrm{T}} S(M) x_M - x_N^{\mathrm{T}} S(N) x_N + \sum_{k=M}^{N-1} \|(B^{\mathrm{T}}S(k+1)B + F^{\mathrm{T}}F)u_k$$

$$+ (B^{\mathrm{T}}S(k+1)A + F^{\mathrm{T}}D)x_k\|^2_{(B^{\mathrm{T}}S(k+1)B+F^{\mathrm{T}}F)^{-1}}.$$

PROOF Using (B.4) we have

$$x_N^{\mathrm{T}} S(N) x_N - x_M^{\mathrm{T}} S(M) x_M$$

$$= \sum_{k=M}^{N-1} (x_{k+1}^{\mathrm{T}} S(k+1) x_{k+1} - x_k^{\mathrm{T}} S(k) x_k)$$

$$= \sum_{k=M}^{N-1} ((Ax_k + Bu_k)^T S(k+1)(Ax_k + Bu_k) - x_k^{\mathrm{T}} S(k) x_k).$$

Thus

$$\sum_{k=M}^{N-1} \|Dx_k + Fu_k\|^2 + x_N^{\mathrm{T}} S(N) x_N - x_M^{\mathrm{T}} S(M) x_M$$

$$= \sum_{k=M}^{N-1} \{(x_k^{\mathrm{T}} D^{\mathrm{T}} + u_k^{\mathrm{T}} F^{\mathrm{T}})(Dx_k + Fu_k)$$

$$+ (x_k^{\mathrm{T}} A^{\mathrm{T}} + u_k^{\mathrm{T}} B^{\mathrm{T}}) S(k+1)(Ax_k + Bu_k) - x_k^{\mathrm{T}} S(k) x_k\}.$$

The kth term in the sum can be written, after some rearrangement, as

$$x_k^{\mathrm{T}}[A^{\mathrm{T}}S(k+1)A + D^{\mathrm{T}}D - (A^{\mathrm{T}}S(k+1)B + D^{\mathrm{T}}F)$$

$$\cdot (B^{\mathrm{T}}S(k+1)B + F^{\mathrm{T}}F)^{-1}(B^{\mathrm{T}}S(k+1)A + F^{\mathrm{T}}D - S(k)]x_k$$

$$+ \|(B^{\mathrm{T}}S(k+1)B + F^{\mathrm{T}}F)u_k$$

$$+ (B^{\mathrm{T}}S(k+1)A + F^{\mathrm{T}}D)x_k\|^2_{(B^{\mathrm{T}}S(k+1)B+F^{\mathrm{T}}F)^{-1}}.$$

This gives the result in view of (B.1). □

Corollary B.4

Let S_0 be a symmetric non-negative definite matrix and x an n-vector. Define $S(-1), S(-2), \ldots$ by (B.1) with $S(0) = S_0$. Then for $j = 0, 1, \ldots,$

$$x^{\mathrm{T}} S(-j) x = \min \left(\sum_{k=0}^{j-1} \|Dx_k + Fu_k\|^2 + x_j^{\mathrm{T}} S_0 x_j \right)$$

where the minimum is taken over all sequences $\{x_k, u_k, k = 0, 1, \ldots, j\}$ satisfying (B.4).

Lemma B.5

Suppose that (A, B) is stabilizable and let $S(-1), S(-2), \ldots$ be defined by (B.1) with $S(0) = 0$. Then, as $k \to -\infty$, $S(k) \to S$ for some matrix S satisfying (B.2).

PROOF Since $S_0 = 0$ it follows from (B.1) that all the $S(k)$ are symmetric, and it is evident from Corollary B.4 that $S(k-1) \geq S(k)$ for $k = 0, -1, \ldots$ It will be shown presently that there exists a constant c such that

$$x^{\mathrm{T}} S(k) x \leq c \|x\|^2 \tag{B.5}$$

for all $x \in \mathbb{R}^n$, $k = 0, -1, \ldots$

Thus $y_k := x^{\mathrm{T}} S(k) x$ is a sequence of numbers which is increasing as $k \to -\infty$ in that $y_{k-1} \geq y_k$, and $y_k \leq c \|x\|^2$ for all k. Any such sequence converges to a limit; call it $\alpha(x)$. According to Proposition D.1.4 in Appendix D this implies that $\alpha(x) = x^{\mathrm{T}} S x$ for some symmetric non-negative definite matrix S and that $S(k) \to S$. Taking the limit as $k \to -\infty$ on both sides of (B.1), we conclude that S satisfies (B.2). It remains to prove (B.5).

Let L be a matrix such that $A - BL$ is stable (such a matrix exists by the stabilizability hypothesis). Take $x \in \mathbb{R}^n$ and define

$$\bar{x}_0 = x$$
$$\bar{x}_{i+1} = A\bar{x}_i + B\bar{u}_i, \qquad \bar{u}_i = -L\bar{x}_i \qquad i = 0, 1, \ldots$$

Since $(A - BL)$ is stable there exist constants c_1, c_2, not depending on x, such that

$$\sum_{i=0}^{\infty} \|D\bar{x}_i + F\bar{u}_i\|^2 \leq c_1 \|x\|^2$$

and

$$\bar{x}_i^{\mathrm{T}} S_0 \bar{x}_i \leq c_2 \|x\|^2, \qquad i = 0, 1, \ldots$$

Now apply Lemma B.3 with $N = 0$, $x_k = \bar{x}_{k-M}$, $u_k = \bar{u}_{k-M}$. This gives

$$\begin{aligned} x^{\mathrm{T}} S(M) x &= \sum_{k=0}^{-M-1} \|D\bar{x}_k + F\bar{u}_k\|^2 + \bar{x}_M^{\mathrm{T}} S_0 \bar{x}_M \\ &\quad - \sum_{k=0}^{-M-1} \|(B^{\mathrm{T}} S(k+1-M)B + F^{\mathrm{T}}F)\bar{u}_k \\ &\quad - (B^{\mathrm{T}} S(k+1-M)A + F^{\mathrm{T}}D)\bar{x}_k\|^2_{(B^{\mathrm{T}}S(k+1-M)B + F^{\mathrm{T}}F)} \\ &\leq (c_1 + c_2)\|x\|^2. \end{aligned}$$

□

Lemma B.6

Let $(\hat{D}, \hat{A})$ be detectable and suppose that S satisfies (B.2). Then $(A - BK)$ is stable, where K is given by (B.3).

PROOF Let L be a matrix such that $\hat{A} - L\hat{D}$ is stable. Such a matrix exists by the detectability hypothesis. Let $x \in \mathbb{R}^n$ be arbitrary. Now apply Lemma B.1 with $M = 0$, $S(k) = S$, $x_0 = x$ and $u_k = -Kx_k$ (so that $x_k = (A - BK)^k x$). This gives

$$x^{\mathrm{T}} S x = \|(A - BK)^N x\|_S^2 + \sum_{k=0}^{N-1} \|(D - FK)x_k\|^2. \tag{B.6}$$

Denote $M = F\Theta F^{\mathrm{T}}$ (this is the projection onto the range of F). Then

$$\begin{aligned} (D - FK)x_k &= (I - M + M)Dx_k - FKx_k \\ &= \hat{D}x_k + F(\Theta F^{\mathrm{T}} D - K)x_k. \end{aligned}$$

The two terms are orthogonal since $\hat{D}x_k$ is orthogonal to the range of F. Thus (B.6) becomes

$$\begin{aligned} x^{\mathrm{T}} S x = \|(A - BK)^N x\|_S^2 + \sum_{k=0}^{N-1} (\|\hat{D}x_k\|^2 \\ + \|(\Theta F^{\mathrm{T}} D - K)x_k\|^2_{F^{\mathrm{T}}F}). \end{aligned} \tag{B.7}$$

Since (a) the left-hand side of (B.7) is independent of N; (b) the terms on the right are all positive; and (c) $\|u\|^2 \leq \delta^{-1} \|u\|^2_{F^{\mathrm{T}}F}$ for any $u \in \mathbb{R}^m$, this shows that

$$\sum_{k=0}^{\infty} (\|\hat{D}x_k\|^2 + \|(\Theta F^{\mathrm{T}}D - K)x_k\|^2) \leq k_1 < \infty \tag{B.8}$$

for some constant k_1 depending on x.

Next, note that, since $\hat{A} = A - B\Theta F^{\mathrm{T}}D$,

$$\begin{aligned} A - BK &= \hat{A} + B(\Theta F^{\mathrm{T}}D - K) \\ &= (\hat{A} - L\hat{D}) + B(\Theta F^{\mathrm{T}}D - K) + L\hat{D}. \end{aligned}$$

The first term on the right will be denoted $\tilde{A}$ and is stable by hypothesis. In view of this identity, if a sequence z_k is generated by

$$\begin{aligned} z_0 &= x \\ z_{k+1} &= \tilde{A}z_k + (B(\Theta F^{\mathrm{T}}D - K) + L\hat{D})z_k, \end{aligned} \tag{B.9}$$

then $z_k = x_k$. On the other hand, writing down the solution of (B.9) as a difference equation with x_k replacing the last z_k gives

$$z_k = \tilde{A}^k x + \sum_{i=0}^{k-1} \tilde{A}^{k-i}(B(\Theta F^{\mathrm{T}}D - K) + L\hat{D})x_i.$$

We therefore have the identity

$$(A - BK)^k x = \tilde{A}^k x + \sum_{i=0}^{k-1} \tilde{A}^{k-i}(B(\Theta F^{\mathrm{T}}D - K) + L\hat{D})(A - BK)^i x.$$

In view of the properties of matrix norms, this shows that

$$a_k \leq \tilde{a}_k + \sum_{i=0}^{k-1} g_{k-i}h_i =: \tilde{a}_k + l_k \tag{B.10}$$

where

$$a_k = \|(A - BK)^k x\|, \qquad \tilde{a}_k = \|\tilde{A}^k x\|, \qquad g_j = \|\tilde{A}^j\|^{\dagger}$$

and

$$h_i = c_1(\|(\Theta F^{\mathrm{T}}D - K)x_k\|^2 + \|\hat{D}x_k\|^2)^{1/2}$$

where

$$c_1 = \max\{\|B\|, \|L\|\}.$$

Now regard the terms in (B.10) as the kth components of

† Matrix norms in this appendix are taken to be the spectral norms.

$(N+1)$-vectors $\mathbf{a}$, $\tilde{\mathbf{a}}$, $\mathbf{l}$ and use the triangle inequality to give

$$\left(\sum_0^N a_k^2\right)^{1/2} \leq \left(\sum_0^N \tilde{a}_k^2\right)^{1/2} + \|\mathbf{l}\|. \tag{B.11}$$

Examining l_k we see that $\mathbf{l}$ can be written

$$\mathbf{l} = \sum_{i=1}^N g_i h_i$$

where $\mathbf{h}_i^{\mathrm{T}} = (0, \ldots, 0, h_0, h_1, \ldots, h_{N-i})$. Since $g_i \geq 0$ this shows that

$$\|\mathbf{l}\| \leq \sum_{i=1}^N g_i \|\mathbf{h}_i\| = \sum_{i=1}^N g_i \left(\sum_{k=0}^{N-i} h_k^2\right).$$

Since $\tilde{A}$ is stable, $g_i \leq c_2 \lambda^i$ for some $\lambda < 1$. Using this together with (B.8) we see that

$$\|\mathbf{l}\| \leq k_1^{1/2} c_2 \lambda (1-\lambda)^{-1}.$$

Now

$$\sum_0^\infty \tilde{a}_k^2 < \infty$$

since $\tilde{A}$ is stable, and hence from (B.11),

$$\sum_{k=0}^\infty \|(A - BK)^k x\|^2 < \infty. \tag{B.12}$$

If $(A - BK)$ were not stable, the real symmetric matrix $(A - BK)^{\mathrm{T}}(A - BK)$ would have a real eigenvalue λ with $\lambda \geq 1$. If x is a corresponding eigenvector then

$$x^{\mathrm{T}}(A^{\mathrm{T}} - K^{\mathrm{T}} B^{\mathrm{T}})^k (A - BK)^k x \geq \|x\|^2 \qquad k = 0, 1, \ldots$$

But this contradicts (B.12). Therefore $(A - BK)$ must be stable. □

Lemma B.7

Suppose that S satisfies (B.2) and $(A - BK)$ is stable, where K is given by (B.3). Then S is the unique solution to (B.2) with this property, and for every $x \in \mathbb{R}^n$,

$$x^{\mathrm{T}} S x = \sum_{k=0}^\infty \|D\bar{x}_k + F\bar{u}_k\|^2 \tag{B.13}$$

where $\{\bar{x}_k, \bar{u}_k, k=0,1,2,\dots\}$ are defined by

$$\bar{x}_0 = x$$
$$\bar{x}_{k+1} = A\bar{x}_k + B\bar{u}_k$$
$$\bar{u}_k = -K\bar{x}_k.$$

PROOF For every $x \in \mathbb{R}^n$ define

$$\eta^0(x) = \inf\left\{\sum_{k=0}^{\infty} \|Dx_k + Fu_k\|^2\right\} \tag{B.14}$$

where the infimum is taken over sequences $\{x_k, u_k\}$ such that the sum in (B.13) is finite, (B.4) is satisfied, and $\|x_k\| \to 0$. (The set of such sequences is non-empty since it includes $\{\bar{x}_k, \bar{u}_k\}$.) Consider such a sequence $\{x_k, u_k\}$. Application of Lemma B.1 with $S(k) = S$ for all k and $M = 0$ and passage to the limit as $N \to \infty$ yields

$$\sum_{k=0}^{\infty} \|Dx_k + Fu_k\|^2$$
$$= x^{\mathrm{T}}Sx + \sum_{k=0}^{\infty} \|(B^{\mathrm{T}}SB + F^{\mathrm{T}}F)u_k + (B^{\mathrm{T}}SA + F^{\mathrm{T}}D)x_k\|^2_{(B^{\mathrm{T}}SB + F^{\mathrm{T}}F)^{-1}}.$$

Since $S \geq 0$ it follows that (B.13) holds and that $\eta^0(k) = x^{\mathrm{T}}Sx$. Now let Q be any other solution to (B.2) for which $A - B(B^{\mathrm{T}}QB + F^{\mathrm{T}}F)^{-1}(B^{\mathrm{T}}QA + F^{\mathrm{T}}F)$ is stable. Bearing in mind that S did not enter the definition of η^0 we conclude that $x^{\mathrm{T}}Qx = x^{\mathrm{T}}Sx$, and hence that $S = Q$, since S and Q are symmetric and x is arbitrary. It follows that S is the unique solution to (B.2) such that $A - BK$ is stable.

Lemma B.8

Suppose that (A, B) is stabilizable and $(\hat{D}, \hat{A})$ is detectable. Let S be the solution to (B.2). Then given $x \in \mathbb{R}^n$,

$$x^{\mathrm{T}}Sx \leq \sum_{i=0}^{\infty} \|Dx_i + Fu_i\|^2$$

for every pair of sequences $\{x_i, u_i\}$ satisfying (B.4).

PROOF The hypotheses imply that (B.2) has a unique non-negative definite solution S. Define $S(-1), S(-2), \dots$ by (B.1) with $S(0) = 0$ and apply Lemma B.3 with $N = 0$, $M < 0$. Since $S(M)$ is non-negative definite, we conclude that

$$x^{\mathrm{T}} S(M) x \leq \sum_{k=0}^{-M-1} \|Dx_k + Fu_k\|^2.$$

Now take the limit as $M \to -\infty$. By Lemma B5, $S(M) \to S$. The result follows. □

Lemma B.9

Suppose that (A, B) is stabilizable and $(\hat{D}, \hat{A})$ is detectable. Let S_0 be an arbitrary symmetric non-negative definite $n \times n$ matrix and let $Q(-1), Q(-2), \ldots$ be defined by (B.1) with $Q(0) = S_0$. Then $Q(k) \to S$ as $k \to -\infty$, where S is the solution to (B.2).

PROOF Take $x \in \mathbb{R}^n$ and let $S(-1), S(-2), \ldots$ satisfy (B.1) with $S(0) = 0$. By Corollary B.4,

$$x^{\mathrm{T}} Q(k) x \geq x^{\mathrm{T}} S(k) x \qquad k = -1, -2, \ldots$$

since these are the minimal costs for the k-stage control problems with terminal cost matrices S_0 and 0 respectively. By Lemma B.5, $S(k) \to S$, so that

$$\liminf_{k \to -\infty} x^{\mathrm{T}} Q(k) x \geq x^{\mathrm{T}} S x. \tag{B.15}$$

On the other hand, again by Corollary B.4, for each $j \geq 0$

$$x^{\mathrm{T}} Q(-j) x \leq \sum_{i=0}^{j-1} \|Dx_i + Fu_i\|^2 + x_j^{\mathrm{T}} S_0 x_j$$

where $\{x_i, u_i, i = 0, 1, \ldots, j\}$ satisfy (B.4) with $u_i = -Kx_i$, since this stationary control is sub-optimal for the j-stage problem. By Lemma B.6, $x_j^{\mathrm{T}} S_0 x_j \to 0$ as $j \to \infty$. By Lemma B.7, the sum converges to $x^{\mathrm{T}} S x$. It follows that

$$\limsup_{j \to -\infty} x^{\mathrm{T}} Q(j) x \leq x^{\mathrm{T}} S x \tag{B.16}$$

Now (B.15) and (B.16) imply that $x^{\mathrm{T}} Q(j) x \to x^{\mathrm{T}} S x$ as $j \to -\infty$ for arbitrary x. In view of Proposition D.1.4, Appendix D, this implies that $Q(j) \to S$. □

APPENDIX C

Proof of Theorem 7.4.2

In this appendix we provide a proof of Theorem 7.4.2, showing that when an ARMAX system is controlled by the Unit Delay Algorithm 7.4.1, performance is 'asymptotically optimal' under the stated conditions. All notation is that of Section 7.4. We start by making the following definitions:

$$e_k := y_k - y_k^* = y_k - \phi_{k-1}^{\mathrm{T}} \hat{\theta}_{k-1}$$
$$s_{k-1} := e_k - w_k = E[y_k | \mathcal{Y}_{k-1}] - y_k^*$$
$$b_k := -\phi_k^{\mathrm{T}} \tilde{\theta}_k.$$

where

$$\tilde{\theta}_k = \hat{\theta}_k - \theta_0.$$

Note that all these processes are adapted to $\mathcal{Y}_k$ in that at each time k they are functions of the output and initial conditions $\{y_1, \ldots, y_k, w_0, \ldots, w_{m^*}\}$ (in the case of s_k, recall that y_{k+1}^* is deterministic). The 'true system' equation in predictor form (7.1.3) is

$$C(z^{-1})(y_{k+1} - y_{k+1}^*) = [B(z^{-1})u_k + D(z^{-1})y_k - C(z^{-1})y_{k+1}^*] + C(z^{-1})w_{k+1}.$$

This can be expressed as

$$C(z^{-1})(e_{k+1} - w_{k+1}) = \phi_k^{\mathrm{T}} \theta_0 - y_{k+1}^*$$

or alternatively as

$$C(z^{-1})s_k = \phi_k^{\mathrm{T}} \theta_0 - y_{k+1}^* \tag{C.1}$$

where θ_0 is as given by (7.4.4), (7.4.5) but extended to dimension $(n_A^0 + n_B^0 + n_C^0 - 1)$ by the addition of zero coefficients. This is where the degree condition on n_A, etc., is required. We shall prove two lemmas, from the second of which the assertions of the Theorem follow quickly. The proofs of these lemmas require some technical results which are collected together at the end of the section.

We write $\bar{a}^0 = \bar{a}$ throughout for convenience.

Lemma C.1

With probability one, if $r_N \to \infty$ as $N \to \infty$ then

$$\lim_{N\to\infty} \frac{1}{r_N} \sum_{k=1}^{N} s_k^2 = 0.$$

PROOF Define

$$V_k = \tilde{\theta}_k^{\mathrm{T}} \tilde{\theta}_k.$$

From (7.4.3a), $\tilde{\theta}_k$ satisfies

$$\tilde{\theta}_k = \tilde{\theta}_{k-1} + \frac{\bar{a}}{r_{k-1}} \phi_{k-1} e_k.$$

A little algebra using the fact that $y_k - E[y_k | \mathscr{Y}_{k-1}] = w_k$ shows that

$$V_k = V_{k-1} - 2\frac{\bar{a}}{r_{k-1}} b_{k-1} s_{k-1} + 2\frac{\bar{a}}{r_{k-1}} b_{k-1} w_k$$
$$+ \left(\frac{\bar{a}}{r_{k-1}}\right)^2 \phi_{k-1}^{\mathrm{T}} \phi_{k-1} (s_{k-1} + w_k)^2.$$

Taking the conditional expectation of both sides and using the properties that $E[w_k | \mathscr{Y}_{k-1}] = 0$ and that s_{k-1} is a function of $\mathscr{Y}_{k-1}$, we obtain

$$E[V_k | \mathscr{Y}_{k-1}] = V_{k-1} - 2\frac{a}{r_{k-1}} b_{k-1} s_{k-1}$$
$$+ \frac{\bar{a}^2}{r_{k-1}^2} \phi_{k-1}^{\mathrm{T}} \phi_{k-1} s_{k-1}^2 + \frac{\bar{a}^2}{r_{k-1}^2} \phi_{k-1}^{\mathrm{T}} \phi_{k-1} \sigma^2.$$

From (7.4.3b), $\phi_{k-1}^{\mathrm{T}} \phi_{k-1} \le r_{k-1}$ and hence the third term on the right is, for any $\rho > 0$, less than or equal to

$$\frac{\bar{a}^2}{r_{k-1}} s_{k-1}^2 = \left(\frac{2\bar{a}}{r_{k-1}} \left(\frac{\bar{a} + \rho}{2} \right) - \frac{\rho \bar{a}}{r_{k-1}} \right) s_{k-1}^2,$$

Thus

$$E[V_k | \mathscr{Y}_{k-1}] \le V_{k-1} - \frac{2\bar{a}}{r_{k-1}} h_{k-1} s_{k-1} - \frac{\rho \bar{a}}{r_{k-1}} s_{k-1}^2$$
$$+ \frac{\bar{a}^2}{r_{k-1}^2} \phi_{k-1}^{\mathrm{T}} \phi_{k-1} \sigma^2 \qquad \text{(C.2)}$$

where

$$h_k = b_k - \frac{\bar{a}+\rho}{2} s_k.$$

Now using (C.1) and (7.4.3c) we see that

$$C(z^{-1})s_{k-1} = \phi_{k-1}^{\mathrm{T}}\theta_0 - y_k^* = \phi_{k-1}^{\mathrm{T}}\theta_0 - \phi_{k-1}^{\mathrm{T}}\hat{\theta}_{k-1} = b_{k-1}.$$

The process h_k is therefore the following moving average of s_k:

$$h_k = \left[C(z^{-1}) - \frac{\bar{a}+\rho}{2} \right] s_k \qquad k \geq m^*.$$

Since $C(z^{-1}) - \frac{1}{2}\bar{a}$ is strictly positive real, $C(z^{-1}) - \frac{1}{2}(\bar{a}+\rho)$ is positive real for some $\rho > 0$. Choose such a ρ and define

$$S_k := 2\bar{a} \sum_{j=1}^{k} h_{j-1}s_{j-1} + K.$$

It follows from Lemma C.3 below that there exists $K > 0$ such that $S_k \geq 0$ for all k. Now define

$$Z_k := V_k + \frac{1}{r_{k-1}} S_k.$$

Substituting $V_k = Z_k - S_k/r_{k-1}$ in (C.2), we find that

$$E[Z_k | \mathscr{Y}_{k-1}] \leq Z_{k-1} - \frac{\rho\bar{a}}{r_{k-1}} s_{k-1}^2 + \frac{\bar{a}^2}{r_{k-1}^2} \phi_{k-1}^{\mathrm{T}} \phi_{k-1} \sigma^2.$$

Note that

(a) $Z_k \geq 0$ for all k;
(b) The second term on the right is non-positive;
(c) For the third term we have

$$\sum_{k=1}^{\infty} \frac{\bar{a}^2}{r_k^2} \phi_k^{\mathrm{T}} \phi_k \sigma^2 < \infty.$$

Part (c) follows from (7.4.3b) since, denoting $\alpha_k := \phi_k^{\mathrm{T}} \phi_k$, we have

$$r_k = r_{k-1} + \alpha_k, \qquad k > m^*$$

so that

$$\sum_{k=m^*}^{N} \frac{\alpha_k}{r_k^2} \leq \sum_{k=m^*}^{N} \frac{\alpha_k}{r_k r_{k-1}} = \sum_{k=m^*}^{N} \frac{r_k - r_{k-1}}{r_k r_{k-1}}$$

$$= \sum_{k=m^*}^{N} \frac{1}{r_{k-1}} - \frac{1}{r_k} = 1 - \frac{1}{r_N} \leq 1.$$

The result follows. In view of (a), (b), (c), the martingale convergence theorem (Lemma C.4 below) implies that with probability one,

$$Z_k \to Z_\infty$$

where Z_∞ is some random variable with $Z_\infty \geq 0$, $EZ_\infty < \infty$; and that

$$\sum_{k=1}^{\infty} \frac{\rho \bar{a}}{r_{k-1}} s_{k-1}^2 < \infty .$$

The conclusion of the lemma now follows from the Kronecker Lemma C.5 (which requires $r_k \uparrow \infty$). □

Lemma C.2

With probability one,

$$\lim_{N\to\infty} \frac{1}{N} \sum_{k=1}^{N} s_k^2 = 0.$$

PROOF First consider a realization of the process such that $r_k \uparrow r_\infty < \infty$. From (7.4.3b) this implies $\|\phi_k\| \to 0$. Since $Z_k = \tilde{\theta}_k^{\mathrm{T}}\tilde{\theta}_k + S_k/r_{k-1} \to Z < \infty$ and $S_k/r_{k-1} \geq 0$, there exists n_0 such that $\|\tilde{\theta}_k\| < 2Z$ for $k > n_0$. Now s_k is generated by $C(z^{-1})s_k = b_k = -\phi_k^{\mathrm{T}}\tilde{\theta}_k$ and C is stable; thus $s_k \to 0$ since $|\phi_k^{\mathrm{T}}\tilde{\theta}_k| \to 0$, and hence

$$\frac{1}{N} \sum_{k=1}^{N} s_k^2 \to 0.$$

For the remainder of the proof we take a realization of the process such that $r_k \uparrow \infty$, and that (7.4.2) and the conclusion of Lemma C.1 hold. Together with the case considered above, this covers all possible realizations except a set of probability zero.

Think of system (7.4.1) as a stable linear system with inputs w_k, y_k and output u_k. This can be realized in state-space form in the standard way, and it then follows from the bounded input/bounded output stability Lemma C.6 that there are constants K_1, K_2, N_0 such that

$$\frac{1}{N} \sum_{k=1}^{N} u_k^2 \leq \frac{K_1}{N} \sum_{k=1}^{N} y_{k+1}^2 + K_2 \qquad \text{for } N > N_0.$$

Using the fact that

$$r_{m^*+k} = 1 + \sum_{m^*}^{m^*+k} \phi_j^{\mathrm{T}} \phi_j$$

and the definition of ϕ_j, this implies that

$$\frac{r_N}{N} \leq \frac{K_3}{N} \sum_{j=1}^{N} y_{k+1}^2 + K_4 \qquad N > N_0 \tag{C.3}$$

for some constants K_3, K_4. Now

$$y_k = y_k^* + w_k + s_{k-1}. \tag{C.4}$$

It follows from (7.4.2) that

$$\limsup_{N \to \infty} \frac{1}{N} \sum_{k=1}^{N} w_k^2 < \infty$$

and hence from (C.4), since $|y_k^*|$ is bounded,

$$\frac{1}{N} \sum_{k=1}^{N} y_{k+1}^2 \leq \frac{K_5}{N} \sum_{k=1}^{N} s_k^2 + K_6 \qquad N \geq N_1$$

for some K_5, K_6, N_1. Combining this with (C.3) we have

$$\frac{1}{N} r_N \leq \frac{K_7}{N} \sum_{j=1}^{N} s_k^2 + K_8. \tag{C.5}$$

We can use this relation to show that

$$\frac{1}{N} \sum_{k=1}^{N} y_{k+1}^2$$

must be bounded. Indeed, suppose

$$\frac{1}{N} \sum_{k=1}^{N} y_{k+1}^2$$

is *not* bounded; then

$$\limsup_{N \to \infty} \frac{1}{N} r_N = \infty$$

since, by the definition of r_N

$$r_N \geq \sum_{k=1}^{N} y_{k+1}^2.$$

In view of (C.5) this implies that

$$\limsup_{N \to \infty} \frac{1}{N} \sum_{k=1}^{N} s_k^2 = \infty. \tag{C.6}$$

Write

$$\bar{s}_N^2 = \frac{1}{N}\left(\sum_1^N s_k^2\right);$$

then (C.5) states that $N/r_N \geq 1/(K_7\bar{s}_N^2 + K_8)$ or that

$$\frac{N}{r_N}\bar{s}_N^2 \geq \frac{\bar{s}_N}{K_7\bar{s}_N + K_8}. \tag{C.7}$$

From (C.6), there exists a sequence $n_1 < n_2 < \ldots$ such that $\lim_k \bar{s}_{n_k}^2 = \infty$ Thus $\lim_k \bar{s}_{n_k}^2/(K_7\bar{s}_{n_k} + K_8) = 1/K_7$, and from (C.7)

$$\liminf_{N\to\infty} \frac{N}{r_N}\bar{s}_N^2 \geq \frac{1}{K_7}.$$

However, this contradicts Lemma C.1; so it must be the case that

$$\frac{1}{N}\sum_1^N y_{k+1}^2$$

is bounded. Returning to (C.3) this means that

$$\limsup_{N\to\infty} \frac{1}{N} r_N < \infty$$

or, equivalently, that

$$\liminf_{N\to\infty} \frac{N}{r_N} \geq \delta > 0.$$

Combining this with Lemma C.1 we conclude that

$$\lim_{N\to\infty} \frac{1}{N}\sum_{k=1}^N s_k^2 = 0.$$

This completes the proof. □

Proof of Theorem 7.4.2

This is almost immediate from Lemma C.2. Indeed, (7.4.8) and (7.4.9) have already been shown in the proof of Lemma C.2. For (7.4.10), denote $\hat{y}_{k|k-1} := E[y_k|\mathcal{Y}_{k-1}]$. Then

$$y_k - y_k^* = [y_k - \hat{y}_{k|k-1}] + s_{k-1}.$$

Now s_{k-1} is a function of $\mathscr{Y}_{k-1}$, so that

$$E[(y_k - \hat{y}_{k|k-1})s_{k-1}|\mathscr{Y}_{k-1}] = s_{k-1}E[y_k - \hat{y}_{k|k-1}|\mathscr{Y}_{k-1}] = 0.$$

Hence

$$\begin{aligned} E[(y_k - y_k^*)^2|\mathscr{Y}_{k-1}] &= s_{k-1}^2 + E[(y_k - y_{k|k-1})^2|\mathscr{Y}_{k-1}] \\ &= s_{k-1}^2 + \sigma^2. \end{aligned}$$

Thus using Lemma C.2 we see that

$$\lim_{N\to\infty} \frac{1}{N} \sum_{k=1}^{N} E[(y_k - y_k^*)^2|\mathscr{Y}_{k-1}] = \sigma^2.$$

which is the result claimed. □

The following lemmas are used in the proof.

Lemma C.3 (Positive real lemma)

Suppose

$$G(z^{-1}) = 1 + g_1 z^{-1} + \cdots + g_p z^{-p}$$

is positive real, i.e. $\mathrm{Re}\{G(e^{-i\omega})\} \geq 0$ for $0 \leq \omega \leq 2\pi$. Suppose doubly infinite sequences $\{h_k, s_k\}$ satisfy

$$h_k = G(z^{-1})s_k \qquad \text{for } k \geq m^* \geq p.$$

Then there exists a constant K such that, for all $N \geq m^*$,

$$\sum_{k=0}^{N} h_k s_k + K \geq 0.$$

REMARK Positive realness is a kind of 'passivity' condition. If one thinks of G as a transfer function relating (sampled) current h_k and voltage s_k in a network then $\Sigma h_k s_k$ is the energy dissipation and this is, apart from initial conditions, positive, corresponding to a passive network. The result is true if G is a rational function (rather than a polynomial) but the proof is then somewhat more complicated.

PROOF Fix $N \geq m^*$ and suppose, without loss of generality, that $h_k = s_k = 0$ for $k < 0$ and $s_k = 0$ for $k > N$. Define

$$H(\omega) = \sum_{k=-\infty}^{\infty} h_k e^{-i\omega k}, \qquad S(\omega) = \sum_{k=-\infty}^{\infty} s_k e^{-i\omega k}$$

(these are finite sums). By the generalized Parseval equality,

$$\frac{1}{2\pi}\int_0^{2\pi} S^*(\omega)H(\omega)d\omega = \sum_{k=0}^{\infty} s_k^* h_k. \tag{C.8}$$

Let $\bar{h}_k$ be given by

$$\bar{h}_k = G(z^{-1})s_k, \qquad \text{for all } k.$$

Thus $\bar{h}_k = h_k$ for $k \geq m^*$, and

$$\sum_{k=-\infty}^{\infty} \bar{h}_k e^{-i\omega k} = \bar{H}(\omega) = G(e^{-i\omega})S(\omega),$$

so that

$$H(\omega) = G(e^{-i\omega})S(\omega) + R(\omega)$$

where

$$R(\omega) = \sum_{k=0}^{m^*-1} (h_k - \bar{h}_k)e^{-i\omega k}.$$

Now s_k and h_k are real, so (C.8) becomes

$$\sum_{k=0}^{N} s_k h_k = \operatorname{Re}\left\{\frac{1}{2\pi}\int_0^{2\pi} S^*(\omega)G(e^{i\omega})S(\omega)d\omega + \frac{1}{2\pi}\int_0^{2\pi} S^*(\omega)R(\omega)d\omega\right\}$$

$$\geq \operatorname{Re}\frac{1}{2\pi}\int_0^{2\pi} S^*(\omega)R(\omega)d\omega$$

(invoking the positive real condition). This last expression is equal to

$$\sum_{k=0}^{m^*-1} s_k(h_k - \bar{h}_k)$$

and is independent of N. This completes the proof.

Lemma C.4 (Martingale convergence theorem)

Let $x_0, y_1, y_2, \ldots$ be a sequence of random variables and denote $\mathscr{Y}_k = \{x_0, y_1, \ldots, y_k\}$. A sequence of random variables $\{T_k\}$ is *adapted to* $\mathscr{Y}_k$ if, for each $k \geq 1$, $T_k = g_k(x_0, y_1, \ldots, y_k)$ for some function g_k.

Let $\{T_k\}$, $\{\alpha_k\}$, $\{\beta_k\}$ be sequences of non-negative random variables adapted to $\mathscr{Y}_k$ such that

$$E[T_k|\mathscr{Y}_{k-1}] \leq T_{k-1} - \alpha_{k-1} + \beta_{k-1}.$$

If

$$\sum_{k=1}^{\infty} \beta_k < \infty$$

with probability 1, then, also with probability 1, T_k converges to a finite random variable T and

$$\sum_{k=1}^{\infty} \alpha_k < \infty.$$

PROOF We cannot give a self-contained proof of this result here. See Neveu (1975) or Goodwin *et al.* (1981) listed in the references to Chapter 7.

Lemma C.5 (Kronecker lemma)

Let s, $\{x_n, b_n, n=1,2,\ldots\}$ be real numbers such that $0 < b_n \uparrow \infty$ and

$$s_{n+1} := \sum_{k=1}^{n} x_k \to s \quad \text{as } n \to \infty.$$

Then

$$\lim_{n\to\infty} \frac{1}{b_n} \sum_{k=1}^{n} b_k x_k = 0.$$

PROOF Define $b_0 := 0$ and $a_k := b_k - b_{k-1}$. Then

$$\frac{1}{b_n} \sum_{k=1}^{n} b_k x_k = (s_{n+1} - s) - \frac{1}{b_n} \sum_{k=1}^{n} a_k (s_k - s).$$

Thus it suffices to show that the second term on the right converges to 0 as $n \to \infty$. For any $\varepsilon > 0$ there exists n_ε such that $|s_k - s| < \varepsilon$ for $k > n_\varepsilon$, so that for $n > n_\varepsilon$

$$\left|\frac{1}{b_n} \sum_{k=1}^{n} a_k (s_k - s)\right| \leq \left|\frac{1}{b_n} \sum_{k=1}^{n_\varepsilon} a_k (s_k - s)\right| + \left|\frac{1}{b_n} \sum_{k=n_\varepsilon+1}^{n} a_k (s_k - s)\right|$$
$$\leq \left|\frac{1}{b_n} \sum_{k=1}^{n_\varepsilon} a_k (s_k - s)\right| + \varepsilon.$$

The result follows on letting $n \uparrow \infty$, $\varepsilon \downarrow 0$ (in that order).

Lemma C.6 (Bounded input/bounded output stability)

Let ξ_k, u_k be respectively the m-vector input and scalar output of the stable linear system

$$x_{k+1} = Ax_k + B\xi_k$$
$$u_k = c^{\mathrm{T}} x_k + d^{\mathrm{T}} \xi_k.$$

Then there exist constants c_1, c_2 independent of N such that

$$\sum_{k=1}^{N} u_k^2 \leq c_1 \sum_{k=0}^{N} |\xi_k|^2 + c_2.$$

PROOF The output is given explicitly by

$$u_k = c^{\mathrm{T}} A^k x_0 + d^{\mathrm{T}} \xi_k + \sum_{j=1}^{k} c^{\mathrm{T}} A^j B \xi_{k-j}.$$

Let $\| A \|$ denote the spectral norm $\| A \| = \max_{|x|=1} |Ax|$. Since A is stable, there exists $\lambda, 0 \leq \lambda < 1$, and $K < \infty$ such that $\| A^j \| \leq K\lambda^j$. Thus

$$\| u_k \|^2 \leq 3\left[\| c \|^2 \| A^k \|^2 \| x_0 \|^2 + \| d \|^2 \| \xi_k \|^2 + \left(\sum_{j=1}^{k} \| c \| \| A^j \| \| B \| \| \xi_{t-j} \| \right)^2 \right]$$

$$\leq K_1 \lambda^{2k} + K_2 \| \xi_k \|^2 + K_3 \left(\sum_{j=1}^{k} \lambda^j \| \xi_{k-j} \| \right)^2.$$

Using the Schwarz inequality we have

$$\left(\sum_{j=1}^{k} \lambda^{j/2} \lambda^{j/2} \| \xi_{k-j} \| \right)^2 \leq \sum_{j=1}^{k} \lambda^j \sum_{j=1}^{k} \lambda^j \| \xi_{k-j} \|^2$$

and hence

$$\sum_{k=1}^{N} \| u_k \|^2 \leq K_4 + K_2 \sum_{k=1}^{N} \| \xi_k \|^2 + K_6 \sum_{k=1}^{N} \sum_{j=1}^{k} \lambda^j \| \xi_{k-j} \|^2.$$

On introducing the variables $l = k - j$ and interchanging the order of summation, the last term becomes

$$K_6 \sum_{l=0}^{N} \sum_{k=l+1}^{N} \lambda^{k-l} \| \xi_l \|^2 \leq \frac{K_6}{1-\lambda} \sum_{l=0}^{N} \| \xi_l \|^2.$$

The proof is complete. □

APPENDIX D

Some properties of matrices

In this appendix we collect together various facts about matrices which are used in this book. There are four sections. The first covers properties of symmetric non-negative definite matrices. In the second, various matrix norms and the relations between them are discussed. The third section is devoted to establishing a uniform bound on $\|A^k\|$ for stable matrices A; this is needed in connection with the analysis of identification algorithms in Chapter 5. In the final section, some identities of matrix calculus are presented.

D.1 Symmetric non-negative definite matrices

Symmetric non-negative definite matrices play an important role in this book. This section establishes their main properties.

Throughout, we consider only matrices with real (as opposed to complex) entries. First some definitions: an $n \times n$ matrix A is

symmetric if $A^T = A$;
non-negative definite if $x^T A x \geq 0$ for all $x \in \mathbb{R}^n$;
positive definite if $x^T A x > 0$ for all $x \in \mathbb{R}^n$, $x \neq 0$;
orthogonal if $A^T A = I$ (the $n \times n$ identity matrix).

From the definition, an orthogonal matrix A is non-singular and $A^{-1} = A^T$.

Lemma D.1.1

A real symmetric matrix has real eigenvalues. With every eigenvalue can be associated a real eigenvector, and the (real) eigenvectors corresponding to distinct eigenvalues are orthogonal.

PROOF In the following, an overbar denotes complex conjugate and a star denotes complex conjugate transpose. Suppose that λ, x are respectively an eigenvalue and an eigenvector of a real symmetric matrix A, so that

$$Ax = \lambda x \qquad \text{and} \qquad x \neq 0. \tag{D.1}$$

Then

$$x^* A x = \lambda x^* x.$$

The left-hand side of this equality is, however, real in view of the symmetry, since

$$\overline{x^* A x} = x^{\mathrm{T}} A \bar{x} = x^* A x.$$

Thus λ is equal to the real quantity $x^* A x / x^* x$. If $x = x_1 + i x_2$ then

$$A x_1 + i A x_2 = \lambda x_1 + i \lambda x_2$$

so that both x_1 and x_2 are real eigenvectors corresponding to λ; at least one must be non-zero. If μ is another eigenvalue, $\mu \neq \lambda$, with real eigenvector y, then

$$Ay = \mu y. \tag{D.2}$$

Premultiplying (D.1) and (D.2) by y^{T} and x^{T} respectively and subtracting, we see that

$$(\lambda - \mu) x^{\mathrm{T}} y = 0$$

and hence that $x \perp y$ since $\lambda - \mu \neq 0$. This completes the proof. □

Suppose A is a symmetric matrix with distinct eigenvalues $\lambda_1, \ldots, \lambda_n$. Then the eigenvectors $u_1, \ldots, u_n$ are mutually orthogonal and hence form a basis of $\mathbb{R}^n$. We suppose the u_i are normalized: $u_i^{\mathrm{T}} u_i = 1$. Let U be the $n \times n$ matrix with columns $u_1, \ldots, u_n$. We then have

$$U^{\mathrm{T}} U = I \tag{D.3}$$

$$AU = U\Lambda \tag{D.4}$$

where Λ is the diagonal matrix with diagonal entries $\lambda_1, \ldots, \lambda_n$. Thus U is orthogonal, and, premultiplying (D.4) by U^{T} we see that $U^{\mathrm{T}} A U = \Lambda$, i.e. A can be diagonalized by means of the orthogonal matrix U. It is important that a similar result holds even when the eigenvalues are not distinct.

Proposition D.1.2

Let A be a symmetric matrix with eigenvalues $\lambda_1, \ldots, \lambda_n$ (not necessarily distinct) and form the diagonal matrix Λ as above. Then there exists an orthogonal matrix U such that $U^{\mathrm{T}} A U = \Lambda$.

PROOF The proof is by induction on the order n. Suppose the result holds for $n = k - 1$ and let A be a $k \times k$ matrix and λ, x be an eigenvalue/eigenvector pair. Let P be the orthogonal matrix which rotates x so as to align with the first coordinate vector e_1, i.e. such that

$$Px = le_1 \tag{D.5}$$

where $l = \|x\|$. Now $Ax = \lambda x$ so that

$$PAx = P\lambda x,$$

or

$$PAP^{\mathrm{T}}(Px) = \lambda(Px).$$

In view of (D.5) this shows that

$$PAP^{\mathrm{T}}e_1 = \lambda e_1.$$

Now the left-hand side is just the first column of PAP^{T}, so that PAP^{T} takes the form

$$PAP^{\mathrm{T}} = \left[\begin{array}{c|c} \lambda & b^{\mathrm{T}} \\ \hline 0 & B \end{array}\right].$$

However, PAP^{T} is symmetric, so $b = 0$. By the induction hypothesis, B can be written $B = V^{\mathrm{T}}MV$ where M is diagonal and V is a $(k-1) \times (k-1)$ orthogonal matrix. Thus $A = U^{\mathrm{T}}\Lambda U$, where

$$\Lambda = \left[\begin{array}{c|c} \lambda & 0 \\ \hline 0 & M \end{array}\right], \qquad U = \left[\begin{array}{c|c} 1 & 0 \\ \hline 0 & V \end{array}\right]P.$$

Since U is orthogonal, this establishes the result for $n = k$. The result is trivially true when $n = 1$, so that the induction argument is complete. □

Non-negative definite symmetric matrices have non-negative eigenvalues since if A is such a matrix and λ, x an eigenvalue/eigenvector pair, then

$$0 \leq x^{\mathrm{T}}Ax = \lambda x^{\mathrm{T}}x.$$

In view of the representation $A = U^{\mathrm{T}}\Lambda U$ it is evident that the rank deficiency of A is equal to the number of zero eigenvalues and that A is positive definite if and only if all its eigenvalues are strictly greater than zero. The following results on the existence of 'square root' matrices are used in several places in the book.

Proposition D.1.3

Let A be a symmetric non-negative definite matrix.

(a) If A has rank k, there exists an $n \times k$ matrix B such that $A = BB^{\mathrm{T}}$;
(b) If A is positive definite, i.e. has rank n, then there exists a positive definite symmetric matrix $A^{1/2}$ such that $A = (A^{1/2})^2$.

PROOF Write $A = U^{\mathrm{T}}\Lambda U$ and suppose, without loss of generality, that Λ takes the form

$$\begin{bmatrix} \lambda_1 & & & & & & \bigcirc \\ & \cdot & & & & & \\ & & \lambda_k & & & & \\ & & & 0 & & & \\ & & & & \cdot & & \\ \bigcirc & & & & & & 0 \end{bmatrix}.$$

Now let C be the the $n \times k$ matrix

$$C = \begin{bmatrix} \lambda_1^{1/2} & & \bigcirc \\ & \ddots & \\ \bigcirc & & \lambda_k^{1/2} \\ \hline & \bigcirc & \end{bmatrix}$$

and define $B = U^{\mathrm{T}}C$. Then $A = BB^{\mathrm{T}}$. If $k = n$ we can define $A^{1/2} = U^{\mathrm{T}}CU$. This is symmetric and positive definite, and $(A^{1/2})^2 = U^{\mathrm{T}}\Lambda U = A$. □

Finally, we need a result on convergence of sequences of symmetric non-negative definite matrices.

Proposition D.1.4

Let $P(k)$, $k = 1, 2, \ldots$ be a sequence of $n \times n$ symmetric non-negative definite matrices and suppose that for each $x \in \mathbb{R}^n$, the scalar sequence $x^{\mathrm{T}}P(k)x$ converges to some number $\alpha(x)$. Then there exists a non-negative definite symmetric matrix P such that $\alpha(x) = x^{\mathrm{T}}Px$.

PROOF Let e_i be the unit vector in the ith coordinate direction of $\mathbb{R}^n$ and define

$$P_{ii} = \alpha(e_i) = \lim_{k \to \infty} e_i^{\mathrm{T}} P(k) e_i.$$

Now note the identity

$$(e_i + e_j)^{\mathrm{T}} P(k)(e_i + e_j) = e_i^{\mathrm{T}} P(k) e_i + e_j^{\mathrm{T}} P(k) e_j + 2e_i^{\mathrm{T}} P(k) e_j$$

where we have used the symmetry of $P(k)$. Taking the limit as $k \to \infty$ of this identity shows that

$$e_i^{\mathrm{T}} P(k) e_j \to \tfrac{1}{2}(\alpha(e_i + e_j) - \alpha(e_i) - \alpha(e_j)), \qquad k \to \infty.$$

Denote this limit P_{ij} and let P be the symmetric matrix with i,jth entry P_{ij}.

Then for arbitrary x,

$$x = \sum_1^n x_i e_i,$$

we have,

$$\lim_{k \to \infty} x^{\mathrm{T}} P(k) x = \lim_{k \to \infty} \sum_{i,j=1} x_i x_j e_i^{\mathrm{T}} P(k) e_j$$

$$= \sum_{i,j=1} x_i x_j P_{ij} = x^{\mathrm{T}} P x.$$

Thus $x^{\mathrm{T}}Px \geq 0$ since $x^{\mathrm{T}}P(k)x \geq 0$ for all k, so that P is non-negative definite. □

D.2 Matrix norms

Consider first of all the space of n-vectors (over the real or complex field). A real-valued function on the space is called a *norm*, and is written $\|\cdot\|$, if it possesses the following properties:

(a) $\|\cdot\| \geq 0$ and $\|x\| = 0$ if and only if $x = 0$;
(b) $\|\alpha x\| = |\alpha| \; \|x\|$ for all scalars α;
(c) $\|x + y\| \leq \|x\| + \|y\|$.

These axiomatize the notion of 'length' of a vector. (If the field is complex, $|\cdot|$ here indicates the modulus and, if real, the absolute value).

An important example is

$$\| x \| = \left(\sum_{i=1}^{n} |x_i|^2 \right)^{1/2}$$

where the x_i are the components of x. This is the Euclidean norm. Others are

$$\| x \| = \sum_{i=1}^{n} |x_i|$$

and

$$\| x \| = \max_i |x_i|.$$

We can define a norm also on the space on $n \times m$ matrices. This is a function, again written $\| \cdot \|$, which satisfies axioms analogous to (a)–(c) above, namely:

(a′) $\| A \| \geq 0$ with equality if and only if A is the zero matrix;
(b′) $\| \alpha A \| = |\alpha| \; \| A \|$ for all scalars α;
(c′) $\| A + B \| \leq \| A \| + \| B \|$.

There are many possible choices of matrix norm. The Euclidean (matrix) norm of a matrix is simply the Euclidean norm of the vector assembled from the entries of the matrix. It is a useful fact that this norm can be expressed as

$$\| A \|^2 = \text{trace}\{A^* A\} \quad (\text{or trace}\{A A^*\}).$$

(Here A^* denotes the 'simple' transpose or complex conjugate transpose of A, depending on whether the field is real or complex). Other possible choices are

$$\| A \| = \sum_{i=j}^{n} \sum_{j=1}^{m} |a_{ij}|$$

(where the a_{ij} are the entries of A) or

$$\| A \| = \max_{i,j} |a_{ij}|.$$

A particularly important class of norms are those which take the form

$$\| A \| = \max_{\|x\| = 1} \| Ax \|, \tag{D.6}$$

the definition of which depends, of course, on our choice of norms on the domain and range spaces of A. These are called *induced* matrix norms ('induced' by our choice of norms on the domain and range spaces). The reason for their importance is that, if a norm is defined according to (D.6) then the norm satisfies the inequality

$$\| AB \| \leq \| A \| \cdot \| B \|. \tag{D.7}$$

(Of course it is assumed here that the same norm is adopted for the range of B and the domain space of A for the purposes of defining the induced norms $\| A \|$ and $\| B \|$). We frequently need to bound the magnitude of the product of several matrices; noting inequality (D.7) we see that, if induced norms are used, a bound is provided simply by the product of the norms of the matrices involved.

An induced matrix norm which crops up particularly frequently is that which results from the choice of the Euclidean norm on both the domain and range spaces. It is named the *spectral norm*. It can be shown that, if $\| \cdot \|$ is the spectral norm, then for any matrix A we have that $\| A \|^2$ is the maximum eigenvalue (the eigenvalues will all be real) of the matrix AA^*, or equivalently of the matrix A^*A. It is the relationship of the spectral norm A with the eigenvalues ('spectrum', as the set of eigenvalues is called) of the associated matrix A^*A which gives rise to the terminology 'spectral norm'.

All the norms considered in this book are defined on matrices of arbitrary dimension and satisfy (in addition to the norm axioms) the condition

$$\| A \| = \| A^* \|$$

for arbitrary A.

There is a sense in which all matrix norms are equivalent: if $\| \cdot \|$ and $\| \cdot \|'$ are two norms on the space of $n \times m$ matrices, it can be shown that there exist real numbers $c_0(n, m)$ and $c_1(n, m)$ such that

$$\| A \| \leq c_0(n, m) \| A \|'$$

and

$$\| A \|' \leq c_1(n, m) \| A \|$$

for any matrix A. This means that we can pass from bounds (above or below) on one matrix norm to another by simple scaling. This device is extremely useful when we require a bound with respect to one particular norm, but calculations are much more easily carried out with some other norm. Often in analysis we do not need the actual numerical values of $c_0(n, m)$ and $c_1(n, m)$, but merely the fact of their existence; for example, the inequalities imply that if members of a set of norm matrices are uniformly bounded in magnitude (above or below) with respect to one matrix norm then they will be uniformly bounded with respect to any other matrix norm.

We utter a word of caution here, though. We can expect matrix norms to be equivalent only if we limit attention to matrices of fixed dimension. Indeed, if the members of a set of matrices, not of fixed dimension, are uniformly bounded with respect to one matrix norm, it does not necessarily follow that the same is true when we substitute another matrix norm.

Often in this book it will not matter what choice of norm or matrix norm is made. The reader should assume, for concreteness, that the vector norm is the Euclidean norm and the matrix norm is the spectral norm unless explicitly told to the contrary. This convention is consistent since for a vector, interpreted as a matrix with one column, the Euclidean and spectral norms coincide.

D.3 A uniform bound for stable matrices and applications

It is a well-known property of (real) $n \times n$ matrices A, which are 'stable' in the sense that all the eigenvalues lie in the open unit disc, that the numbers $\|A^k\|$, $k = 1, 2, \ldots$, decay exponentially. (Here, and for the rest of this appendix $\|\cdot\|$ denotes the spectral norm). This property has the important implication for linear dynamical systems that a solution $\{x_k\}$ to the dynamical system equations

$$x_{k+1} = Ax_k, \qquad k = 0, 1, \ldots \tag{D.8}$$

(for x_0 a given n-vector) has exponential decay.

With applications to system identification in mind, we now consider a family of $n \times n$ matrices, in place of a single matrix. We give conditions under which the exponential decay is uniform over the family.

Proposition D.3.1

Let $\mathscr{P}$ be a compact subset of $n \times n$ matrices. Suppose that there exists $\varepsilon \in (0, 1)$ such that, for each matrix $A \in \mathscr{P}$ the eigenvalues of

A are contained in the disc $\{s \in \mathbb{C} : |s| \leq 1 - \varepsilon\}$. Then, corresponding to any $\lambda > 1 - \varepsilon$, there exists $c > 0$ such that

$$\|A^k\| \leq c\lambda^k \qquad \text{for all } A \in \mathscr{P}, \text{ and } k = 0, 1, \ldots$$

PROOF The proof is in several steps.
Let $\bar{\lambda}$ and λ be numbers such that

$$1 - \varepsilon < \bar{\lambda} < \lambda < 1.$$

Step 1 Take A to be a fixed element in $\mathscr{P}$. We shall show that there exists a number c_A (which depends on A) such that

$$\|A^k\| \leq c_A \bar{\lambda}^k, \qquad k = 1, 2, \ldots \tag{D.9}$$

Jordan decomposition of the matrix A gives

$$A = M^{-1} J M.$$

Here M is a (possibly complex) non-singular $n \times n$ matrix. The matrix J can be partitioned as follows:

$$J = \begin{bmatrix} J_1 & & & & \bigcirc \\ & J_2 & & & \\ & & \cdot & & \\ & & & \cdot & \\ \bigcirc & & & & J_d \end{bmatrix}.$$

Here, each J_i is a square matrix (of dimension n_i) which takes the form

$$J_i = \begin{bmatrix} \lambda_i & 1 & & & \bigcirc \\ & \lambda_i & 1 & & \\ & & \cdot & \cdot & \\ & & & \cdot & 1 \\ \bigcirc & & & & \lambda_i \end{bmatrix}$$

for λ_i some eigenvalue of A.

Powers of A can be expressed in terms of J as follows

$$A^k = M^{-1}JMM^{-1}JM \dots M^{-1}JM = M^{-1}J^kM$$

$$= M^{-1}\begin{bmatrix} J_1^k & & & \bigcirc \\ & \cdot & & \\ & & \cdot & \\ \bigcirc & & & \cdot \\ & & & J_d^k \end{bmatrix} M.$$

It follows from properties of matrix norms that there exists a number $c_1 > 0$ such that

$$\|A^k\| \le c_1 \max_{i \in \{1,2,\dots,d\}} \|J_i^k\|, \qquad k = 0, 1, \dots \tag{D.10}$$

Next we note that, for $i = 1, 2, \dots, d$, $J_i = \lambda_i I + S_i$, where

$$S_i = \begin{bmatrix} 0 & 1 & & & \bigcirc \\ & 0 & 1 & & \\ & & \cdot & \cdot & \\ & & & \cdot & \cdot \\ \bigcirc & & & \cdot & 1 \\ & & & & 0 \end{bmatrix}.$$

Observe that $S_i^k = 0$ for $k \ge n_i$. We deduce that there exists a number $c_2 > 0$ (which does not depend on i) such that

$$\|S_i^k\| \le c_2(\bar{\lambda} - 1 + \varepsilon)^k, \qquad k = 0, 1, \dots \tag{D.11}$$

Using the binomial expansion, we calculate, for $i = 1, \dots, d$,

$$\|J_i^k\| = \|(\lambda_i I + S_i)^k\| = \left\| \sum_{l=0}^{k} c_{lk} \lambda_i^l S_i^{k-l} \right\|,$$

where $c_{lk} = k!/(k-l)!l!$,

$$\le \sum_{l=0}^{k} c_{lk} |\lambda_i|^l \|S_i^{k-l}\|$$

$$\le c_2 \sum_{l=0}^{k} c_{lk} |\lambda_i|^l (\bar{\lambda} - 1 + \varepsilon)^{k-l}$$

by (D.11)

$$= c_2(|\lambda_i| + \bar{\lambda} - 1 + \varepsilon)^k$$

(we have appealed once again to the binomial expansion)

$$\leq c_2\bar{\lambda}^k, \qquad i=1,2,\ldots,d,\ k=0,1,\ldots$$

in view of our assumptions on the eigenvalues of A. It follows from (D.10) that (D.9) is true with $c_A = c_1c_2$.

Step 2 Let A be a fixed element in $\mathscr{P}$. We shall show that

$$\|\tilde{A}^k\| \leq c_A\lambda^k$$

for all $\tilde{A}$ in an open ball about A of radius $\lambda - \bar{\lambda}$ ('ball' here is understood in the sense of the induced norm).

Suppose that B is an $n \times n$ matrix such that $\|B\| < \lambda - \bar{\lambda}$. Then

$$\|(A+B)^k\| = \left\| \sum_{l=0}^{k} c_{lk}A^lB^{k-l} \right\| \leq \sum_{l=0}^{k} c_{lk}\|A^l\|\,\|B^{k-l}\|$$
$$\leq c_A\sum_{l=0}^{k} c_{lk}\bar{\lambda}^l(\lambda-\bar{\lambda})^{k-l}$$

by (D.9) (once again the c_{lk} are the 'binomial' coefficients),

$$= c_A\lambda^k, \qquad k=1,2,\ldots$$

This is the required inequality.

Conclusion of the proof The collection of sets $\{\tilde{A}: \|\tilde{A}-A\| < \lambda - \bar{\lambda}\}$, $A \in \mathscr{P}$, forms an open covering of $\mathscr{P}$. Since $\mathscr{P}$ is compact there is a finite subcovering; in other words there exist matrices $A_1, \ldots, A_p$ in $\mathscr{P}$ such that

$$\mathscr{P} \subset \bigcup_{i=1}^{p} \{\tilde{A}: \|\tilde{A}-A_i\| < \lambda - \bar{\lambda}\}.$$

We set $c = \max_i c_{A_i}$. Given any $A \in \mathscr{P}$, A will lie in the set

$$\{\tilde{A}: \|\tilde{A}-A_i\| < \lambda-\bar{\lambda}\} \quad \text{for some value of } i.$$

By the results of step 2,

$$\|A^k\| \leq c_{A_i}\lambda^k \qquad k=1,2,\ldots$$
$$\leq c\lambda^k, \qquad k=1,2,\ldots$$

The proposition is proved. □

Given the close relationship between state-space system descriptions and descriptions expressed in terms of matrices of rational functions in the delay operator (see Section 2.3 of Chapter 2), one would expect Proposition D.3.1, which concerns the state-space equation (D.8), to find a counterpart governing rational functions. Obtaining such a result is the goal of the rest of this appendix. To this end we prove a preliminary lemma.

Lemma D.3.2

Let $\mathscr{D} \subset \mathbb{R}^q$ be a compact set. Consider the polynomial in σ:

$$p_\theta(\sigma) = 1 + \alpha_1(\theta)\sigma + \cdots + \alpha_n(\theta)\sigma^n$$

whose coefficients $\alpha_1(\theta), \ldots, \alpha_n(\theta)$ are continuous (real-valued) functions of the parameter θ on $\mathscr{D}$. Suppose that for each $\theta \in \mathscr{D}$, all the zeros of $\sigma \to p_\theta(\sigma)$ lie in the set $\{\sigma \in \mathbb{C} : |\sigma| > r_0\}$ for some fixed $r_0 > 1$. Then corresponding to any number $\lambda > r_0^{-1}$ there exists a number c such that the coefficients $d_k(\theta)$, $k = 1, 2, \ldots$ in the formal expansion

$$[p_\theta(\sigma)]^{-1} = 1 + d_1(\theta)\sigma + d_2(\theta)\sigma^2 + \cdots$$

satisfy

$$|d_k(\theta)| \leq c\lambda^k, \qquad \text{for all } \theta \in \mathscr{D},\ k = 1, 2, \ldots$$

PROOF We note the following identity: for arbitrary θ,

$$[p_\theta(\sigma)]^{-1} = h^{\mathrm{T}}[I - \sigma A(\theta)]^{-1} b \tag{D.12}$$

in which

$$F(\theta) = \begin{bmatrix} 0 & & & & -\alpha_n(\theta) \\ 1 & \bigcirc & & & \cdot \\ & \cdot & & & \cdot \\ & & \cdot & & \cdot \\ & & & \cdot & -\alpha_2(\theta) \\ & \bigcirc & & 1 & -\alpha_1(\theta) \end{bmatrix} \qquad b = \begin{bmatrix} 0 \\ \cdot \\ \cdot \\ \cdot \\ \\ 1 \end{bmatrix} \qquad h^{\mathrm{T}} = [0, \ldots, 0, 1].$$

To see this we have merely to observe that $[p_\theta(z^{-1})]^{-1}$ is the transfer function of the system described by

$$p_\theta(z^{-1})y_k = u_k \tag{D.13}$$

and $h^{\mathrm{T}}[I - z^{-1}A(\theta)]^{-1}b$ is that of the system described by

$$\begin{aligned} x_{k+1} &= A(\theta)x_k + bu_{k+1} \\ y_k &= h^{\mathrm{T}}x_k. \end{aligned} \tag{D.14}$$

The responses of the two systems (D.13) and (D.14) are the same for zero initial conditions (i.e. $y_k = 0$, $k \leq 0$ and $x_0 = 0$) and arbitrary inputs $\{u_k\}$. This can be deduced from Proposition 2.4.2. It follows that the transfer functions are the same, which amounts to (D.12).

Expanding the right-hand side of (D.12) about $\sigma = 0$, we obtain

$$[p_\theta(\sigma)]^{-1} = \sum_{k=0}^{\infty} d_k(\theta)\sigma^k$$

in which

$$d_k(\theta) = h^{\mathrm{T}} A^k(\theta) b, \qquad k = 0, 1, \ldots$$

It follows that

$$|d_k(\theta)| \leq \|h\| \, \|b\| \, \|A^k(\theta)\|. \tag{D.15}$$

Now the characteristic polynomial of $A(\theta)$ is $s^n + \alpha_1(\theta)s^{n-1} + \cdots + \alpha_n(\theta)$. Bearing in mind that $p_\theta(\sigma)$ cannot vanish at $\sigma = 0$, we deduce that $\bar{\sigma}$ is a zero of $\sigma \to p_\theta(\sigma)$ if $\bar{\sigma} \neq 0$ and $\bar{\sigma}^{-1}$ is an eigenvalue of $A(\theta)$. It follows from our assumptions about the zeros of $p_\theta(\sigma)$ that the eigenvalues of $A(\theta)$ are contained in $\{s \in \mathbb{C} : |s| < r_0^{-1}\}$. Note also that $\{A(\theta) : \theta \in \mathscr{D}\}$ is a compact set of $n \times n$ matrices since $\mathscr{D}$ is compact and $A(\theta)$ depends continuously on θ.

Take a real number $\lambda > r_0^{-1}$. By Proposition D.3.1 there exists a number c_1 such that

$$\|A^k(\theta)\| \leq c_1 \lambda^k, \qquad k = 0, 1, \ldots$$

It follows now from (D.15) that

$$|d_k(\theta)| \leq c\lambda^k, \qquad \text{for } k = 0, 1, \ldots \text{ and } \theta \in \mathscr{D}$$

where $c = \|h\| \, \|b\| c_1$. The lemma is proved. □

Proposition D.3.3

Let $\mathscr{D} \subset \mathbb{R}^q$ be a compact subset. Consider an $r \times l$ matrix of rational functions $T_\theta(\sigma)$ in σ which can be represented

$$T_\theta(\sigma) = [g_\theta(\sigma)]^{-1} G(\sigma)$$

in which

$$g_\theta(\sigma) = 1 + \alpha_1(\theta)\sigma + \cdots + \alpha_n(\theta)\sigma^n$$

and

$$G_\theta(\sigma) = Q_0(\theta) + Q_1(\theta)\sigma + \cdots + Q_n(\theta)\sigma^n.$$

Here $\alpha_1(\theta), \ldots, \alpha_n(\theta)$ are (real-valued) continuous functions of θ (on $\mathscr{D}$) and $Q_0(\theta), \ldots, Q_n(\theta)$ are continuous $r \times l$ matrix-valued functions of θ. We suppose that there exists $r_0 > 1$ such that the zeros of $\sigma \rightarrow g_\theta(\sigma)$ are contained in $\{\sigma \in \mathbb{C}: |\sigma| > r_0\}$ for all $\theta \in D$. Let $H_0(\theta)$, $H_1(\theta), \ldots$ be the coefficients in the formal expansion of $T_\theta(\sigma)$ about $\sigma = 0$:

$$T_\theta(\sigma) = H_0(\theta) + H_1(\theta)\sigma + \cdots \tag{D.16}$$

Then, corresponding to any $\lambda > r_0^{-1}$, there exists $c > 0$ such that

$$\|H_k(\theta)\| \leq c\lambda^k, \qquad \text{for } \theta \in \mathscr{D},\ k = 0, 1, \ldots$$

PROOF Fix $\lambda > r_0^{-1}$. By Lemma D.3.2 there exists $c_1 > 0$ such that

$$|d_k(\theta)| \leq c_1 \lambda^k \qquad \text{for } \theta \in \mathscr{D},\ k = 0, 1, \ldots \tag{D.17}$$

where $d_1(\theta)$, $d_2(\theta), \ldots$ are the coefficients in the formal expansion of $[g_\theta(\sigma)]^{-1}$ about $\sigma = 0$:

$$[g_\theta(\sigma)]^{-1} = 1 + d_1(\theta)\sigma + d_2(\theta)\sigma^2 + \cdots$$

Now the $H_k(\theta)$, given by (D.16), are related to the $d_k(\theta)$ by the formula:

$$H_k(\theta) = \sum_{j=0}^{\min\{k,n\}} d_{k-j}(\theta) Q_j(\theta), \qquad \theta \in \mathscr{D},\ k = 0, 1, \ldots$$

It follows now from (D.17) and properties of matrix norms that

$$\begin{aligned} \|H_k(\theta)\| &\leq c_1 \left(\sum_{j=0}^{n} \|Q_j(\theta)\| \lambda^{k-j} \right) \\ &\leq c\lambda^k, \qquad \theta \in \mathscr{D},\ k = 0, 1, \ldots \end{aligned}$$

Here the constant c is given by

$$c = \frac{c_1}{\lambda^n} \max_{\theta \in \mathscr{D}} \left[\sum_{j=0}^{n} \|Q_j(\theta)\| \right].$$

This completes the proof. □

D.4 Some matrix calculus identities

We collect together in this appendix a number of identities of importance in identification. Let $F(t)$ be a matrix-valued function of a scalar parameter t and let $m(S)$ be a scalar-valued function on a space of matrices S. In what follows we shall interpret

$$\frac{\partial}{\partial t}F(t) \quad \text{and} \quad \frac{\partial}{\partial S}m(S)$$

as having components

$$\left(\frac{\partial}{\partial t}F(t)\right)_{ij}=\frac{\partial}{\partial t}[F(t)]_{ij} \quad \text{and} \quad \left(\frac{\partial}{\partial S}m(S)\right)_{ij}=\frac{\partial}{\partial s_{ji}}m(S).$$

This interpretation is consistent with the convention, adhered to elsewhere in this book, that the gradient of a scalar-valued function of a column vector is a row vector.

Lemma D.4.1

Let $M(t)$ and $N(t)$ be continuously differentiable functions of the scalar parameter t. Suppose that $M(t)$ is $p \times r$ matrix valued and $N(t)$ is $r \times q$ matrix valued. Then

$$\frac{\partial}{\partial t}(M(t)N(t))=\left(\frac{\partial}{\partial t}M(t)\right)N(t)+M(t)\left(\frac{\partial}{\partial t}N(t)\right).$$

PROOF The (i, j)th component of

$$\frac{\partial}{\partial t}(M(t)N(t))$$

is

$$\begin{aligned}\frac{\partial}{\partial t}[M(t)N(t)]_{ij}&=\frac{\partial}{\partial t}\sum_k m_{ik}(t)n_{kj}(t)\\&=\sum_k\left(\frac{\partial}{\partial t}m_{ik}(t)\right)n_{kj}(t)+\sum_k m_{ik}(t)\left(\frac{\partial}{\partial t}n_{kj}(t)\right).\\&=\left[\left(\frac{\partial}{\partial t}M(t)N(t)\right)\right]_{ij}+\left[M(t)\left(\frac{\partial}{\partial t}N(t)\right)\right]_{ij}.\end{aligned}$$ □

Lemma D.4.2

Let $F(t)$ be a continuously differentiable $n \times n$ matrix-valued function of the scalar parameter t. Suppose that $F(t)$ is non-singular at $t=\bar{t}$. Then

$$\frac{\partial}{\partial t}F^{-1}(t)=-F^{-1}(t)\left(\frac{\partial}{\partial t}F(t)\right)F^{-1}(t)$$

at $t=\bar{t}$.

PROOF Since $F(\bar{t})$ is non-singular and F is continuous, $F(t)$ is non-singular on some neighbourhood $\mathcal{N}$ of $\bar{t}$ and we can write

$$F(t)F^{-1}(t) = I \text{ on } \mathcal{N}. \tag{D.18}$$

Now in consequence of the implicit function theorem, the neighbourhood $\mathcal{N}$ can be so chosen that the function $F^{-1}(t)$ is continuously differentiable on $\mathcal{N}$. Differentiating both sides of equation (D.18) we deduce from Lemma D.4.1 that

$$\left(\frac{\partial}{\partial t}F^{-1}(t)\right)F(t) + F^{-1}(t)\left(\frac{\partial}{\partial t}F(t)\right) = 0 \quad \text{on } \mathcal{N}.$$

It follows that

$$\frac{\partial}{\partial t}F^{-1}(t) = -F^{-1}(t)\left(\frac{\partial}{\partial t}F(t)\right)F^{-1}(t) \qquad \square$$

at $t = \bar{t}$.

Lemma D.4.3

Let D be an $r \times q$ matrix. Then

$$\frac{\partial}{\partial S}\text{trace}\,\{SD\} = D \qquad \text{on the space of } p \times r \text{ matrices.}$$

PROOF The (i, j)th component of d/dS trace$\{SD\}$ is

$$\frac{\partial}{\partial s_{ji}}\text{trace}\{SD\} = \frac{\partial}{\partial s_{ji}}\sum_{k,l} s_{kl}d_{lk} = d_{ij}. \qquad \square$$

Lemma D.4.4

Let $\bar{S}$ be a non-singular $n \times n$ matrix. Then

$$\frac{\partial}{\partial S}\log\det S = S^{-1}$$

on a neighbourhood of $\bar{S}$ in the space of $n \times n$ matrices.

PROOF Let $\mathcal{N}$ be a neighbourhood of $\bar{S}$ on which $\det S \neq 0$. Fix a pair of indices (i, j). By Cramér's rule,

$$(\det S)I = S \operatorname{Adj} S$$

(Adj denotes the adjugate matrix of S). Equating the (j, j)th components of the matrices in this equation, we obtain

$$\det S = \sum_k s_{jk}[\text{Adj } S]_{kj}.$$

It follows that on $\mathcal{N}$,

$$\left[\frac{\partial}{\partial S}\log\det S\right]_{ij} = \frac{\partial}{\partial s_{ji}}\log\det S = (\det S)^{-1}\frac{\partial}{\partial s_{ji}}\det S$$
$$= (\det S)^{-1}[\text{Adj } S]_{ij}.$$

(We have used the fact that $[\text{Adj } S]_{kj}$ does not depend on s_{ji} for any k).

□

Lemma D.4.5

Let $\bar{S}$ be a non-singular $n \times n$ matrix and let a be an n-vector. Then

$$\frac{\partial}{\partial S} a^{\mathrm{T}}S^{-1}a = S^{-1}aa^{\mathrm{T}}S^{-1}$$

on a neighbourhood of $\bar{S}$ in the space of $n \times n$ matrices.

PROOF Let $\mathcal{N}$ be a neighbourhood of $\bar{S}$ on which $\det S \neq 0$. On $\mathcal{N}$ we have, by Lemma D.4.2,

$$\left[\frac{\partial}{\partial S} a^{\mathrm{T}}S^{-1}a\right]_{ij} = \frac{\partial}{\partial s_{ji}} a^{\mathrm{T}}S^{-1}a = a^{\mathrm{T}}S^{-1}O(i, j)S^{-1}a$$

(here $O(i, j)$ denotes the matrix with 1 in the (j, i)th entry and zeros elsewhere)

$$= \text{trace}\{O(i, j)S^{-1}aa^{\mathrm{T}}S^{-1}\}$$
$$= \sum_{k,l} [O(i, j)]_{k,l}[S^{-1}aa^{\mathrm{T}}S^{-1}]_{l,k}$$
$$= [S^{-1}aa^{\mathrm{T}}S^{-1}]_{i,j}.$$

□

APPENDIX E

Some inequalities of Hölder type

We collect together in this appendix a number of useful inequalities which centre around the Hölder inequality for finite sequences of real numbers.

Theorem E.1 (The Hölder inequality)

Let p and q be numbers (possibly infinite) such that $1 \leq p \leq \infty$, $1 \leq q \leq \infty$, and $1/p + 1/q = 1$. Then for any positive integer n and numbers $\alpha_1, \alpha_2, \ldots, \alpha_n$ and $\beta_1, \beta_2, \ldots, \beta_n$, we have

$$\sum_{i=1}^{n} |\alpha_i \beta_i| \leq \left[\sum_{i=1}^{n} |\alpha_i|^p \right]^{1/p} \left[\sum_{i=1}^{n} |\beta_i|^q \right]^{1/q}. \tag{E.1}$$

(when $p = \infty$ the relationship $1/p + 1/q = 1$ is taken to indicate $q = 1$ and

$$\left[\sum_{i=1}^{n} |\alpha_i|^p \right]^{1/p}$$

is interpreted as $\max_i |\alpha_i|$).

PROOF The inequality is obviously true if all the α_i, or all the β_i, are zero. It is obviously true also in the cases $p = 1$ (when $q = \infty$) and $p = \infty$ (when $q = 1$). We need consider then only the case when the α_i are not all zero, the β_i are not all zero, and $1 < p < \infty$, $1 < q < \infty$.

The proof hinges on an auxiliary inequality:

$$x^{\lambda} y^{1-\lambda} \leq \lambda x + (1 - \lambda) y, \tag{E.2}$$

valid for any numbers $x \geq 0, y \geq 0, 0 < \lambda < 1$. To show (E.2) we consider the function $r: [0, \infty) \rightarrow \mathbb{R}$ given by

$$r(t) = t^{\lambda} - \lambda t, \qquad 0 \leq t < \infty. \tag{E.3}$$

Note that the derivative $r'(t)$ $(= \lambda(t^{\lambda-1} - 1))$ is positive for $t < 1$ and negative for $t > 1$. It follows that r achieves its maximum over $[0, \infty)$ at $t = 1$, so

$$r(t) \le r(1), \qquad t \ge 0.$$

From (E.3),

$$t^{\lambda} \le \lambda t + 1 - \lambda, \qquad t \ge 0 \tag{E.4}$$

We are now ready to prove (E.2). Clearly we can limit attention to the case $y \ne 0$ since, otherwise, the inequality is trivial. But if $y \ne 0$ the inequality follows from substitution of $t = x/y$ into (E.4).

Next, for $i = 1, \ldots, n$ we apply (E.2) when x, y and λ are taken as follows:

$$x = \frac{|\alpha_i|^p}{\sum_{j=1}^{n} |\alpha_i|^p}, \quad y = \frac{|\beta_i|^q}{\sum_{j=1}^{n} |\beta_i|^q}, \quad \lambda = \frac{1}{p} \left(\text{whence } 1 - \lambda = \frac{1}{q} \right).$$

This choice of x and y makes sense, since by assumption neither all the α_i nor all the β_i are zero. There results

$$\frac{|\alpha_i \beta_i|}{\left(\sum_j |\alpha_j|^p \right)^{1/p} \left(\sum_j |\beta_i|^q \right)^{1/q}} \le \frac{1}{p} \frac{|\alpha_i|^p}{\left(\sum_j |\alpha_j|^p \right)} + \frac{1}{q} \frac{|\beta_i|^q}{\left(\sum_j |\beta_j|^q \right)}.$$

Summing over i, we obtain

$$\frac{\sum_i |\alpha_i \beta_i|}{\left(\sum_j |\alpha_j|^p \right)^{1/p} \left(\sum_j |\beta_j|^q \right)^{1/q}} \le \frac{1}{p} + \frac{1}{q} = 1.$$

This is the Hölder inequality. □

Undoubtedly the most frequently used case of this inequality is that which results when we take $p = q = 2$. Here the inequality takes the form

$$\sum_i |\alpha_i \beta_i| \le \left(\sum_i |\alpha_i|^2 \right)^{1/2} \left(\sum_i |\beta_i|^2 \right)^{1/2}. \tag{E.5}$$

This is the Schwarz inequality. An alternative direct proof can be given along the lines of the proof of the similarly-named inequality in Proposition 1.1.2.

We remark that the names ‘Hölder’ and ‘Schwarz’ are given to inequalities similar in character to (E.1) and (E.5) but when infinite sequences, functions or random variables replace $\{\alpha_i\}_{i=1}^n$, $\{\beta_i\}_{i=1}^n$.

The Hölder inequality is the source of a variety of inequalities, obtained by consideration of special classes of numbers $\{\alpha_1,\ldots,\alpha_n\}$ $\{\beta_1,\ldots,\beta_n\}$. One which is particularly useful in stability analysis in the following.

Corollary E.2

Let p and q be numbers (possibly infinite) such that $1 \le p < \infty$, $1 \le q \le \infty$ and $1/p + 1/q = 1$. Then for any positive integer n and numbers $\lambda_1,\ldots,\lambda_n$, $\mu_1,\ldots,\mu_n$, we have

$$\sum_{i=1}^{n} |\lambda_i \mu_i| \le \left(\sum_{i=1}^{n} |\lambda_i|\right)^{1/p} \left(\sum_{i=1}^{n} |\lambda_i||\mu_i|^q\right)^{1/q}. \tag{E.6}$$

PROOF Apply the Hölder inequality with $\alpha_i = |\lambda_i|^{1/p}$, $\beta_i = |\lambda_i|^{1/q}\mu_i$ for $i = 1,\ldots,n$. □

Author index

Subject index

Page numbers given in **bold type** refer to the definition or principle discussion of the corresponding entry.

GPSR Compliance

The European Union's (EU) General Product Safety Regulation (GPSR) is a set of rules that requires consumer products to be safe and our obligations to ensure this.

If you have any concerns about our products, you can contact us on ProductSafety@springernature.com

In case Publisher is established outside the EU, the EU authorized representative is:

Springer Nature Customer Service Center GmbH
Europaplatz 3
69115 Heidelberg, Germany

Batch number: 10607194

Printed by Printforce, the Netherlands